Stem Cells and Revascularization Therapies

BIOTECHNOLOGY AND BIOPROCESSING SERIES

Series Editor
Anurag Rathore

1. Membrane Separations in Biotechnology, *edited by W. Courtney McGregor*
2. Commercial Production of Monoclonal Antibodies: A Guide for Scale-Up, *edited by Sally S. Seaver*
3. Handbook on Anaerobic Fermentations, *edited by Larry E. Erickson and Daniel Yee-Chak Fung*
4. Fermentation Process Development of Industrial Organisms, *edited by Justin O. Neway*
5. Yeast: Biotechnology and Biocatalysis, *edited by Hubert Verachtert and René De Mot*
6. Sensors in Bioprocess Control, *edited by John V. Twork and Alexander M. Yacynych*
7. Fundamentals of Protein Biotechnology, *edited by Stanley Stein*
8. Yeast Strain Selection, *edited by Chandra J. Panchal*
9. Separation Processes in Biotechnology, *edited by Juan A. Asenjo*
10. Large-Scale Mammalian Cell Culture Technology, *edited by Anthony S. Lubiniecki*
11. Extractive Bioconversions, *edited by Bo Mattiasson and Olle Holst*
12. Purification and Analysis of Recombinant Proteins, *edited by Ramnath Seetharam and Satish K. Sharma*
13. Drug Biotechnology Regulation: Scientific Basis and Practices, *edited by Yuan-yuan H. Chiu and John L. Gueriguian*
14. Protein Immobilization: Fundamentals and Applications, *edited by Richard F. Taylor*
15. Biosensor Principles and Applications, *edited by Loï'efc J. Blum and Pierre R. Coulet*
16. Industrial Application of Immobilized Biocatalysts, *edited by Atsuo Tanaka, Tetsuya Tosa, and Takeshi Kobayashi*
17. Insect Cell Culture Engineering, *edited by Mattheus F. A. Goosen, Andrew J. Daugulis, and Peter Faulkner*
18. Protein Purification Process Engineering, *edited by Roger G. Harrison*
19. Recombinant Microbes for Industrial and Agricultural Applications, *edited by Yoshikatsu Murooka and Tadayuki Imanaka*
20. Cell Adhesion: Fundamentals and Biotechnological Applications, *edited by Martin A. Hjortso and Joseph W. Roos*
21. Bioreactor System Design, *edited by Juan A. Asenjo and José C. Merchuk*
22. Gene Expression in Recombinant Microorganisms, *edited by Alan Smith*
23. Interfacial Phenomena and Bioproducts, *edited by John L. Brash and Peter W. Wojciechowski*

CRC Press
Taylor & Francis Group
6000 Broken Sound Parkway NW, Suite 300
Boca Raton, FL 33487-2742

First issued in paperback 2019

© 2012 by Taylor & Francis Group, LLC
CRC Press is an imprint of Taylor & Francis Group, an Informa business

No claim to original U.S. Government works

ISBN-13: 978-1-4398-0323-3 (hbk)
ISBN-13: 978-0-367-38203-2 (pbk)

Visit the Taylor & Francis Web site at
http://www.taylorandfrancis.com

and the CRC Press Web site at
http://www.crcpress.com

Stem Cells and Revascularization Therapies

EDITED BY

HYUNJOON KONG
ANDREW J. PUTNAM
LAWRENCE B. SCHOOK

CRC Press
Taylor & Francis Group
Boca Raton London New York

CRC Press is an imprint of the
Taylor & Francis Group, an **informa** business

Contents

PART I Defining, Isolating, and Characterizing Various Stem and Progenitor Cell Populations for Neovascularization

PART II In Vitro Studies for Angiogenesis, Vasculogenesis, and Arteriogenesis

PART III   Stem Cell Mobilization Strategies

PART IV   Stem Cell Transplantation Strategies

Preface

Blood vessels are responsible for the convective delivery of oxygen, nutrients, and other large macromolecules, as well as immune cells, to all tissues in the human body. New vessels form either via angiogenesis, the sprouting of new vessels from preexisting ones, or via vasculogenesis, the de novo formation of vascular networks from cellular building blocks. Understanding and controlling these two vascularization processes are critical to understanding normal development, wound healing, and tissue remodeling, and are of critical importance to the treatment of various diseases. For the past three to four decades, strategies to impede vascularization have been explored in both preclinical and clinical research, primarily in the context of treating a variety of cancers. Fundamental insights gained from this type of work have, in turn, enabled the development of new approaches to promote vascularization, a limiting factor in the development of engineered tissues for regenerative medicine applications.

Revascularization therapy aims to rebuild vascular networks in ischemic tissues and has emerged as a promising strategy to treat cardiovascular patients who may not be candidates for, or benefit from, traditional surgical or medical device interventions. Revascularization therapies have consistently proven to be powerful tools to treat acute and chronic wounds, ischemic tissues (including the brain, the heart, and endocrine organs), and promote regeneration within tissue defects. Promising strategies have employed the use of various pro-angiogenic factors to promote revascularization. However, clinical trials relying on bolus injection of individual factors have been somewhat disappointing, perhaps due to the limited half-life of most protein growth factors, the lack of temporal and spatial control over growth factor release, and the inability of single factors to properly regulate neovascularization. Newer strategies involving sustained delivery of pro-angiogenic factors or genes from biodegradable scaffolds have helped to overcome protein stability issues. Furthermore, delivery of multiple pro-angiogenic factors in a time-dependent fashion to mimic the process of natural vessel development has shown enhanced potential versus single-factor delivery alone, yielding more mature vascular networks that are both functional and stable. However, because even combinations of multiple factors may not fully recapitulate the complex milieu of pro-angiogenic signals presented to cells *in vivo*, these approaches are now increasingly complemented by the use of stem cells as vascular medicines.

The theoretical regenerative power of stem cells, whether embryonic stem cells or progenitor cells from a variety of postnatal tissues, is due in part to their pleiotropic effects. Not only can stem cells from a variety of sources offer the potential to differentiate into the cellular building blocks of blood vessels, they can also endogenously express multiple pro-angiogenic factors subject to endogenous physiologic controls. Therefore, they may be more potent than molecular-based medicines. Significant advancements in stem cell biology and engineering over the last few decades have enabled progress toward the clinical use of stem cells in revascularization therapies.

Some strategies involved the mobilization of endogenous stem cell populations, while others have explored cell transplantation. In both cases, biomaterials and other biomedical tools have been designed via multidisciplinary efforts to improve and control the fate of stem cells. However, there are significant hurdles with respect to our understanding of stem cell biology that, once overcome, will further enhance their utility in revascularization therapies.

In this context, the goal of *Stem Cells and Revascularization Therapies* is to provide a concise and topically focused volume of fundamental and applied studies on both stem cell biology and perspectives associated with the development of revascularization strategies. To facilitate readers' understanding of the multidisciplinary issues associated with this topic, we have divided the book into four parts. Part I: (Chapters 1 through 3) explores the topics of defining, isolating, and characterizing various stem and progenitor cell populations for neovascularization. Part II (Chapters 4 through 7) summarizes some especially useful model systems and approaches to regulate angiogenesis, vasculogenesis, and arteriogenesis, and explores their implications for forming functional vessels *in vivo*. Part III (Chapters 8 and 9) focuses on stem cell homing to sites of injury and inflammation, and strategies to exploit this mobilization phenomenon. Part IV covers topics related to stem cell transplantation, including recreating features of endogenous stem cell niches in order to maintain the multipotency of transplanted cells (Chapter 10) and combinatorial delivery of cells and molecular factors (Chapter 11). We hope this integrated collection of contributions from some of the most prominent scientists and engineers working in this emerging field of research will help both novices to the field and seasoned researchers alike to better comprehend emergent findings and remaining challenges involving the use of stem cells to promote new blood vessel formation. Ultimately, we hope to inspire researchers to make new contributions to improve the therapeutic efficacy of stem cell–based revascularization therapies.

Contributors

Patrick Allen
Department of Surgery
Children's Hospital Boston
and
Department of Biomedical Engineering
Boston University
Boston, Massachusetts

James Ankrum
Harvard-MIT Division of Health
   Sciences and Technology
Department of Medicine
Brigham and Women's Hospital
Harvard Medical School
Harvard Stem Cell Institute
Cambridge, Massachusetts

Suk Ho Bhang
School of Chemical and Biological
   Engineering
Seoul National University
Seoul, Republic of Korea

Joyce Bischoff
Department of Surgery
Children's Hospital Boston
and
Department of Surgery
Harvard Medical School
Boston, Massachusetts

Christopher S. Chen
Department of Bioengineering
University of Pennsylvania
Philadelphia, Pennsylvania

Limor Chen-Konak
Faculty of Biomedical Engineering
Technion
Haifa, Israel

Ross J. DeVolder
Department of Chemical and
   Biomolecular Engineering
University of Illinois at
   Urbana-Champaign
Urbana-Champaign, Illinois

Amir Fine
Faculty of Biomedical Engineering
Technion
Haifa, Israel

Ji Woong Han
Division of Cardiology
Department of Medicine
Emory University School of Medicine
Atlanta, Georgia

Jeffrey J.D. Henry
Department of Bioengineering
University of California, Berkeley
Berkeley, California
and
University of California, San Francisco
San Francisco, California

Jeffrey M. Karp
Harvard-MIT Division of Health
   Sciences and Technology
Department of Medicine
Brigham and Women's Hospital
Harvard Medical School
Harvard Stem Cell Institute
Cambridge, Massachusetts

Byung-Soo Kim
School of Chemical and Biological
   Engineering
Seoul National University
Seoul, Republic of Korea

Justin T. Koepsel
Departments of Biomedical
    Engineering
University of Wisconsin
Madison, Wisconsin

Hyunjoon Kong
Department of Chemical and
    Biomolecular Engineering

and

Institute of Genomic Biology
University of Illinois at
    Urbana-Champaign
Urbana-Champaign, Illinois

Namit Kumar
Harvard-MIT Division of Health
    Sciences and Technology
Department of Medicine
Brigham and Women's Hospital
Harvard Medical School
Harvard Stem Cell Institute
Cambridge, Massachusetts

Shulamit Levenberg
Faculty of Biomedical Engineering
Technion
Haifa, Israel

Rebecca Diane Levit
Division of Cardiology
Department of Medicine
Emory University School of Medicine
Atlanta, Georgia

Song Li
Department of Bioengineering
University of California, Berkeley
Berkeley, California

and

University of California, San Francisco
San Francisco, California

David J. Mooney
School of Engineering and Applied
    Science
Harvard University
Cambridge, Massachusetts

William L. Murphy
Department of Biomedical Engineering

and

Department of Pharmacology

and

Department of Materials Science and
    Engineering
University of Wisconsin
Madison, Wisconsin

Andrew J. Putnam
Department of Biomedical Engineering
University of Michigan
Ann Arbor, Michigan

Debanjan Sarkar
Harvard-MIT Division of Health
    Sciences and Technology
Department of Medicine
Brigham and Women's Hospital
Harvard Medical School
Harvard Stem Cell Institute
Cambridge, Massachusetts

Lawrence B. Schook
Department of Animal Sciences

and

Institute for Genomic Biology

and

Division of Biomedical Research
University of Illinois at
    Urbana-Champaign
Urbana-Champaign, Illinois

Colette J. Shen
Department of Bioengineering
University of Pennsylvania
Philadelphia, Pennsylvania

Dmitry Shvartsman
School of Engineering and Applied
    Science
Wyss Institute for Biologically Inspired
    Engineering
Harvard University
Cambridge, Massachusetts

Wei Suong Teo
Harvard-MIT Division of Health
    Sciences and Technology
Department of Medicine
Brigham and Women's Hospital
Harvard Medical School
Harvard Stem Cell Institute
Cambridge, Massachusetts

Gregory Timp
Institute for Genomic Biology
and
Department of Electrical and Chemical
    Engineering
University of Illinois at
    Urbana-Champaign
Urbana-Champaign, Illinois

Lisa R. Trump
Department of Animal Sciences
and
Institute for Genomic Biology
University of Illinois at
    Urbana-Champaign
Urbana-Champaign, Illinois

Young-sup Yoon
Division of Cardiology
Department of Medicine
Emory University School of Medicine
Atlanta, Georgia

Liang Youyun
Department of Chemical and
    Biomolecular Engineering
University of Illinois at
    Urbana-Champaign
Urbana-Champaign, Illinois
and
Department of Chemical and
    Biomolecular Engineering
National University of Singapore
Singapore, Singapore

Weian Zhao
Harvard-MIT Division of Health
    Sciences and Technology
Department of Medicine
Brigham and Women's Hospital
Harvard Medical School
Harvard Stem Cell Institute
Cambridge, Massachusetts

Part I

Defining, Isolating, and Characterizing Various Stem and Progenitor Cell Populations for Neovascularization

1 Embryonic Stem Cells

*Limor Chen-Konak, Amir Fine,
and Shulamit Levenberg*

CONTENTS

1.1   INTRODUCTION

Stem cells are characterized by their abilities to indefinitely self-renew and to differentiate into a diverse range of specialized cell types. The last decade has witnessed significant progress and achievements in the field of stem cell applications, due to landmarks, including the isolation of murine embryonic stem (ES) cells in 1981,[1,2] which paved the way for basic research of the subsequent landmark, the isolation of human ES cells, in 1998.[3] Human ES cells bear the capacity to massively generate virtually any cell type and tissue of the human body, rendering them an ideal cell supply for tissue engineering therapies, which carry the potential of curing a variety of human diseases. However, a thorough understanding of stem cell origin, characteristics, and the molecular basis underlying their differentiation potential are prerequisites to their effective application in regenerative medicine. This chapter will focus on the characterization, derivation, and multilineage differentiative potential of human ES cells and will also touch upon their vascular lineage pathway and its potential implications in neovascularization applications.

1.2   DEFINING AND CHARACTERIZING STEM CELLS

1.2.1   STEM CELLS: DEFINITION AND CLASSIFICATION

Two unique properties define a stem cell: self-renewal—the ability to replicate itself and multipotency—the ability to differentiate into one or more specialized cell types.[1] Stem cells can be derived from multiple stages of embryonic development, as well as from numerous adult tissues, as is reflected by their classification. ES cells are derived from the embryonic blastocyst and, due to their pluripotency, can give rise to cells derived from any of the three germ layers, thus theoretically can serve as an inexhaustible supply of cells. Similarly, embryonic germ (EG) cells are derived from an embryonal origin, from the primordial fetal germ cells (mouse EGs, for example, are derived from E8.5 to E12.5 mouse embryos), and they are also pluripotent stem cells. However, adult or somatic stem cells are categorized by the various differentiated tissues from which they are derived, such as bone marrow, brain, peripheral blood, neural tissues, liver, muscle, or skin.[4] While these cells can self-renew, their differentiative potential is limited to the repertoire of cell types found in the tissue of origin. Although adult stem cell therapies using patient-specific cells may bypass the immunorejection problem, factors such as their restricted lineage potential and poor growth rates *in vitro*, coupled with challenges in access and isolation of such stem cell pools, make their use less practical. However, a novel, artificially created pluripotent stem cell that also bypasses the need for immunosuppression therapies has been recently introduced to the therapeutic arena. Induced pluripotent stem cells (iPS), which is discussed in more detail in Chapter 3, are derived from nonpluripotent cells, typically adult somatic cell, by specifically inducing expression of a group of requisite genes.[5-7] In many respects, they are believed to be similar to natural pluripotent stem cells and are, therefore, expected to provide a more suitable substitute cell source for ES cells. However, the full extent of their therapeutic potential in relation to natural pluripotent stem cells is still being assessed.

1.2.2 Characterization of Embryonic Stem Cells

The basic characteristics of ES cells include unlimited self-renewal, multilineage differentiation capacity (pluripotency) *in vitro* and *in vivo*, clonogenicity, and normal karyotype. Their remarkable capacity to differentiate into virtually any cell type in the body sets ES cells apart from other cells, in general and from adult stem cells, in particular. ES cells can be maintained and expanded indefinitely as pure populations of undifferentiated cells, so long as they are supplied with medium supplemented with either leukemia inhibitory factor (LIF) for mouse ES cells or basic fibroblast growth factor (bFGF) for human ES cells. In an *in vitro* environment, ES cells are cultured in colonies, where their growth is neither retarded by contact inhibition nor by proliferative senescence. Under well-regulated culture conditions, ES cells maintain their self-renewing capacity over extended culture periods while retaining a considerably normal karyotype throughout. These unique characteristics endow ES cells with enormous prospects for advancing medicine and science.

ES cell pluripotency manifests itself via the capacity to differentiate into cells derived from any of the three primary germ layers, namely, endoderm, mesoderm, and ectoderm. This characteristic can be assessed by (1) allowing for spontaneous differentiation in of ES cells inculture,[8,9] (2) their implantation into adult mice resulting in teratoma formation,[10] and (3) their introduction into a morula or blastocyst resulting in chimeras.[11,12] While *in vivo* differentiation of ES cells can be assessed in animal species via germ line chimerism production, this method cannot be applied in humans. Thus, monitoring of teratomas formed from *in vivo* differentiation of human ES cells is a typical mode of determining their differentiative potential.

There are certain genetic characteristics specific to all stem cells that represent their cell stemness. Critical regulators of the stem cell character include transcription factors such as the POU domain octamer-4 (Oct4), Nanog, gastrulation brain homeobox 2 (Gbx2), the homeobox domain transcription factor *Sry*-type HMG box 2 (SOX-2), the zinc finger RNA editing exouridylylase 1 (REX-1), the transcriptional coactivator undifferentiated embryonic cell transcription factor 1 (UTF-1), and the CP2-related transcriptional repressor 1 (CRTR-1).[13–19] Specific cell surface molecules serve as additional molecular markers for stem cell characterization including the stage-specific embryonic antigen (SSEA-1 in mice and SSEA-3,4 in humans) and embryonal stem cell–specific gene 1 (Esg-1) exclusively associated with pluripotency. Other typical markers highly expressed in ES cells are alkaline phosphatase activity and telomerase maintenance molecules.[20–23]

1.2.3 Isolation and Characterization of Human Embryonic Stem Cells

The first isolation of human ES cells[3] led to a dramatic elevation in global interest in the therapeutic potential of ES cells, moving the theory one step closer to reality. It was quickly understood that human ES cells can provide a renewable source of replacement cells for the treatment of medical disorders including Parkinson's and Alzheimer's diseases, diabetes, stroke, burn wounds, spinal cord damage, and heart disease. The following sections summarize human ES cell derivation and expansion, their unique properties, and the extracellular factors regulating their self-renewal.

1.2.3.1   Isolation, Derivation, and Expansion of Human Embryonic Stem Cells

Human ES cells were first derived from the inner cell mass (ICM) of blastocyst-stage human embryos[3] basically in the same manner as the rhesus monkey ES cells.[24] Cleavage-stage human embryos formed for clinical purposes by *in vitro* fertilization techniques were donated and cultured to produce the blastocyst. The blastocyst contains the ICM, the part rich in epiblast cells, which can further differentiate to form the three embryonic germ layers. The ICM is isolated using antisera or microdissection and plated onto mitotically inactivated murine embryonic fibroblast (MEF) feeder layers. ICM-derived cells are cultured in the presence of serum, and colonies demonstrating characteristic undifferentiated morphology are subsequently selected and expanded. Following the initial cell derivation stage, human ES cell lines can be maintained and propagated on MEF feeder layers in medium containing serum alone or serum replacement (SR) supplemented with bFGF. While the murine LIF cytokine is responsible for supporting mouse ES cell derivation, it fails to maintain human ES cells.[25,26]

Concerns about exposure of human ES cells to animal-derived products have precluded their widespread use in human clinical applications. Such considerations were substantiated by a study showing that human ES cells cultured in the presence of animal-derived products express the nonhuman sialic acid, N-glycolylneuraminic acid (NeuGc).[27] Thus, efforts have been made to find alternatives for potent support of human ES cell self-renewal while avoiding potential contamination with xenoproteins and xenogeneic tissues. A number of cell sources have been tested for their efficacy as feeder cells, including human embryo-derived fibroblasts, foreskin fibroblasts, fetal epithelial cells, adult bone marrow cells, fallopian tube–derived cells, and placental cells.[28–32] However, cell and tissue therapies demand maintenance of large quantities of undifferentiated human ES cells, and therefore, usage of feeder cells that are extremely labor intensive to prepare may not provide the optimal solution. An alternative solution is the feeder-free systems that avoid the use of feeder cells in maintaining human ES cells. Xu et al. reported the first successful feeder-free culture, in which human ES cells were grown on culture dishes coated with biologically active materials such as laminin and Matrigel™ in MEF-conditioned medium supplemented with serum replacement and bFGF.[33] Later studies further supported these findings and successfully upgraded culturing protocols by determining conditions independent of feeder-conditioned media by either applying high concentrations of bFGF or in combination with other growth factors, together with usage of culture dishes coated with natural extracellular matrix (ECM)-based biomaterials.[34–38] Additionally, in order to avoid contamination of cultures with animal products, growth medium can be enriched with human serum in place of animal serum, or alternatively, with growth factors under serum-free conditions.[38,39] The defined conditions system provides an additional method of culturing human ES cells while avoiding exposure to animal-derived contaminants or yet undetermined influences of components from growth factor cocktails.[40,41] However, the use of cell-based ECM components poses threats of pathogen transmission. Yet, although the use of human laminin in place of Matrigel may provide a partial solution, significant variability exists between lots of commercial laminin sources. Thus, synthetic ECM-based culture support offers numerous advantages including a pathogen/allergen-free environment, ease of scale-up, and tight control of biochemical and biomechanical properties.

1.2.3.2 Properties of Human Embryonic Stem Cells

Human ES cell morphology resembles that of nonhuman primate ES colonies, exhibiting high nucleus–cytoplasm ratios. They form relatively flat and compact colonies but retain distinct cell borders. The proliferative rate of human ES cells is slower than that of mouse ES cells, with a population doubling time of ~36 h compared to ~12 h in mouse ES cells.[42] In addition, human ES cells differ from their murine counterparts with regard to their cell-surface antigen phenotype. Human ES cells express SSEA-3 and SSEA-4, high molecular weight glycoproteins tumor rejection antigen (TRA)-1-60 and TRA-1-81, and alkaline phosphatase.[3,43] Human ES cells also express transcriptional factors characteristic of the undifferentiated ES cell state, including Oct4 (POU5F1), Nanog, UTF1, Sox2, TDGF1 (Cripto), and DPPA5.[13,14,44,45] Additionally, human ES cells demonstrate remarkably stable karyotypes and high telomerase activity even after >300 population doublings and passages spanning beyond one year in culture.[42] The self-renewal of human ES cells is critically dependent upon a stoichiometric balance between various signaling molecules, where any shift can cause loss of ES cell identity as elaborated in the following.

1.2.3.3 Extracellular Factors Regulating Human ES Cell Self-Renewal

While the molecular mechanisms underlying self-renewal of human ES cells remain to be elucidated, it is believed that ES propagation is most likely maintained by a highly coordinated signaling network. Although several signaling pathways are known to be involved in the self-renewal of mouse ES cells, they are still poorly understood in human ES cells.[26,46] Gene-expression analyses of stem cells using microarray technology have provided further insight into the signaling pathways involved in human ES cells. Such analyses have shown that the genes associated with bFGF, TGFβ1/ bone morphogenetic protein (BMP), and Wnt signaling are expressed in human ES cells and play a role in their self-renewal mechanisms.[21,22,44,47–49]

The fibroblast growth factor (FGF) signaling pathway is thought to be central to regulation of human ES cell self-renewal. The FGF family consists of 22 ligands signaling through four cell surface FGF receptors (FGFR1–4) equipped with intrinsic tyrosine kinase activity. Gene expression analyses have indicated active FGF signaling in undifferentiated human ES cells.[44,48] Moreover, excessive levels of bFGF, or alternatively, physiological bFGF concentrations in combination with other factors can effectively maintain undifferentiated human ESC viability in the absence of MEF-conditioned medium.[38] While bFGF can be replaced by TGFβ1 or activin in the unconditioned medium state, human ES cells demonstrated retarded proliferation rates, suggesting the critical role of bFGF in human ES cells mitogenesis.[34] In direct correlation with these findings, FGF receptor 1 (FGFR1), the cognate bFGF receptor, has been shown to be exceedingly expressed in undifferentiated human ES cells.[14,44,50] In addition, expression of other FGF receptors has been detected in human ES cells, including those which bind stem cell factor (SCF) and fetal liver tyrosine kinase-3 ligand (Flt3L).

A constant struggle exists between cell propagation vs. differentiation-inducing factors in determining ES cell fate. Addition of BMP-2, BMP-4, or BMP-7 to culture media leads to differentiation of human ES cells to primitive endoderm, whereas suppression of BMP-mediated signaling is necessary for human ES cell maintenance. The undifferentiated state of human ES cells is characterized by high levels of phosphorylated Smad2/3, indicating a functional role of the TGFβ/activin/nodal signaling pathway in regulating human ES cells self-renewal.[38,51,52] In line with these findings, noggin and gremlin, two BMP antagonists, have been detected in MEF-conditioned medium enhancing human ES cell culture maintenance. However, addition of noggin to human ES cell cultures grown on feeder layers resulted in elevation of neuroectodermal markers. Therefore, a delicate balance between BMP signaling and BMP antagonists is required to regulate the undifferentiated state of human ES cells.[38,52,53]

Wnt/β-catenin signaling has been implicated in control of various animal developmental processes, including embryonic induction, generation of cell polarity, and cell fate.[54] Several Wnt signaling pathway components have been detected in human ES cells,[44,47,55] where the levels of different receptors varied between undifferentiated and differentiated populations. Dravid et al. described Wnt-dependent stimulation of both human ES cell proliferation and differentiation, but minimal β-catenin-mediated transcriptional activity in the undifferentiated state. Furthermore, it was found that factors secreted by feeders supporting human ES cells are not Wnt ligands based on blocking experiments.[56] Therefore, the regulation of human ES cells by Wnt signaling is still inconclusive.

1.3 DIFFERENTIATION OF HUMAN EMBRYONIC STEM CELLS

As pluripotent cells, human ES cells can give rise to tissues derived from all three dermal layers, namely endoderm, mesoderm, and ectoderm. Differentiation of human ES cells can be performed by either *in vivo* formation of teratomas or by either spontaneously or directed differentiation *in vitro*.

1.3.1 *IN VIVO* DIFFERENTIATION: TERATOMA FORMATION

Injection of human ES cells into immunodeficient mice results in the formation of teratomas, the benign tumors composed of multiple tissue-type lineages cells.[3,46] This assay constitutes the most rigorous technique available for the assessment of the ability of human ES cells to differentiate into each of the three germ layers. Human ES cells can be injected beneath the testis capsules, under the kidney capsule, or directly into the hindlimb muscle of severe combined immunodeficiency (SCID) mice. Approximately 6–8 weeks after injection, tumors are formed and teratomas are histologically analyzed for the presence of tissues representative of all three germ layers.

1.3.2 *In Vitro* Three-Dimensional Differentiation

One of the most common methods used to induce human ES cell differentiation *in vitro* is via embryoid body (EB) formation, which is a three-dimensional (3D) aggregate consisting of multiple differentiated tissue lineages as a recapitulation of early embryonic development.[57,58] In order to form EBs, human ES cells are removed from the feeder layer and cultured in nonadherent plates. This results in spherical EBs in which the cells undergo spontaneous differentiation to yield cells of the endoderm, ectoderm, and mesoderm layers.[3,46,59] EB differentiation in 3D culture provides enhanced cell–cell interactions and allows for both entrapment of secreted extracellular matrix components and maintenance of spherical cellular morphologies.[60,61] In addition, the 3D culture environment offers structural support for higher order tissue organization and remodeling. However, the heterogeneous nature of EBs may result in inconsistent cell responses to exogenous factors. In addition, since initial differentiation within the EB occurs in its extraembryonic endoderm portion (yolk sac, endoderm, and visceral endoderm), signaling molecules produced from these tissues keep regulating the differentiation of other parts of the EB and, in this manner, may eliminate the effect of exogenous cytokines and growth factors in inducing differentiation within the EB. Thus, further optimization of EB culture strategies aimed at regulating stem cell behavior and differentiation is still required.

Spontaneous EB-based 3D differentiation is considered less efficient for the induction of specific ES-derived cell types, resulting in small percentages of each cell type.[62] For example, only 3% of human ES-derived EB cultures express the hematopoietic-related differentiation CD34 protein[63] and less than 2% demonstrated CD31 (PECAM1) expression,[64] correlated with the endothelial lineage. Other protocols inducing differentiation in two-dimensional (2D) systems without requiring EB formation have resulted in more controlled differentiation and increased yield of cells of interest (detailed in Section 1.3.6.1).

Human EBs are commonly formed as a heterogeneous mixture of cell aggregates with wide variability both between and within individual aggregates.[58,65] The heterogeneous size and shape of human EBs resulting from suspension cultures often influence their differentiation potential.[66] Recently, methods such as those using round-bottom 96-well plates, conical tubes, and microwells have been adopted to form EBs from predetermined numbers of human ES cells.[67–69] This method results in generation of EBs of uniform size and cell distribution, which can then be used to promote formation of monodisperse EBs. Recently, Ungrin et al. reported the development of a microwell culture system that produces morphologically organized human ES cell aggregates from a single-cell EB capable of multilineage differentiation.[70] The size of human EBs plays a significant role in establishing differentiation outcomes; thus, the ability to control this parameter is fundamental to its diverse applications.

Differentiation of human ES cells in 3D cultures can also be performed using scaffolds as applied in tissue engineering techniques. Scaffolds provide a 3D environment that closely mimics the natural ECM and serves as a physical support to the cells embedded within. Various natural or synthetic biomaterials have been used to form scaffolds supporting human ES cell differentiation (for reviews see

Refs. [71,72]). Levenberg et al. demonstrated application of a biodegradable poly-L-lactic acid (PLLA)/ polylactic-co-glycolic acid (PLGA) scaffold,[73] in which early differentiated human ES cells (day-eight EBs) cultured with specific growth factors, generated complex structures with features of various committed embryonic tissues. Cell structures with characteristics of neural, cartilage and liver tissues, as well as a vessel-like network were observed in the human ES cell–embedded scaffolds.

1.3.3 TWO-DIMENSIONAL DIFFERENTIATION OF HUMAN ES CELLS

Human ES cell differentiation can also be induced in a 2D setting, namely, by culturing human ES cell monolayers on ECM protein-coated plates or directly on supportive stromal layers.[74] While both 2D and 3D culturing methods can yield a broad spectrum of cell types from human ES cells, 2D cultures do not closely mimic physiological environments and may result in inefficient and heterogeneous differentiation. However, in certain cases, 2D techniques offer conditions advantageous to particular application sets. More specifically, Wang et al. reported on a 2D culture system used to direct differentiation of human ES cells to hematopoietic and endothelial lineages. This process aimed at avoiding the 3D-EB spontaneous differentiation method found to be inefficient for such cell lineages. The 2D differentiation performed by coculturing of human ES cells with MEF cells in addition to growth factors resulted in a successful and efficient differentiation into hematopoietic and endothelial progenitors. The authors reported a five to eightfold increase in the number of CD34+ cells differentiated in the 2D system than in the 3D system.[75] Significant differences were also found in culture environment profiles required for ES differentiation in 3D compared to 2D.[76,77] When comparing the conditions required for the induction of hematopoietic lineage differentiation in a 2D system of human ES cells cocultured with stromal cells versus a 3D EBs forming setup, addition of the SCF, thrombopoietin (Tpo), and Flt3L growth factors to serum-free media supported hematopoiesis in the 2D but not in the EB system.[78] However, supplementing the EB system with BMP-4 and vascular endothelial growth factor (VEGF) improved hematopoietic differentiation. Thus, promotion of human ES differentiation to specific cell types requires distinct cytokines and growth factor in the different systems.

1.3.4 GROWTH FACTOR–INDUCED DIFFERENTIATION

Various morphogenetic factors including hedgehog proteins, Wnt proteins, notch ligands, members of TGFβ superfamily of growth factors, or FGFs, have been implicated to play important roles in controlling the regulation of embryo development, and thus were studied in human ES differentiation. Schuldiner et al.[79] were the first group to question the role of a number of different growth factors in the differentiation of human ES cells and the induction of EBs into mesodermal, endodermal, or ectodermal lineages. Their study reported a wide variety of receptors expressed at relatively high levels in human ES cells, including members of the protein-tyrosine phosphatase (PTP)-, fibroblast growth factor (FGF)-, insulin growth factor (IGF)-, BMP-, activin-, and tumor necrosis factor (TNF)-receptor families, all of which are known to participate in a range of developmental pathways such as gastrulation, mesendoderm commitment, and neural maturation.[80] However, these studies did not result in homogeneous differentiation of ES cells. Hence, the current challenge is to find an optimized combination of the various cytokines and growth factors that would bias differentiation specifically toward a desired lineage or self-organization into tissues. The exact role of the various growth factors in the differentiation of human ES cells was demonstrated in the multiple protocols used in each of a cell lineage protocol differentiation, as mentioned in the few examples of representative lineages described in Section 1.3.6 and as described in detail in Murry and Keller's recent review.[81]

1.3.5 SMALL MOLECULE–INDUCED DIFFERENTIATION

In addition to growth factors and cell-secreted morphogenetic factors, the fate of stem cells can be regulated by small cell–permeable molecules such as dexametha-sone, ascorbic acid, sodium pyruvate, thyroid hormones, prostaglandin E2, dibutryl cAMP, concanavalin A, vanadate, and retinoic acid (RA).[82,83] Such small molecules play important roles during embryogenesis and may be utilized to direct or control the differentiation process of ES cells. For example, RA enhances expression of neural crest cells while reducing mesodermal differentiation.[84,85] ES cell exposure to sodium butyrate resulted in production of hepatocyte-like cells exhibiting a gly-colytic phenotype.[86] Thyroid hormones have also been implicated as potent differ-entiation factors.[87,88] These findings stress the significance of cell-permeable small molecule–mediated biological signals in directing differentiation of ES cells. With increased understanding of the identity and nature of these molecules along with their roles in stem cell biology, they can be incorporated into tissue engineering scaf-fold design so as to harness their beneficial effects for lineage specific differentiation and tissue development.

1.3.6 *IN VITRO* DIFFERENTIATION INTO THE THREE GERM LAYER LINEAGES

Formation of the three embryonic germ layers during gastrulation—the ectoderm, mesoderm, and endoderm—constitutes one of the first and primary stages of the *in vivo* embryogenic process. This process is to some extent recapitulated during the *in vitro* human ES cell differentiation.[80] Many culture protocols have been developed to induce generation and propagation of specific cell types from human ES cells. Studies of directed human ES cell differentiation largely rely on developmental biol-ogy disciplines, studies of mouse ES cells, and also from empirical work. This sec-tion summarizes a number of examples describing human ES differentiation into each of the germ layer lineages.

1.3.6.1 Differentiation into Mesoderm: Endothelial Cells

Protocols to direct human ES cell differentiation toward mesoderm, which gives rise to blood, hematopoietic, endothelial cells, cardiomyocytes, and other cell types, have been established. Herein, we have chosen to describe the derivation of endothelial cells (ECs) from human ES cells as this lineage will be discussed later with regard to its therapeutic potential. Two main strategies have been proposed to differentiate human ES cells into endothelial cells or their progenitors: spontaneous differentiation through EBs[64,89–94] and directed differentiation of 2D culture.[74,75,95–97] Both methods carried out isolation of human ES-derived ECs via fluorescence-activated cell sorting (FACS) or magnetic cell sorting (MACS), followed by both *in vitro* and *in vivo* characterization. Levenberg et al. successfully isolated an endothelial population from spontaneously differentiated 13-day-old EBs by CD31-targeted FACS.[64] The isolated ECs demonstrated endothelial characteristics, such as expression of CD31, CD34, fetal liver kinase 1 (Flk1), vascular endothelial cadherin (VE-cad), as well as the mature endothelial Von Willebrand factor (vWF) marker. Moreover, the ability to incorporate Dil-Ac-LDL and to form blood vessel-like structures were observed. Other groups reported similar EB-derived EC isolation, although ECs generated from 9- to 10-day-old EBs did not yet express detectable levels of mature endothelial proteins, such as vWF and eNOS, and were therefore termed "primitive endothelial-like cells."[90] Further differentiation of these endothelial-like precursors resulted in the rise of either mature endothelial cells or hematopoietic cells, depending on culture conditions. Stimulation of human EBs by specific growth factors led to enhanced endothelial precursor differentiation.[92,98] Hemangioblast precursor cells were identified from EBs stimulated with BMP4, VEGF, and bFGF and were found to express KDR and CD117, but neither CD31 nor CD34.[98] Goldman et al. reported that a CD144(+)/KDR(+) population derived from BMP4-stimulated EBs represented an immature endothelial cell population that could further differentiate in culture to yield mature endothelial cells.[92]

The low efficiency (~2%) of EB-based endothelial cell isolation[64,91] represents a significant disadvantage of this method. The harsh digestion step applied when generating single cells out of EBs is believed to lead to a substantial reduction in cell viability. 2D models have been suggested in an attempt to improve EC differentiation yields. Various groups have demonstrated efficient differentiation of human ES cells into ECs from cultures using feeder layers or ECM component-coated plates. Growth of human ES cells on collagen IV in conjunction with VEGF treatment resulted in endothelial differentiation determined by the expression of endothelial progenitor markers such as CD31, CD34, AC133, Tie2, and GATA3.[95] Kaufman et al. cocultured human ES cells with murine yolk sac ECs or bone marrow stromal cells and reported isolation of CD34-expressing endothelial and hematopoietic precursors, where approximately 50% coexpressed CD31.[74] Wang et al. employed MEF feeder cells in 10 day cocultures with human ES cells to induce human ES differentiation toward hematopoietic precursors. Between 5% and 10% of CD34+ progenitor cells were generated, presenting a potential to differentiate into both hematopoietic and endothelial cells.[75]

However, other groups tackled the drawbacks of EB-based EC isolation by combining the 3D-EB system with other 2D continuing differentiation procedures.[89,99–101] A two-step differentiation technique was designed to generate blast precursor cells from human ES cells.[89,99] EBs were first generated from human ES cells and cultured for 3.5 days while being supplied with a mixture of cytokines, including BMP4, VEGF, SCF, Tpo, and Flt3L. Single cells from dissociated EBs were then cultured in semisolid culture medium containing methylcellulose for an additional 4–6 days. This method efficiently produced hemangioblast progenitors from human ES cells

$(400 \times 10^6$ blast cells derived from approximately 12×10^6 MA01 human ES cell line). An additional different two-step method combining EBs and 2D culture settings was established.[100,101] The technique is based on findings describing a rich endothelial-like EB core in the center region of attached EB in comparison to the outgrowth region. Therefore, the central regions of attached human EBs were either mechanically[100] or enzymatically[101] isolated after a 5–9 day incubation. Subsequent FACS analysis of the isolated cells defined more than 50% as vWF positive and the expression of other endothelial-specific markers as CD31, Flk-1, and Tie2.

1.3.6.2 Differentiation into Endoderm: Pancreatic Cell Differentiation

The first derivation of human ES cells in 1998[3] introduced the possibility of using human ES- derived insulin-producing cells as a potential cell source for the therapy of diabetes. The various studies aimed at differentiating human ES cells into insulin-producing cells using different approaches, mostly involving the induction of growth factors and growth conditions.

Spontaneous differentiation of human ES cells into insulin-producing cells was first described by Assady et al. using either gelatin-based adherent or EB-based suspension culture conditions. Differentiated cells featured characteristics typical of insulin-producing beta cells and maintained the capacity to express low levels of insulin even after 30 days in culture.[102] Similarly, Segev et al.[103] presented a method using human ES cell–derived EBs for the formation of islet-like clusters of insulin-producing cells. Their protocol included several stages, starting with EB culturing in insulin-transferrin-selenium-fibronectin-enriched medium, followed by bFGF supplementation and addition of nicotinamide. However, the described methods remained inadequate in their differentiation efficiency, secreted insulin levels, and in the difficulty to determine whether the detected insulin originated from the media or from de novo synthesis by insulin-producing cells.[104–106] These studies and others from murine ES cell–based models yielded ES-derived cells expressing endocrinic pancreatic lineage markers but failed to demonstrate mature pancreatic beta cell characteristics such as glucose regulation *in vitro* and correction of blood glucose levels *in vivo*.[102,103,107–109]

A substantial breakthrough was made while inducing direct differentiation into early developmental intermediate steps, thereby bypassing the EB stage. As the pancreas develops from the ectoderm layer, induction of a mesendoderm and definitive endoderm differentiation using activin A was the first step to be established.[110,111] Differentiation of human ES cells in the presence of activin A and low-serum concentrations for 4–5 days produced cultures highly enriched for definitive endoderm.[111] The same group later developed a multistep protocol recapitulating pancreatic organogenesis *in vivo*, in which human ES cells underwent a series of intermediate stages through mesendoderm, definitive endoderm, primitive gut tube, posterior foregut, pancreatic endoderm, and, finally, insulin-producing endocrine cells. Each sequential stage was triggered using specific culture conditions, including activin A treatment in a serum-free medium, FGF, and RA signaling with inhibition of sonic hedgehog homolog (SHH) signaling, notch inhibitor, and a cocktail of other growth factors. Seven percent of the resulting insulin-producing cells also expressed high levels of proinsulin but failed to secret insulin in response to heightened glucose levels.[112,113] Despite the progress described for multistep differentiation protocols, the resulting insulin-producing cells still failed in glucose-sensitivity assays,[112–115] although some reports describe a partial correction of hyperglycemia *in vivo*.[116,117]

Remarkable progress was achieved by the Novocell Inc. research team demonstrating production of functional insulin-producing cells capable of correcting hyperglycemia *in vivo*.[118] Their reports describe generation of ES cell–derived pancreatic endocrine precursor cells via a four-stage protocol involving treatment with activin A, Wnt3a, KGF, RA, noggin, and cyclopamine. The resultant endocrine precursor cells were implanted into immunodeficient mice and allowed to further differentiate *in vivo*. After several months, the implanted cells function was examined in their capacity to protect hyperglycemia induced animals. Through effective regulation of blood glucose levels, the transplanted cells corrected diabetic phenotypes and exhibited morphological and functional similarity to natural pancreatic islets, including insulin and c-peptide expression. Further confirmation of the role of human ES-derived insulin-producing cells in preventing diabetes was demonstrated upon their removal, which led to immediate hyperglycemia. This pioneering work indicates the vast potential borne by ES-derived cells for the advancement of clinical regenerative therapy.[118]

Melton and colleagues recently introduced a model exploiting specific small molecules for prompting endodermal lineages via directed induction of human ES cell differentiation.[119,120] The described system led to formation of definitive endoderm by 80% of the induced ES cells.[119] Expression of pancreatic markers including pancreatic and duodenal homeobox 1 (Pdx-1) was elevated upon exposure of human ES cell to (-)-indolactam V.[120] Yet, further studies will be required to confirm the resemblance of chemically induced endoderm cells to pancreatic beta cells.

1.3.6.3 Differentiation into Ectoderm: Neural Differentiation

Due to its great potential and value for treatment of neurodegenerative disorders such as Parkinson's disease, Alzheimer's disease, and spinal cord injuries, extensive work has been dedicated to study of neuronal differentiation from human ES cells.

Human ES cells can differentiate in culture to form neural cells including functional neurons, glial cells, and oligodendrocytes.[121–124] Although various protocols describe such induction, development of an optimal protocol generating a homogenous neural progenitor population remains a major challenge.

To date, protocols for directing ES cell differentiation to neuroectoderm lineages rely on signaling pathways established as central regulators of neural cell fate in early embryogenetic stages. Such pathways include notch (reviewed in Refs. [125–127]), SHH,[128] Wnt,[129,130] and members of the FGF[131] and TGFβ superfamilies.[132] Suspension and adherent culture, in the form of EBs or monolayers, respectively,[133,134] are employed in the various reported protocols. To trigger proneural differentiation and improve survival of neural progenitors, growth factors or morphogenetic factors are often included in culture media. The use of noggin[135] and RA[135,136] have been described, as have alternation of ES cell culture conditions to include serum and serum-free culture,[137] or coculture of ES cells with specific stromal cell lines such as PA6.[138] Using Sox1-targeted fluorescently labeled cDNA,[49] 60% of the ES cells grown in monolayer cultures were shown to form neuroectoderm in serum-free cultures, which proved to be highly dependent on FGF signaling.[134] Addition of BMP4 to serum-free cultures shifted developmental trends from neuro-ectoderm development to induction of mesoderm.[133,134,139] Further evidence support-ing the reported neural inhibitory role of BMP was demonstrated by the elevated expression of noggin, an established BMP antagonist, in neurally differentiation ES cells.[135] Similarly, Pera et al. reported that noggin-induced inhibition of endog-enous BMP signaling leads to efficient differentiation of human ES cells into neural progenitors through the upregulation of neural transcription factors Pax6 and Sox2 and other neural markers such as Nestin.[52] These results correlate with findings in Xenopus, where BMP4 inhibitors—noggin and chordin—were shown to induce neu-ral development.[140,141] The Wnt signaling pathway also appears to play an inhibitory role in neuroectoderm development in ES cell cultures. Aubert et al demonstrated that inhibition of Wnt by the expression of the Wnt inhibitor Sfrp2 led to enhanced neural development in EBs, whereas overexpression of Wnt1 inhibited neural devel-opment.[142] The notch pathway is also a key player in controlling neural differentia-tion. Hitoshi et al. reported that differentiated ES cells could not be maintained in the neural progenitor state and were quickly lost to differentiation in the absence of notch signaling.[126] Lowell et al. reported that notch activation exclusively promoted neural differentiation, whereas notch inhibition blocked formation of neural pro-genitors. He also reported that promotion of neural progenitor formation via notch ligands is mediated by FGF receptor.[127] Taken together, these data implicate that notch signaling is necessary in establishment of neural progenitor cells by regulating cell survival and promoting expansion of the neural progenitors through blocking their differentiation.

1.4 THERAPEUTIC POTENTIAL OF HUMAN EMBRYONIC STEM CELLS

Human ES cells represent a theoretically unlimited source of precursor cells that can be differentiated into any cell type. This feature provides a basis for treatment of degenerative, malignant, or genetic diseases by repairing or replacing damaged tis-sues with those of artificial sources. Various lineages have been derived from human ES cells, including neurons,[143–145] cardiomyocytes,[146–148] smooth muscle cells, hema-topoietic cells,[74,98,149] osteogenic cells,[150] hepatocytes,[151] insulin-producing cells,[118] keratinocytes,[152] and EC.[64,75] In addition to their pluripotency, human ES cells bear low immunogenic potential,[153,154] which presents a significant advantage in cell-ther-apy applications. Challenges that remain to be addressed before full effectuation of human ES cells' potential in cell replacement therapies will be discussed in the following.

1.4.1 VASCULAR APPLICATIONS OF HUMAN EMBRYONIC STEM CELLS

The increasing interest in use of endothelial cells for therapeutic purposes has led to the development of methods to isolate endothelial progenitor cells from human ES cells, as described in Section 1.3.6.1. Human ES-derived ECs or endothelial pro-genitor cells (EPCs) can be applied in cell transplantation for ischemic tissue repair, engineering of artificial blood vessels and heart valves, repair of damaged vessels, and the formation of blood vessel networks in engineered tissues.[155]

One of the critical challenges in tissue engineering is providing the implanted tissue construct sufficient oxygen and nutrient transport for its cells to survive, which can be achieved by a vascular network. Vascularization of engineered tissues before transplantation has been shown to be essential to construction of complex and thick tissues *in vitro*. Vascularization *in vitro* enhances cell viability during tissue growth, enhances structural organization, and promotes integration upon implantation. Levenberg et al.[156] demonstrated the importance of vascularization in a 3D skeletal muscle tissue construct by coculturing myoblasts, embryonic fibroblasts, and endothelial cells on porous biodegradable scaffolds. Incorporation of human ES-derived endothelial cells allowed the construct to undergo neovascularization prior to *in vivo* implantation. This prevascularization was then demonstrated to be directly responsible for improved blood perfusion, survival, and integration of the engineered muscle construct after transplantation to SCID mice. A different multicellular scaffold construct demonstrating the necessity of neovascularized engineered tissue was demonstrated by Caspi et al.[157] This study described the formation of contracting engineered cardiac tissue construct containing endothelial vessel networks. *In vitro* tissue vascularization was promoted using the combination of human ES-derived ECs and embryonic fibroblasts together with the human ES-derived cardiomyocytes. The multicellular tissue engineered construct enabled the generation of highly vascularized engineered cardiac tissue with cardiac-specific ultrastructural, molecular, and functional properties. The presence of the fibroblasts found to decrease EC death and increased their proliferation. Recent report[158] confirmed the feasibility of *in vivo* implantation of the engineered prevascularized constructs showing the formation of viable perfused grafts in immunosuppressed rat hearts. A similar work was reported by Stevens et al.[159] using a scaffold-free prevascularized human heart tissue, which survived *in vivo* transplantation and integrated with the host coronary circulation. In this study, cardiomyocytes derived from human ES cells were placed into a rotating orbital shaker to create a scaffold-free cardiac tissue patch. Only patches containing cardiomyocytes together with endothelial cells (both hESC-derived endothelial cells and human umbilical vein) and fibroblasts could survive to form significant grafts after implantation *in vivo* and formed anastomoses with the rat host coronary circulation. These studies demonstrate the importance of prevascularization by including endothelial and stromal elements in engineered tissues.

Animal experiments describing transplantation of human ES cell–derived ECs or their progenitors have shown encouraging results in repair of ischemic tissues and restoration of blood flow.[89,96,100] Wang et al.[75] demonstrated that human ES cell–derived ECs transplanted into SCID mice contributed to efficient blood vessel formation upon their integration into the host vascular networks, where they served as functional blood vessels for 150 days. Nakao's group[96,160] examined the potential of human ES-derived vascular cells in promoting vascular regeneration in mouse ischemic models. As vascular smooth muscle cells and pericytes play an important role in supporting and stabilizing blood vessels, human ES-derived vascular progenitor cells (VPCs) were induced to differentiate into both VE-cadherin+ endothelial cells and alpha smooth muscle actin+ mural cells (MCs) by addition of VEGF or PDGF-B, respectively.

The VE-cadherin+ cells, which were also confirmed to be CD34+ and VEGF-R2+, were transplanted to the hindlimb of immunodeficient ischemic mouse model and were shown to contribute to both neovascularization and improvement of blood flow. Moreover, upon transplantation of both human ES-VPC-derived endothelial cells (ECs) and MCs, effective incorporation into host circulating vessels and maintenance of long-term vascular integrity were recorded. Furthermore, blood flow recovery and capillary density were significantly improved when compared to other sources of human endothelial progenitor cells or to transplantation of ECs alone. In a more recent study,[161] the vascular potential of human ES-derived vascular cells was examined in a mouse ischemic brain model. In this model, the transplanted ECs were also successfully incorporated into the host vessels and MCs were detected surrounding the endothelial tubes. At 28 days post-occlusion, the cerebral blood flow and the vascular density were significantly improved and a major neurological recovery was observed in the treated mice. This effective combined transplantation of human ES-derived ECs and MCs may present a novel strategy for potential vascular applications.

1.4.2 CHALLENGES IN USING HUMAN ES CELLS FOR CLINICAL APPLICATIONS

Although human ES cells bear tremendous potential for advancing clinical medicine, there remain several challenges yet to be surmounted before effectively curing human diseases. A comprehensive review describing these challenges has been recently published by Murry and Keller.[81] A major question to be considered is with reference to the pretransplantation differentiated state of human ES cells. As undifferentiated cells can form teratomas *in vivo*, samples must be cleared of all undifferentiated cells to eliminate any tumor-forming potential. However, when considering the less proliferative character of differentiated cells, progenitor cell populations are often preferred for transplantation purposes. In addition, progenitor populations may lead to a more effective differentiation in the *in vivo* environment rather than any *in vitro* environment, resulting in the rise of the most physiologically appropriate cell type or in some cases to few related cell types having the same precursors. This may present an advantage when a number of related lineages are required, as in cases of repair of spinal cord injury.[162,163] However, progenitor cells still pose tumorigenic threats as progenitors still retain proliferative potential *in vivo*, as observed upon transplantation of neuronal progenitors in the rat brain.[164]

Aside from safety concerns, purification of the desired differentiated cells is necessary to assure that no undesired cell type appear in an inappropriate location. Purification of differentiated cells can be performed by microdissection of cellular aggregates, density-gradient centrifugation, FACS or MACS sorting, or via genetic selection. While microdissection is the less efficient method for purification, FACS and genetic selections yield highly homogeneous populations.

Immune rejection presents a major concern when considering transplantation of human ES-derived cells. While human ES cells express weak immunogenecity due to low levels of class I major histocompatibility antigens and absence of class II–binding molecules,[165,166] they can induce an immune response when transplanted, exhibit increased expression of histocompatibility antigens, and, thus, may be rejected in an allogeneic setting.[166] Several approaches have been suggested to address this obstacle,[167,168] including immunosuppression, banking of human ES cell lines, developing a "universal donor cell," such as blood-type O cells or cells with suppressed HLA molecules, or engineering human ES cells that locally secrete immunosuppressive molecules such as soluble interleukin-1 receptor.[81] A promising strategy suggests to generate tolerogenic human ES cell–derived hematopoietic cells that will be transplanted before or with the therapeutically differentiated cells derived from the same human ES cell line, thus ensuring their acceptance.[169] Additional models attempt to overcome the immune responses by using patient-specific somatic cell nuclear transfer (SCNT)[170] or iPS cells,[5–7] which would be recognized as "self" by the immune system, thereby preventing cell transplant rejection.

In order to develop a clinically appropriate therapeutic product, human ES cells must conform to highly regulated standards. Firstly, they should be maintained in animal-free reagents, including culturing on human feeders or in feeder-free systems.[32,35,40] In addition, cell populations must be analyzed for viral or bacterial contaminations, including murine viruses or retroviruses. Genetic stability of human ES cells must be monitored and can include testing for karyotypic aberrations that can occur with passaging. Therefore, a careful investigation of the karyotypic, genetic, and epigenetic characteristics of the cells must be performed before transplantations in clinical settings.

1.5 CONCLUSIONS

As human ES cells feature the capacity to generate any cell type in the body, these cells bear tremendous potential in regenerative therapy. Although recent years have witnessed progress in induced differentiation and isolation of many cell lineages from human ES cells, a number of significant challenges remain as obstacles to effective cure of human diseases. However, as this discipline attracts global interest, in efforts to advance the understanding of the basic biology and characteristics of these cells, much hope remains for the clinical future of human ES cell applications. Such insight will enhance the full exploitive potential of ES cells and will accelerate their shift from the bench to the clinic. The encouraging results reported in a gamut of animal vascular models plant further hope that this vision of utilizing stem cell potential in the clinic is approaching reality.

ACKNOWLEDGMENT

The authors would like to acknowledge the financial support of the Marie-Curie Reintegration Grant.

REFERENCES

1. Evans, M. J.; Kaufman, M. H., Establishment in culture of pluripotential cells from mouse embryos. *Nature* 1981, 292 (5819), 154–156.
2. Martin, G. R., Isolation of a pluripotent cell line from early mouse embryos cultured in medium conditioned by teratocarcinoma stem cells. *Proc Natl Acad Sci USA* 1981, 78 (12), 7634–7638.

3. Thomson, J. A.; Itskovitz-Eldor, J.; Shapiro, S. S.; Waknitz, M. A.; Swiergiel, J. J.; Marshall, V. S.; Jones, J. M., Embryonic stem cell lines derived from human blastocysts. *Science* 1998, *282* (5391), 1145–1147.

4. Zandstra, P. W.; Nagy, A., Stem cell bioengineering. *Annu Rev Biomed Eng* 2001, *3*, 275–305.

5. Maherali, N.; Sridharan, R.; Xie, W.; Utikal, J.; Eminli, S.; Arnold, K.; Stadtfeld, M.; Yachechko, R.; Tchieu, J.; Jaenisch, R.; Plath, K.; Hochedlinger, K., Directly reprogrammed fibroblasts show global epigenetic remodeling and widespread tissue contribution. *Cell Stem Cell* 2007, *1* (1), 55–70.

6. Okita, K.; Ichisaka, T.; Yamanaka, S., Generation of germline-competent induced pluripotent stem cells. *Nature* 2007, *448* (7151), 313–317.

7. Wernig, M.; Meissner, A.; Foreman, R.; Brambrink, T.; Ku, M.; Hochedlinger, K.; Bernstein, B. E.; Jaenisch, R., In vitro reprogramming of fibroblasts into a pluripotent ES-cell-like state. *Nature* 2007, *448* (7151), 318–324.

8. Doetschman, T. C.; Eistetter, H.; Katz, M.; Schmidt, W.; Kemler, R., The in vitro development of blastocyst-derived embryonic stem cell lines: Formation of visceral yolk sac, blood islands and myocardium. *J Embryol Exp Morphol* 1985, *87*, 27–45.

9. Guan, K.; Rohwedel, J.; Wobus, A. M., Embryonic stem cell differentiation models: Cardiogenesis, myogenesis, neurogenesis, epithelial and vascular smooth muscle cell differentiation in vitro. *Cytotechnology* 1999, *30* (1–3), 211–226.

10. Kaufman, M. H.; Robertson, E. J.; Handyside, A. H.; Evans, M. J., Establishment of pluripotential cell lines from haploid mouse embryos. *J Embryol Exp Morphol* 1983, *73*, 249–261.

11. Beddington, R. S.; Robertson, E. J., An assessment of the developmental potential of embryonic stem cells in the midgestation mouse embryo. *Development* 1989, *105* (4), 733–737.

12. Wood, S. A.; Allen, N. D.; Rossant, J.; Auerbach, A.; Nagy, A., Non-injection methods for the production of embryonic stem cell-embryo chimaeras. *Nature* 1993, *365* (6441), 87–89.

13. Bhattacharya, B.; Miura, T.; Brandenberger, R.; Mejido, J.; Luo, Y.; Yang, A. X.; Joshi, B. H.; Ginis, I.; Thies, R. S.; Amit, M.; Lyons, I.; Condie, B. G.; Itskovitz-Eldor, J.; Rao, M. S.; Puri, R. K., Gene expression in human embryonic stem cell lines: Unique molecular signature. *Blood* 2004, *103* (8), 2956–2964.

14. Carpenter, M. K.; Rosler, E. S.; Fisk, G. J.; Brandenberger, R.; Ares, X.; Miura, T.; Lucero, M.; Rao, M. S., Properties of four human embryonic stem cell lines maintained in a feeder-free culture system. *Dev Dyn* 2004, *229* (2), 243–258.

15. Mitsui, K.; Tokuzawa, Y.; Itoh, H.; Segawa, K.; Murakami, M.; Takahashi, K.; Maruyama, M.; Maeda, M.; Yamanaka, S., The homeoprotein Nanog is required for maintenance of pluripotency in mouse epiblast and ES cells. *Cell* 2003, *113* (5), 631–642.

16. Chapman, G.; Remiszewski, J. L.; Webb, G. C.; Schulz, T. C.; Bottema, C. D.; Rathjen, P. D., The mouse homeobox gene, Gbx2: Genomic organization and expression in pluripotent cells in vitro and in vivo. *Genomics* 1997, *46* (2), 223–233.

17. Niwa, H.; Miyazaki, J.; Smith, A. G., Quantitative expression of Oct-3/4 defines differentiation, dedifferentiation or self-renewal of ES cells. *Nat Genet* 2000, *24* (4), 372–376.

18. Pelton, T. A.; Sharma, S.; Schulz, T. C.; Rathjen, J.; Rathjen, P. D., Transient pluripotent cell populations during primitive ectoderm formation: Correlation of in vivo and in vitro pluripotent cell development. *J Cell Sci* 2002, *115* (Pt 2), 329–339.

19. Tomioka, M.; Nishimoto, M.; Miyagi, S.; Katayanagi, T.; Fukui, N.; Niwa, H.; Muramatsu, M.; Okuda, A., Identification of Sox-2 regulatory region which is under the control of Oct-3/4-Sox-2 complex. *Nucleic Acids Res* 2002, *30* (14), 3202–3213.

20. Anisimov, S. V.; Tarasov, K. V.; Tweedie, D.; Stern, M. D.; Wobus, A. M.; Boheler, K. R., SAGE identification of gene transcripts with profiles unique to pluripotent mouse R1 embryonic stem cells. *Genomics* 2002, *79* (2), 169–176.

21. Ivanova, N. B.; Dimos, J. T.; Schaniel, C.; Hackney, J. A.; Moore, K. A.; Lemischka, I. R., A stem cell molecular signature. *Science* 2002, *298* (5593), 601–604.

22. Ramalho-Santos, M.; Yoon, S.; Matsuzaki, Y.; Mulligan, R. C.; Melton, D. A., "Stemness": Transcriptional profiling of embryonic and adult stem cells. *Science* 2002, *298* (5593), 597–600.

23. Tanaka, T. S.; Kunath, T.; Kimber, W. L.; Jaradat, S. A.; Stagg, C. A.; Usuda, M.; Yokota, T.; Niwa, H.; Rossant, J.; Ko, M. S., Gene expression profiling of embryo-derived stem cells reveals candidate genes associated with pluripotency and lineage specificity. *Genome Res* 2002, *12* (12), 1921–1928.

24. Thomson, J. A.; Kalishman, J.; Golos, T. G.; Durning, M.; Harris, C. P.; Becker, R. A.; Hearn, J. P., Isolation of a primate embryonic stem cell line. *Proc Natl Acad Sci USA* 1995, *92* (17), 7844–7848.

25. Draper, J. S.; Smith, K.; Gokhale, P.; Moore, H. D.; Maltby, E.; Johnson, J.; Meisner, L.; Zwaka, T. P.; Thomson, J. A.; Andrews, P. W., Recurrent gain of chromosomes 17q and 12 in cultured human embryonic stem cells. *Nat Biotechnol* 2004, *22* (1), 53–54.

26. Smith, A. G., Embryo-derived stem cells: Of mice and men. *Annu Rev Cell Dev Biol* 2001, *17*, 435–462.

27. Martin, M. J.; Muotri, A.; Gage, F.; Varki, A., Human embryonic stem cells express an immunogenic nonhuman sialic acid. *Nat Med* 2005, *11* (2), 228–232.

28. Amit, M.; Margulets, V.; Segev, H.; Shariki, K.; Laevsky, I.; Coleman, R.; Itskovitz-Eldor, J., Human feeder layers for human embryonic stem cells. *Biol Reprod* 2003, *68* (6), 2150–2156.

29. Cheng, L.; Hammond, H.; Ye, Z.; Zhan, X.; Dravid, G., Human adult marrow cells support prolonged expansion of human embryonic stem cells in culture. *Stem Cells* 2003, *21* (2), 131–142.

30. Hovatta, O.; Mikkola, M.; Gertow, K.; Stromberg, A. M.; Inzunza, J.; Hreinsson, J.; Rozell, B.; Blennow, E.; Andang, M.; Ahrlund-Richter, L., A culture system using human foreskin fibroblasts as feeder cells allows production of human embryonic stem cells. *Hum Reprod* 2003, *18* (7), 1404–1409.

31. Richards, M.; Fong, C. Y.; Chan, W. K.; Wong, P. C.; Bongso, A., Human feeders support prolonged undifferentiated growth of human inner cell masses and embryonic stem cells. *Nat Biotechnol* 2002, *20* (9), 933–936.

32. Genbacev, O.; Krtolica, A.; Zdravkovic, T.; Brunette, E.; Powell, S.; Nath, A.; Caceres, E.; McMaster, M.; McDonagh, S.; Li, Y.; Mandalam, R.; Lebkowski, J.; Fisher, S. J., Serum-free derivation of human embryonic stem cell lines on human placental fibroblast feeders. *Fertil Steril* 2005, *83* (5), 1517–1529.

33. Xu, C.; Inokuma, M. S.; Denham, J.; Golds, K.; Kundu, P.; Gold, J. D.; Carpenter, M. K., Feeder-free growth of undifferentiated human embryonic stem cells. *Nat Biotechnol* 2001, *19* (10), 971–974.

34. Amit, M.; Shariki, C.; Margulets, V.; Itskovitz-Eldor, J., Feeder layer- and serum-free culture of human embryonic stem cells. *Biol Reprod* 2004, *70* (3), 837–845.

35. Klimanskaya, I.; Chung, Y.; Meisner, L.; Johnson, J.; West, M. D.; Lanza, R., Human embryonic stem cells derived without feeder cells. *Lancet* 2005, *365* (9471), 1636–1641.

36. Levenstein, M. E.; Ludwig, T. E.; Xu, R. H.; Llanas, R. A.; VanDenHeuvel- Kramer, K.; Manning, D.; Thomson, J. A., Basic fibroblast growth factor support of human embryonic stem cell self-renewal. *Stem Cells* 2006, *24* (3), 568–574.

37. Wang, G.; Zhang, H.; Zhao, Y.; Li, J.; Cai, J.; Wang, P.; Meng, S.; Feng, J.; Miao, C.; Ding, M.; Li, D.; Deng, H., Noggin and bFGF cooperate to maintain the pluripotency of human embryonic stem cells in the absence of feeder layers. *Biochem Biophys Res Commun* 2005, *330* (3), 934–942.

38. Xu, C.; Rosler, E.; Jiang, J.; Lebkowski, J. S.; Gold, J. D.; O'Sullivan, C.; Delavan-Boorsma, K.; Mok, M.; Bronstein, A.; Carpenter, M. K., Basic fibroblast growth factor supports undifferentiated human embryonic stem cell growth without conditioned medium. *Stem Cells* 2005, *23* (3), 315–323.

39. Amit, M.; Itskovitz-Eldor, J., Sources, derivation, and culture of human embryonic stem cells. *Semin Reprod Med* 2006, *24* (5), 298–303.

40. Ludwig, T. E.; Levenstein, M. E.; Jones, J. M.; Berggren, W. T.; Mitchen, E. R.; Frane, J. L.; Crandall, L. J.; Daigh, C. A.; Conard, K. R.; Piekarczyk, M. S.; Llanas, R. A.; Thomson, J. A., Derivation of human embryonic stem cells in defined conditions. *Nat Biotechnol* 2006, *24* (2), 185–187.

41. Yao, S.; Chen, S.; Clark, J.; Hao, E.; Beattie, G. M.; Hayek, A.; Ding, S., Long-term self-renewal and directed differentiation of human embryonic stem cells in chemically defined conditions. *Proc Natl Acad Sci USA* 2006, *103* (18), 6907–6912.

42. Amit, M.; Carpenter, M. K.; Inokuma, M. S.; Chiu, C. P.; Harris, C. P.; Waknitz, M. A.; Itskovitz-Eldor, J.; Thomson, J. A., Clonally derived human embryonic stem cell lines maintain pluripotency and proliferative potential for prolonged periods of culture. *Dev Biol* 2000, *227* (2), 271–278.

43. Thomson, J. A.; Marshall, V. S., Primate embryonic stem cells. *Curr Top Dev Biol* 1998, *38*, 133–165.

44. Brandenberger, R.; Wei, H.; Zhang, S.; Lei, S.; Murage, J.; Fisk, G. J.; Li, Y.; Xu, C.; Fang, R.; Guegler, K.; Rao, M. S.; Mandalam, R.; Lebkowski, J.; Stanton, L. W., Transcriptome characterization elucidates signaling networks that control human ES cell growth and differentiation. *Nat Biotechnol* 2004, *22* (6), 707–716.

45. Kim, S. K.; Suh, M. R.; Yoon, H. S.; Lee, J. B.; Oh, S. K.; Moon, S. Y.; Moon, S. H.; Lee, J. Y.; Hwang, J. H.; Cho, W. J.; Kim, K. S., Identification of developmental pluripotency associated 5 expression in human pluripotent stem cells. *Stem Cells* 2005, *23* (4), 458–462.

46. Reubinoff, B. E.; Pera, M. F.; Fong, C. Y.; Trounson, A.; Bongso, A., Embryonic stem cell lines from human blastocysts: Somatic differentiation in vitro. *Nat Biotechnol* 2000, *18* (4), 399–404.

47. Sato, N.; Meijer, L.; Skaltsounis, L.; Greengard, P.; Brivanlou, A. H., Maintenance of pluripotency in human and mouse embryonic stem cells through activation of Wnt signaling by a pharmacological GSK-3-specific inhibitor. *Nat Med* 2004, *10* (1), 55–63.

48. Wei, C. L.; Miura, T.; Robson, P.; Lim, S. K.; Xu, X. Q.; Lee, M. Y.; Gupta, S.; Stanton, L.; Luo, Y.; Schmitt, J.; Thies, S.; Wang, W.; Khrebtukova, I.; Zhou, D.; Liu, E. T.; Ruan, Y. J.; Rao, M.; Lim, B., Transcriptome profiling of human and murine ESCs identifies divergent paths required to maintain the stem cell state. *Stem Cells* 2005, *23* (2), 166–185.

49. Ying, Q. L.; Nichols, J.; Chambers, I.; Smith, A., BMP induction of Id proteins suppresses differentiation and sustains embryonic stem cell self-renewal in collaboration with STAT3. *Cell* 2003, *115* (3), 281–292.

50. Sato, N.; Sanjuan, I. M.; Heke, M.; Uchida, M.; Naef, F.; Brivanlou, A. H., Molecular signature of human embryonic stem cells and its comparison with the mouse. *Dev Biol* 2003, *260* (2), 404–413.

51. Besser, D., Expression of nodal, lefty-A, and lefty-B in undifferentiated human embryonic stem cells requires activation of Smad2/3. *J Biol Chem* 2004, *279* (43), 45076–45084.

52. Pera, M. F.; Andrade, J.; Houssami, S.; Reubinoff, B.; Trounson, A.; Stanley, E. G.; Ward-van Oostwaard, D.; Mummery, C., Regulation of human embryonic stem cell differentiation by BMP-2 and its antagonist noggin. *J Cell Sci* 2004, *117* (Pt 7), 1269–1280.

53. James, D.; Levine, A. J.; Besser, D.; Hemmati-Brivanlou, A., TGFbeta/activin/nodal signaling is necessary for the maintenance of pluripotency in human embryonic stem cells. *Development* 2005, *132* (6), 1273–1282.

54. Cadigan, K. M.; Nusse, R., Wnt signaling: A common theme in animal development. *Genes Dev* 1997, *11* (24), 3286–3305.

55. Walsh, J.; Andrews, P. W., Expression of Wnt and Notch pathway genes in a pluripotent human embryonal carcinoma cell line and embryonic stem cell. *Apmis* 2003, *111* (1), 197–210; discussion 210–211.

56. Dravid, G.; Ye, Z.; Hammond, H.; Chen, G.; Pyle, A.; Donovan, P.; Yu, X.; Cheng, L., Defining the role of Wnt/beta-catenin signaling in the survival, proliferation, and self-renewal of human embryonic stem cells. *Stem Cells* 2005, *23* (10), 1489–1501.

57. Desbaillets, I.; Ziegler, U.; Groscurth, P.; Gassmann, M., Embryoid bodies: An in vitro model of mouse embryogenesis. *Exp Physiol* 2000, *85* (6), 645–651.

58. Itskovitz-Eldor, J.; Schuldiner, M.; Karsenti, D.; Eden, A.; Yanuka, O.; Amit, M.; Soreq, H.; Benvenisty, N., Differentiation of human embryonic stem cells into embryoid bodies compromising the three embryonic germ layers. *Mol Med* 2000, *6* (2), 88–95.

59. Chadwick, K.; Wang, L.; Li, L.; Menendez, P.; Murdoch, B.; Rouleau, A.; Bhatia, M., Cytokines and BMP-4 promote hematopoietic differentiation of human embryonic stem cells. *Blood* 2003, *102* (3), 906–915.

60. Liu, H.; Lin, J.; Roy, K., Effect of 3D scaffold and dynamic culture condition on the global gene expression profile of mouse embryonic stem cells. *Biomaterials* 2006, *27* (36), 5978–5989.

61. Liu, H.; Roy, K., Biomimetic three-dimensional cultures significantly increase hematopoietic differentiation efficacy of embryonic stem cells. *Tissue Eng* 2005, *11* (1–2), 319–330.

62. Mountford, J. C., Human embryonic stem cells: Origins, characteristics and potential for regenerative therapy. *Transfus Med* 2008, *18* (1), 1–12.

63. Bhatia, M., Hematopoiesis from human embryonic stem cells. *Ann NY Acad Sci* 2007, *1106*, 219–222.

64. Levenberg, S.; Golub, J. S.; Amit, M.; Itskovitz-Eldor, J.; Langer, R., Endothelial cells derived from human embryonic stem cells. *Proc Natl Acad Sci USA* 2002, *99* (7), 4391–4396.

65. Weitzer, G., Embryonic stem cell-derived embryoid bodies: An in vitro model of eutherian pregastrulation development and early gastrulation. *Handb Exp Pharmacol* 2006, (174), 21–51.

66. Wartenberg, M.; Gunther, J.; Hescheler, J.; Sauer, H., The embryoid body as a novel in vitro assay system for antiangiogenic agents. *Lab Invest* 1998, *78* (10), 1301–1314.

67. Karp, J. M.; Yeh, J.; Eng, G.; Fukuda, J.; Blumling, J.; Suh, K. Y.; Cheng, J.; Mahdavi, A.; Borenstein, J.; Langer, R.; Khademhosseini, A., Controlling size, shape and homogeneity of embryoid bodies using poly(ethylene glycol) microwells. *Lab Chip* 2007, *7* (6), 786–794.

68. Koike, M.; Kurosawa, H.; Amano, Y., A Round-bottom 96-well Polystyrene Plate Coated with 2-methacryloyloxyethyl Phosphorylcholine as an effective tool for embryoid body formation. *Cytotechnology* 2005, *47* (1–3), 3–10.

69. Kurosawa, H.; Imamura, T.; Koike, M.; Sasaki, K.; Amano, Y., A simple method for forming embryoid body from mouse embryonic stem cells. *J Biosci Bioeng* 2003, *96* (4), 409–411.

70. Ungrin, M. D.; Joshi, C.; Nica, A.; Bauwens, C.; Zandstra, P. W., Reproducible, ultra high-throughput formation of multicellular organization from single cell suspension-derived human embryonic stem cell aggregates. *PLoS One* 2008, *3* (2), e1565.
71. Hanjaya-Putra, D.; Gerecht, S., Vascular engineering using human embryonic stem cells. *Biotechnol Prog* 2009, *25* (1), 2–9.
72. Zoldan, J.; Levenberg, S., Engineering three-dimensional tissue structures using stem cells. *Methods Enzymol* 2006, *420*, 381–391.
73. Levenberg, S.; Huang, N. F.; Lavik, E.; Rogers, A. B.; Itskovitz-Eldor, J.; Langer, R., Differentiation of human embryonic stem cells on three-dimensional polymer scaffolds. *Proc Natl Acad Sci USA* 2003, *100* (22), 12741–12746.
74. Kaufman, D. S.; Hanson, E. T.; Lewis, R. L.; Auerbach, R.; Thomson, J. A., Hematopoietic colony-forming cells derived from human embryonic stem cells. *Proc Natl Acad Sci USA* 2001, *98* (19), 10716–10721.
75. Wang, Z. Z.; Au, P.; Chen, T.; Shao, Y.; Daheron, L. M.; Bai, H.; Arzigian, M.; Fukumura, D.; Jain, R. K.; Scadden, D. T., Endothelial cells derived from human embryonic stem cells form durable blood vessels in vivo. *Nat Biotechnol* 2007, *25* (3), 317–318.
76. Hwang, N. S.; Kim, M. S.; Sampattavanich, S.; Baek, J. H.; Zhang, Z.; Elisseeff, J., Effects of three-dimensional culture and growth factors on the chondrogenic differentiation of murine embryonic stem cells. *Stem Cells* 2006, *24* (2), 284–291.
77. Tanaka, H.; Murphy, C. L.; Murphy, C.; Kimura, M.; Kawai, S.; Polak, J. M., Chondrogenic differentiation of murine Embryonic stem cells: Effects of culture conditions and dexamethasone. *J Cell Biochem* 2004, *93* (3), 454–462.
78. Tian, X.; Morris, J. K.; Linehan, J. L.; Kaufman, D. S., Cytokine requirements differ for stroma and embryoid body-mediated hematopoiesis from human embryonic stem cells. *Exp Hematol* 2004, *32* (10), 1000–1009.
79. Schuldiner, M.; Yanuka, O.; Itskovitz-Eldor, J.; Melton, D. A.; Benvenisty, N., Effects of eight growth factors on the differentiation of cells derived from human embryonic stem cells. *Proc Natl Acad Sci USA* 2000, *97* (21), 11307–11312.
80. Dvash, T.; Mayshar, Y.; Darr, H.; McElhaney, M.; Barker, D.; Yanuka, O.; Kotkow, K. J.; Rubin, L. L.; Benvenisty, N.; Eiges, R., Temporal gene expression during differentiation of human embryonic stem cells and embryoid bodies. *Hum Reprod* 2004, *19* (12), 2875–2883.
81. Murry, C. E.; Keller, G., Differentiation of embryonic stem cells to clinically relevant populations: Lessons from embryonic development. *Cell* 2008, *132* (4), 661–680.
82. Ding, S.; Schultz, P. G., A role for chemistry in stem cell biology. *Nat Biotechnol* 2004, *22* (7), 833–840.
83. Ding, S.; Schultz, P. G., Small molecules and future regenerative medicine. *Curr Top Med Chem* 2005, *5* (4), 383–395.
84. Kawaguchi, J.; Mee, P. J.; Smith, A. G., Osteogenic and chondrogenic differentiation of embryonic stem cells in response to specific growth factors. *Bone* 2005, *36* (5), 758–769.
85. Lee, G. S.; Kochhar, D. M.; Collins, M. D., Retinoid-induced limb malformations. *Curr Pharm Des* 2004, *10* (22), 2657–2699.
86. Sharma, N. S.; Shikhanovich, R.; Schloss, R.; Yarmush, M. L., Sodium butyrate-treated embryonic stem cells yield hepatocyte-like cells expressing a glycolytic phenotype. *Biotechnol Bioeng* 2006, *94* (6), 1053–1063.
87. Siebler, T.; Robson, H.; Shalet, S. M.; Williams, G. R., Dexamethasone inhibits and thyroid hormone promotes differentiation of mouse chondrogenic ATDC5 cells. *Bone* 2002, *31* (4), 457–464.
88. Wakita, R.; Izumi, T.; Itoman, M., Thyroid hormone-induced chondrocyte terminal differentiation in rat femur organ culture. *Cell Tissue Res* 1998, *293* (2), 357–364.

89. Lu, S. J.; Feng, Q.; Caballero, S.; Chen, Y.; Moore, M. A.; Grant, M. B.; Lanza, R., Generation of functional hemangioblasts from human embryonic stem cells. *Nat Methods* 2007, *4* (6), 501–509.

90. Wang, L.; Li, L.; Shojaei, F.; Levac, K.; Cerdan, C.; Menendez, P.; Martin, T.; Rouleau, A.; Bhatia, M., Endothelial and hematopoietic cell fate of human embryonic stem cells originates from primitive endothelium with hemangioblastic properties. *Immunity* 2004, *21* (1), 31–41.

91. Li, Z.; Suzuki, Y.; Huang, M.; Cao, F.; Xie, X.; Connolly, A. J.; Yang, P. C.; Wu, J. C., Comparison of reporter gene and iron particle labeling for tracking fate of human embryonic stem cells and differentiated endothelial cells in living subjects. *Stem Cells* 2008, *26* (4), 864–873.

92. Goldman, O.; Feraud, O.; Boyer-Di Ponio, J.; Driancourt, C.; Clay, D.; Le Bousse-Kerdiles, M. C.; Bennaceur-Griscelli, A.; Uzan, G., A boost of BMP4 accelerates the commitment of human embryonic stem cells into the endothelial lineage. *Stem Cells* 2009, *27* (8), 1750–1759.

93. Zambidis, E. T.; Peault, B.; Park, T. S.; Bunz, F.; Civin, C. I., Hematopoietic differentiation of human embryonic stem cells progresses through sequential hematoendothelial, primitive, and definitive stages resembling human yolk sac development. *Blood* 2005, *106* (3), 860–870.

94. Ferreira, L. S.; Gerecht, S.; Shieh, H. F.; Watson, N.; Rupnick, M. A.; Dallabrida, S. M.; Vunjak-Novakovic, G.; Langer, R., Vascular progenitor cells isolated from human embryonic stem cells give rise to endothelial and smooth muscle like cells and form vascular networks in vivo. *Circ Res* 2007, *101* (3), 286–294.

95. Gerecht-Nir, S.; Ziskind, A.; Cohen, S.; Itskovitz-Eldor, J., Human embryonic stem cells as an in vitro model for human vascular development and the induction of vascular differentiation. *Lab Invest* 2003, *83* (12), 1811–1820.

96. Yamahara, K.; Sone, M.; Itoh, H.; Yamashita, J. K.; Yurugi-Kobayashi, T.; Homma, K.; Chao, T. H.; Miyashita, K.; Park, K.; Oyamada, N.; Sawada, N.; Taura, D.; Fukunaga, Y.; Tamura, N.; Nakao, K., Augmentation of neovascularization [corrected] in hindlimb ischemia by combined transplantation of human embryonic stem cells-derived endothelial and mural cells. *PLoS One* 2008, *3* (2), e1666.

97. Vodyanik, M. A.; Bork, J. A.; Thomson, J. A.; Slukvin, I. I., Human embryonic stem cell-derived CD34+ cells: Efficient production in the coculture with OP9 stromal cells and analysis of lymphohematopoietic potential. *Blood* 2005, *105* (2), 617–626.

98. Kennedy, M.; D'Souza, S. L.; Lynch-Kattman, M.; Schwantz, S.; Keller, G., Development of the hemangioblast defines the onset of hematopoiesis in human ES cell differentiation cultures. *Blood* 2007, *109* (7), 2679–2687.

99. Basak, G. W.; Yasukawa, S.; Alfaro, A.; Halligan, S.; Srivastava, A. S.; Min, W. P.; Minev, B.; Carrier, E., Human embryonic stem cells hemangioblast express HLA-antigens. *J Transl Med* 2009, *7*, 27.

100. Cho, S. W.; Moon, S. H.; Lee, S. H.; Kang, S. W.; Kim, J.; Lim, J. M.; Kim, H. S.; Kim, B. S.; Chung, H. M., Improvement of postnatal neovascularization by human embryonic stem cell derived endothelial-like cell transplantation in a mouse model of hindlimb ischemia. *Circulation* 2007, *116* (21), 2409–2419.

101. Kim, J.; Moon, S. H.; Lee, S. H.; Lee, D. R.; Koh, G. Y.; Chung, H. M., Effective isolation and culture of endothelial cells in embryoid body differentiated from human embryonic stem cells. *Stem Cells Dev* 2007, *16* (2), 269–280.

102. Assady, S.; Maor, G.; Amit, M.; Itskovitz-Eldor, J.; Skorecki, K. L.; Tzukerman, M., Insulin production by human embryonic stem cells. *Diabetes* 2001, *50* (8), 1691–1697.

103. Segev, H.; Fishman, B.; Ziskind, A.; Shulman, M.; Itskovitz-Eldor, J., Differentiation of human embryonic stem cells into insulin-producing clusters. *Stem Cells* 2004, *22* (3), 265–274.

104. Hansson, M.; Tonning, A.; Frandsen, U.; Petri, A.; Rajagopal, J.; Englund, M. C.; Heller, R. S.; Hakansson, J.; Fleckner, J.; Skold, H. N.; Melton, D.; Semb, H.; Serup, P., Artifactual insulin release from differentiated embryonic stem cells. *Diabetes* 2004, *53* (10), 2603–2609.

105. Rajagopal, J.; Anderson, W. J.; Kume, S.; Martinez, O. I.; Melton, D. A., Insulin staining of ES cell progeny from insulin uptake. *Science* 2003, *299* (5605), 363.

106. Sipione, S.; Eshpeter, A.; Lyon, J. G.; Korbutt, G. S.; Bleackley, R. C., Insulin expressing cells from differentiated embryonic stem cells are not beta cells. *Diabetologia* 2004, *47* (3), 499–508.

107. Ku, H. T.; Zhang, N.; Kubo, A.; O'Connor, R.; Mao, M.; Keller, G.; Bromberg, J. S., Committing embryonic stem cells to early endocrine pancreas in vitro. *Stem Cells* 2004, *22* (7), 1205–1217.

108. Lumelsky, N.; Blondel, O.; Laeng, P.; Velasco, I.; Ravin, R.; McKay, R., Differentiation of embryonic stem cells to insulin-secreting structures similar to pancreatic islets. *Science* 2001, *292* (5520), 1389–1394.

109. Xu, X.; Kahan, B.; Forgianni, A.; Jing, P.; Jacobson, L.; Browning, V.; Treff, N.; Odorico, J., Endoderm and pancreatic islet lineage differentiation from human embryonic stem cells. *Cloning Stem Cells* 2006, *8* (2), 96–107.

110. Kubo, A.; Shinozaki, K.; Shannon, J. M.; Kouskoff, V.; Kennedy, M.; Woo, S.; Fehling, H. J.; Keller, G., Development of definitive endoderm from embryonic stem cells in culture. *Development* 2004, *131* (7), 1651–1662.

111. D'Amour, K. A.; Agulnick, A. D.; Eliazer, S.; Kelly, O. G.; Kroon, E.; Baetge, E. E., Efficient differentiation of human embryonic stem cells to definitive endoderm. *Nat Biotechnol* 2005, *23* (12), 1534–1541.

112. D'Amour, K. A.; Bang, A. G.; Eliazer, S.; Kelly, O. G.; Agulnick, A. D.; Smart, N. G.; Moorman, M. A.; Kroon, E.; Carpenter, M. K.; Baetge, E. E., Production of pancreatic hormone-expressing endocrine cells from human embryonic stem cells. *Nat Biotechnol* 2006, *24* (11), 1392–13401.

113. McLean, A. B.; D'Amour, K. A.; Jones, K. L.; Krishnamoorthy, M.; Kulik, M. J.; Reynolds, D. M.; Sheppard, A. M.; Liu, H.; Xu, Y.; Baetge, E. E.; Dalton, S., Activin a efficiently specifies definitive endoderm from human embryonic stem cells only when phosphatidylinositol 3-kinase signaling is suppressed. *Stem Cells* 2007, *25* (1), 29–38.

114. Jiang, J.; Au, M.; Lu, K.; Eshpeter, A.; Korbutt, G.; Fisk, G.; Majumdar, A. S., Generation of insulin-producing islet-like clusters from human embryonic stem cells. *Stem Cells* 2007, *25* (8), 1940–1953.

115. Phillips, B. W.; Hentze, H.; Rust, W. L.; Chen, Q. P.; Chipperfield, H.; Tan, E. K.; Abraham, S.; Sadasivam, A.; Soong, P. L.; Wang, S. T.; Lim, R.; Sun, W.; Colman, A.; Dunn, N. R., Directed differentiation of human embryonic stem cells into the pancreatic endocrine lineage. *Stem Cells Dev* 2007, *16* (4), 561–578.

116. Jiang, W.; Shi, Y.; Zhao, D.; Chen, S.; Yong, J.; Zhang, J.; Qing, T.; Sun, X.; Zhang, P.; Ding, M.; Li, D.; Deng, H., In vitro derivation of functional insulin-producing cells from human embryonic stem cells. *Cell Res* 2007, *17* (4), 333–344.

117. Shim, J. H.; Kim, S. E.; Woo, D. H.; Kim, S. K.; Oh, C. H.; McKay, R.; Kim, J. H., Directed differentiation of human embryonic stem cells towards a pancreatic cell fate. *Diabetologia* 2007, *50* (6), 1228–1238.

118. Kroon, E.; Martinson, L. A.; Kadoya, K.; Bang, A. G.; Kelly, O. G.; Eliazer, S.; Young, H.; Richardson, M.; Smart, N. G.; Cunningham, J.; Agulnick, A. D.; D'Amour, K. A.; Carpenter, M. K.; Baetge, E. E., Pancreatic endoderm derived from human embryonic stem cells generates glucose-responsive insulin-secreting cells in vivo. *Nat Biotechnol* 2008, *26* (4), 443–452.

119. Borowiak, M.; Maehr, R.; Chen, S.; Chen, A. E.; Tang, W.; Fox, J. L.; Schreiber, S. L.; Melton, D. A., Small molecules efficiently direct endodermal differentiation of mouse and human embryonic stem cells. *Cell Stem Cell* 2009, *4* (4), 348–358.

120. Chen, S.; Borowiak, M.; Fox, J. L.; Maehr, R.; Osafune, K.; Davidow, L.; Lam, K.; Peng, L. F.; Schreiber, S. L.; Rubin, L. L.; Melton, D., A small molecule that directs differentiation of human ESCs into the pancreatic lineage. *Nat Chem Biol* 2009, *5* (4), 258–265.

121. Carpenter, M. K.; Inokuma, M. S.; Denham, J.; Mujtaba, T.; Chiu, C. P.; Rao, M. S., Enrichment of neurons and neural precursors from human embryonic stem cells. *Exp Neurol* 2001, *172* (2), 383–397.

122. Keirstead, H. S.; Nistor, G.; Bernal, G.; Totoiu, M.; Cloutier, F.; Sharp, K.; Steward, O., Human embryonic stem cell-derived oligodendrocyte progenitor cell transplants remyelinate and restore locomotion after spinal cord injury. *J Neurosci* 2005, *25* (19), 4694–4705.

123. Shin, S.; Mitalipova, M.; Noggle, S.; Tibbitts, D.; Venable, A.; Rao, R.; Stice, S. L., Long-term proliferation of human embryonic stem cell-derived neuroepithelial cells using defined adherent culture conditions. *Stem Cells* 2006, *24* (1), 125–138.

124. Hong, S.; Kang, U. J.; Isacson, O.; Kim, K. S., Neural precursors derived from human embryonic stem cells maintain long-term proliferation without losing the potential to differentiate into all three neural lineages, including dopaminergic neurons. *J Neurochem* 2008, *104* (2), 316–324.

125. Androutsellis-Theotokis, A.; Leker, R. R.; Soldner, F.; Hoeppner, D. J.; Ravin, R.; Poser, S. W.; Rueger, M. A.; Bae, S. K.; Kittappa, R.; McKay, R. D., Notch signalling regulates stem cell numbers in vitro and in vivo. *Nature* 2006, *442* (7104), 823–826.

126. Hitoshi, S.; Alexson, T.; Tropepe, V.; Donoviel, D.; Elia, A. J.; Nye, J. S.; Conlon, R. A.; Mak, T. W.; Bernstein, A.; van der Kooy, D., Notch pathway molecules are essential for the maintenance, but not the generation, of mammalian neural stem cells. *Genes Dev* 2002, *16* (7), 846–858.

127. Lowell, S.; Benchoua, A.; Heavey, B.; Smith, A. G., Notch promotes neural lineage entry by pluripotent embryonic stem cells. *PLoS Biol* 2006, *4* (5), e121.

128. Maye, P.; Becker, S.; Siemen, H.; Thorne, J.; Byrd, N.; Carpentino, J.; Grabel, L., Hedgehog signaling is required for the differentiation of ES cells into neurectoderm. *Dev Biol* 2004, *265* (1), 276–290.

129. Davidson, K. C.; Jamshidi, P.; Daly, R.; Hearn, M. T.; Pera, M. F.; Dottori, M., Wnt3a regulates survival, expansion, and maintenance of neural progenitors derived from human embryonic stem cells. *Mol Cell Neurosci* 2007, *36* (3), 408–415.

130. Lamba, D. A.; Karl, M. O.; Ware, C. B.; Reh, T. A., Efficient generation of retinal progenitor cells from human embryonic stem cells. *Proc Natl Acad Sci USA* 2006, *103* (34), 12769–12774.

131. Rao, B. M.; Zandstra, P. W., Culture development for human embryonic stem cell propagation: Molecular aspects and challenges. *Curr Opin Biotechnol* 2005, *16* (5), 568–576.

132. Smith, J. R.; Vallier, L.; Lupo, G.; Alexander, M.; Harris, W. A.; Pedersen, R. A., Inhibition of Activin/Nodal signaling promotes specification of human embryonic stem cells into neuroectoderm. *Dev Biol* 2008, *313* (1), 107–117.

133. Wiles, M. V.; Johansson, B. M., Embryonic stem cell development in a chemically defined medium. *Exp Cell Res* 1999, *247* (1), 241–248.

134. Ying, Q. L.; Stavridis, M.; Griffiths, D.; Li, M.; Smith, A., Conversion of embryonic stem cells into neuroectodermal precursors in adherent monoculture. *Nat Biotechnol* 2003, *21* (2), 183–186.

135. Gratsch, T. E.; O'Shea, K. S., Noggin and chordin have distinct activities in promoting lineage commitment of mouse embryonic stem (ES) cells. *Dev Biol* 2002, *245* (1), 83–94.

136. Bain, G.; Kitchens, D.; Yao, M.; Huettner, J. E.; Gottlieb, D. I., Embryonic stem cells express neuronal properties in vitro. *Dev Biol* 1995, *168* (2), 342–357.
137. Okabe, S.; Forsberg-Nilsson, K.; Spiro, A. C.; Segal, M.; McKay, R. D., Development of neuronal precursor cells and functional postmitotic neurons from embryonic stem cells in vitro. *Mech Dev* 1996, *59* (1), 89–102.
138. Kawasaki, H.; Mizuseki, K.; Nishikawa, S.; Kaneko, S.; Kuwana, Y.; Nakanishi, S.; Nishikawa, S. I.; Sasai, Y., Induction of midbrain dopaminergic neurons from ES cells by stromal cell-derived inducing activity. *Neuron* 2000, *28* (1), 31–40.
139. Finley, M. F.; Devata, S.; Huettner, J. E., BMP-4 inhibits neural differentiation of murine embryonic stem cells. *J Neurobiol* 1999, *40* (3), 271–287.
140. Sasai, Y.; Lu, B.; Steinbeisser, H.; De Robertis, E. M., Regulation of neural induction by the Chd and Bmp-4 antagonistic patterning signals in Xenopus. *Nature* 1995, *376* (6538), 333–336.
141. Wilson, P. A.; Lagna, G.; Suzuki, A.; Hemmati-Brivanlou, A., Concentration-dependent patterning of the Xenopus ectoderm by BMP4 and its signal transducer Smad1. *Development* 1997, *124* (16), 3177–3184.
142. Aubert, J.; Dunstan, H.; Chambers, I.; Smith, A., Functional gene screening in embryonic stem cells implicates Wnt antagonism in neural differentiation. *Nat Biotechnol* 2002, *20* (12), 1240–1245.
143. Reubinoff, B. E.; Itsykson, P.; Turetsky, T.; Pera, M. F.; Reinhartz, E.; Itzik, A.; Ben-Hur, T., Neural progenitors from human embryonic stem cells. *Nat Biotechnol* 2001, *19* (12), 1134–1140.
144. Zhang, S. C.; Wernig, M.; Duncan, I. D.; Brustle, O.; Thomson, J. A., In vitro differentiation of transplantable neural precursors from human embryonic stem cells. *Nat Biotechnol* 2001, *19* (12), 1129–1133.
145. Joannides, A. J.; Fiore-Heriche, C.; Battersby, A. A.; Athauda-Arachchi, P.; Bouhon, I. A.; Williams, L.; Westmore, K.; Kemp, P. J.; Compston, A.; Allen, N. D.; Chandran, S., A scaleable and defined system for generating neural stem cells from human embryonic stem cells. *Stem Cells* 2007, *25* (3), 731–737.
146. Kehat, I.; Kenyagin-Karsenti, D.; Snir, M.; Segev, H.; Amit, M.; Gepstein, A.; Livne, E.; Binah, O.; Itskovitz-Eldor, J.; Gepstein, L., Human embryonic stem cells can differentiate into myocytes with structural and functional properties of cardiomyocytes. *J Clin Invest* 2001, *108* (3), 407–414.
147. Xu, C.; Police, S.; Rao, N.; Carpenter, M. K., Characterization and enrichment of cardiomyocytes derived from human embryonic stem cells. *Circ Res* 2002, *91* (6), 501–508.
148. Graichen, R.; Xu, X.; Braam, S. R.; Balakrishnan, T.; Norfiza, S.; Sieh, S.; Soo, S. Y.; Tham, S. C.; Mummery, C.; Colman, A.; Zweigerdt, R.; Davidson, B. P., Enhanced cardiomyogenesis of human embryonic stem cells by a small molecular inhibitor of p38 MAPK. *Differentiation* 2008, *76* (4), 357–370.
149. Davis, R. P.; Ng, E. S.; Costa, M.; Mossman, A. K.; Sourris, K.; Elefanty, A. G.; Stanley, E. G., Targeting a GFP reporter gene to the MIXL1 locus of human embryonic stem cells identifies human primitive streak-like cells and enables isolation of primitive hematopoietic precursors. *Blood* 2008, *111* (4), 1876–1884.
150. Sottile, V.; Thomson, A.; McWhir, J., In vitro osteogenic differentiation of human ES cells. *Cloning Stem Cells* 2003, *5* (2), 149–155.
151. Rambhatla, L.; Chiu, C. P.; Kundu, P.; Peng, Y.; Carpenter, M. K., Generation of hepatocyte-like cells from human embryonic stem cells. *Cell Transplant* 2003, *12* (1), 1–11.
152. Green, H.; Easley, K.; Iuchi, S., Marker succession during the development of keratinocytes from cultured human embryonic stem cells. *Proc Natl Acad Sci USA* 2003, *100* (26), 15625–15630.

153. Rosenstrauch, D.; Poglajen, G.; Zidar, N.; Gregoric, I. D., Stem cell therapy for ischemic heart failure. *Tex Heart Inst J* 2005, *32* (3), 339–347.

154. Drukker, M.; Katchman, H.; Katz, G.; Even-Tov Friedman, S.; Shezen, E.; Hornstein, E.; Mandelboim, O.; Reisner, Y.; Benvenisty, N., Human embryonic stem cells and their differentiated derivatives are less susceptible to immune rejection than adult cells. *Stem Cells* 2006, *24* (2), 221–229.

155. Levenberg, S.; Zoldan, J.; Basevitch, Y.; Langer, R., Endothelial potential of human embryonic stem cells. *Blood* 2007, *110* (3), 806–814.

156. Levenberg, S.; Rouwkema, J.; Macdonald, M.; Garfein, E. S.; Kohane, D. S.; Darland, D. C.; Marini, R.; van Blitterswijk, C. A.; Mulligan, R. C.; D'Amore, P. A.; Langer, R., Engineering vascularized skeletal muscle tissue. *Nat Biotechnol* 2005, *23* (7), 879–884.

157. Caspi, O.; Lesman, A.; Basevitch, Y.; Gepstein, A.; Arbel, G.; Habib, I. H.; Gepstein, L.; Levenberg, S., Tissue engineering of vascularized cardiac muscle from human embryonic stem cells. *Circ Res* 2007, *100* (2), 263–272.

158. Lesman, A.; Habib, M.; Caspi, O.; Gepstein, A.; Arbel, G.; Levenberg, S.; Gepstein, L., Transplantation of a tissue-engineered human vascularized cardiac muscle. *Tissue Eng Part A* 2010, *16* (1), 115–125.

159. Stevens, K. R.; Kreutziger, K. L.; Dupras, S. K.; Korte, F. S.; Regnier, M.; Muskheli, V.; Nourse, M. B.; Bendixen, K.; Reinecke, H.; Murry, C. E., Physiological function and transplantation of scaffold-free and vascularized human cardiac muscle tissue. *Proc Natl Acad Sci USA* 2009, *106* (39), 16568–16573.

160. Sone, M.; Itoh, H.; Yamahara, K.; Yamashita, J. K.; Yurugi-Kobayashi, T.; Nonoguchi, A.; Suzuki, Y.; Chao, T. H.; Sawada, N.; Fukunaga, Y.; Miyashita, K.; Park, K.; Oyamada, N.; Sawada, N.; Taura, D.; Tamura, N.; Kondo, Y.; Nito, S.; Suemori, H.; Nakatsuji, N.; Nishikawa, S.; Nakao, K., Pathway for differentiation of human embryonic stem cells to vascular cell components and their potential for vascular regeneration. *Arterioscler Thromb Vasc Biol* 2007, *27* (10), 2127–2134.

161. Oyamada, N.; Itoh, H.; Sone, M.; Yamahara, K.; Miyashita, K.; Park, K.; Taura, D.; Inuzuka, M.; Sonoyama, T.; Tsujimoto, H.; Fukunaga, Y.; Tamura, N.; Nakao, K., Transplantation of vascular cells derived from human embryonic stem cells contributes to vascular regeneration after stroke in mice. *J Transl Med* 2008, *6*, 54.

162. McDonald, J. W.; Becker, D.; Holekamp, T. F.; Howard, M.; Liu, S.; Lu, A.; Lu, J.; Platik, M. M.; Qu, Y.; Stewart, T.; Vadivelu, S., Repair of the injured spinal cord and the potential of embryonic stem cell transplantation. *J Neurotrauma* 2004, *21* (4), 383–393.

163. Zhang, S. K.; Liu, Y.; Song, Z. M.; Fu, C. F.; Xu, X. X., Green fluorescent protein as marker in chondrocytes overexpressing human insulin-like growth factor-1 for repair of articular cartilage defects in rabbits. *Chin J Traumatol* 2007, *10* (1), 10–17.

164. Roy, N. S.; Cleren, C.; Singh, S. K.; Yang, L.; Beal, M. F.; Goldman, S. A., Functional engraftment of human ES cell-derived dopaminergic neurons enriched by coculture with telomerase-immortalized midbrain astrocytes. *Nat Med* 2006, *12* (11), 1259–1268.

165. Drukker, M.; Katz, G.; Urbach, A.; Schuldiner, M.; Markel, G.; Itskovitz-Eldor, J.; Reubinoff, B.; Mandelboim, O.; Benvenisty, N., Characterization of the expression of MHC proteins in human embryonic stem cells. *Proc Natl Acad Sci USA* 2002, *99* (15), 9864–9869.

166. Nussbaum, J.; Minami, E.; Laflamme, M. A.; Virag, J. A.; Ware, C. B.; Masino, A.; Muskheli, V.; Pabon, L.; Reinecke, H.; Murry, C. E., Transplantation of undifferentiated murine embryonic stem cells in the heart: Teratoma formation and immune response. *Faseb J* 2007, *21* (7), 1345–1357.

167. Cabrera, C. M.; Cobo, F.; Nieto, A.; Concha, A., Strategies for preventing immunologic rejection of transplanted human embryonic stem cells. *Cytotherapy* 2006, *8* (5), 517–518.

168. Drukker, M., Immunogenicity of human embryonic stem cells: Can we achieve tolerance? *Springer Semin Immunopathol* 2004, *26* (1–2), 201–213.
169. Priddle, H.; Jones, D. R.; Burridge, P. W.; Patient, R., Hematopoiesis from human embryonic stem cells: Overcoming the immune barrier in stem cell therapies. *Stem Cells* 2006, *24* (4), 815–824.
170. Lanza, R. P.; Chung, H. Y.; Yoo, J. J.; Wettstein, P. J.; Blackwell, C.; Borson, N.; Hofmeister, E.; Schuch, G.; Soker, S.; Moraes, C. T.; West, M. D.; Atala, A., Generation of histocompatible tissues using nuclear transplantation. *Nat Biotechnol* 2002, *20* (7), 689–696.

2 Building Blood Vessels Using Endothelial and Mesenchymal Progenitor Cells

Patrick Allen and Joyce Bischoff

CONTENTS

2.1 INTRODUCTION: TWO MAJOR CELLULAR CONSTITUENTS OF BLOOD VESSELS

The basic structural unit of a blood vessel is an endothelial cell, which forms the lumen of the vessel and is in contact with the blood, and a pericyte or smooth muscle cell, positioned on the abluminal surface of the endothelium. Here, we will focus on human cells isolated from blood that can form the endothelial lining of vessels in vivo[1–4]—we call these cells endothelial progenitor cells (EPCs); some groups use the term endothelial colony-forming cell, abbreviated ECFC, to emphasize the clonogenic potential of these cells[5–7] while others use the term blood outgrowth endothelial cells (BOECs).[8] EPCs are distinct from blood- and bone marrow–derived cells that are referred to in the literature as early EPCs, CFU-EC (colony-forming unit-endothelial cell, or angiogenic monocytes).[9,10] Cells designated "early EPCs" have been shown to augment capillary tube formation[11] and angiogenesis in vivo[12] but do not directly participate in forming vessel structures. Of note, a murine equivalent of EPCs or ECFCs—highly proliferative, clonogenic cells with vessel building activity in vivo—has been difficult to obtain. One report describes murine BOECs, but the proliferative potential appeared to be modest.[13] The obscurity of the murine counterpart to human EPCs has likely contributed to some of the controversy on the identity of EPCs. The perivascular component of the vessel wall, pericytes and smooth muscle cells, has been studied intensively to determine origins, phenotypic plasticity, and roles in blood vessel formation and stability.[14–16] Here, we will focus on human mesenchymal stem/progenitor cells that have been shown to differentiate into perivascular cells and to support formation of functional blood vessels.[4,17,18]

2.1.1 ENDOTHELIAL CELLS

Endothelial cells are identified by expression of endothelial markers and by functional endothelial properties that can be assayed in vitro and ultimately in vivo. The most reliable cellular markers are CD31, also known as PECAM-1 (platelet–endothelial cell adhesion molecule-1), CD144, also known as VE-cadherin (vascular endothelial-cadherin) or cadherin-5, and von Willebrand Factor (vWF), sometimes referred to as Factor VIII–related antigen. However, there are caveats that must be kept in mind when relying on the presence of these markers to assign an endothelial phenotype. CD31 is expressed on nearly all continuous endothelium, but its expression is variable in fenestrated endothelium such as in the liver sinusoidal endothelial cells.[19] A more pressing issue is that CD31 is expressed on myeloid lineage cells, in particular monocytes, albeit at about 1/10 the level of CD31 present on the surface of endothelial cells. VE-cadherin is also expressed broadly in the adult vasculature and has excellent utility as an endothelial marker, although VE-cadherin transcriptional activity has been detected in hematopoietic cells[20] and in hemogenic endothelial cells.[21,22] vWF is a highly specific marker when localized within rod-shaped organelles known as Weibel–Palade bodies, which are unique to endothelial cells.[23,24] However, vWF is also present in platelets and can be deposited along the endothelium when platelets are activated. Less reliable markers are TIE2, CD34, vascular endothelial growth factor receptors (VEGFRs) and CD105 (also known as endoglin or SH2). These cell surface membrane proteins are expressed to varying degrees in hematopoietic cells,[6,25] such that cultures of adherent cells from blood or bone marrow are often mistakenly identified as EPCs based on expression of these markers.[26] In addition, CD105 is expressed on mesenchymal stem cells (MSCs), and CD146, originally identified on human umbilical vein endothelial cells (HUVECs) using the monoclonal antibody P1H12,[27] has been shown to be a striking marker of human pericytes *in vivo*.[28] Thus, the detection of one or more of these markers is certainly consistent with an endothelial phenotype but does provide sufficient information to designate the cells as endothelial.

Therefore, the expression of endothelial markers is a starting point for establishing endothelial identity, but this must be combined with a demonstration of endothelial function. One assay that is commonly used, although it is not specific, is the endocytosis of fluorescently labeled acetylated low-density lipoprotein (acLDL). Any cell with a scavenger receptor will take up fluorescently labeled acLDL—this includes endothelial cells but also monocytes—and therefore this assay is not useful for assigning endothelial identity or function. Capillary-like tube formation *in vitro* is also widely used to assess endothelial function; the most commonly used assay is to simply plate cells on Matrigel™ and after several hours examine for multicellular, branching cell patterns, or "cords," which are thought to reflect capillaries. However, many non-endothelial cells form such patterns when plated on Matrigel, including MSCs, so the assay lacks specificity for identifying endothelial cells and on its own does not indicate endothelial function. Three-dimensional capillary morphogenesis assays have been developed which, when carefully analyzed, do reveal the ability of human endothelial cells to form lumenal structures, branch points, sprouts, and anastomoses in a culture dish.[29,30] A third functional assay is myeloid cell adhesion to cytokine-stimulated endothelial cells. Cytokines such as tumor necrosis factor (TNF)-α or interleukin(IL)-1 will induce expression of the leukocyte adhesion molecules (e.g., E-selectin, intracellular adhesion molecule-1 (ICAM-1), vascular cell adhesion moleculre-1 (VCAM-1)), which in turn confer calcium-dependent adhesion of myeloid cells.[31,32] As leukocyte trafficking is a critical and unique job of the endothelium, this provides a specific *in vitro* assay for endothelial function. Finally, a definitive test for endothelial function is to demonstrate the ability to form the endothelial lining of blood vessels *in vivo*; this will be discussed in depth below.

2.1.2 PERICYTES

Pericytes, sometimes called microvascular smooth muscle cells, are defined by their anatomical location around capillaries and microvessels. These cells, embedded within the vascular basement membrane, are critical for vessel stability and function, but their origins and plasticity remain mysterious. Evidence has accumulated over many years to show that pericytes are multipotential cells and may even be the origin of MSCs (for reviews, see Refs. [15,33]). The potential for human pericytes to regulate tissue microenvironment and promote regeneration, in addition to their role(s) in the vascular wall, has recently been described.[34] In human

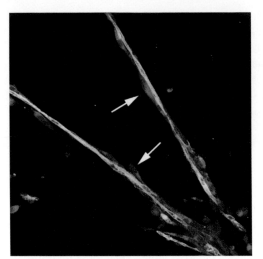

EPCs (green), MPCs (red), nuclei (blue)

FIGURE 2.1 (See color insert.) EPCs and MPCs stained in three-dimensional culture *in vitro*. MPCs were labeled with CellTracker Red (Invitrogen), then suspended with EPCs in Matrigel (BD Biosciences), and cultured overnight in EGM-2, 1% serum, and 3 ng/mL bFGF. EPCs were then stained for CD31 (Dako), green. The figure shows linear, multicellular network segments with MPCs aligned lengthwise along EPC cords.

organs and tissues, pericytes express CD146, neural glial antigen-2 (NG2), platelet-derived growth factor receptor-β (PDGFR-β, CD140b) and are devoid of endothelial and hematopoietic markers. Other pericytic markers are α-smooth muscle actin (α-SMA), desmin, and RGS5.[35,36] In addition, pericytes are positive for CD90 (Thy-1 antigen), CD105 (endoglin), CD73 (5′nucleotidase, SH3), and CD44 (hyaluronate receptor)—each of these cell surface proteins are also expressed on human MSCs. Furthermore, murine pericytes with multi-lineage differentiation potential were shown to express annexin A5 and Sca-1.[37,38] Pericyte functional activity can be measured by increased production of basement membrane constituents, in particular nidogen-1.[38,39] When co-seeded with endothelial cells, pericytes have been shown to align along endothelial cords *in vitro*.[40] An example of this cellular interaction is depicted in Figure 2.1.

2.1.3 Vascular Smooth Muscle Cells

Vascular smooth muscle cells (abbreviated SMCs) surround larger vessels in a circumferential orientation and in multiple layers, providing contractile function and tone to the vessel wall. Often the distinction between pericytes and smooth muscle cells is blurred, and terms such as perivascular or mural cell are used to refer to cells residing on the abluminal side of the endothelium that contribute to vessel stability and function. SMCs are identified by a series of markers, although no single marker is sufficient to establish SMC identity. The markers include α-SMA, h1-calponin, SM22α, smoothelin, caldesmon, and smooth muscle-specific myosin heavy chain

(SM-MHC).[41] The levels of expression of these markers are highly variable during phenotypic modulation of SMCs between contractile and synthetic states. The contractile phenotype is characterized by low proliferation, low synthetic activity, and high expression of the contractile proteins used as SMC markers. However, SMCs can rapidly respond to changes in the local environment, *in vivo* or *in vitro* in culture dishes, and switch to a synthetic phenotype characterized by proliferation, migration, synthesis of extracellular matrix proteins and glycosaminoglycans, and reduced expression of contractile proteins. Besides contractile activity, SMCs influence endothelial phenotype and survival, which can be assayed *in vitro*.[42]

2.2 SOURCES OF ADULT STEM/PROGENITOR CELLS FOR TISSUE VASCULARIZATION

2.2.1 SOURCES OF EPCs

2.2.1.1 Adult Peripheral Blood

A landmark study by Lin and colleagues, which was published in 2000, identified a rare population of endothelial cells in adult peripheral blood with robust proliferative capability.[43] Analysis of these cells in blood samples from patients who had undergone gender-mismatched bone marrow transplantation suggested strongly that the proliferative endothelial cells were derived from the transplanted marrow-derived cells and therefore could be considered "circulating angioblasts." In the culture dish, the cells were called late outgrowth endothelial cells and subsequently BOECs. In the same blood samples, circulating endothelial cells (CECs) with modest proliferative capability were shown to be of recipient origin and were hypothesized to be mature endothelial cells shed from the vessel wall. Precise measurements of CECs in peripheral blood are now sought[44] because CECs have been proposed as a biomarker of anti-angiogenic therapy.[45] For example, it is thought that CECs in the circulation will increase as a result of anti-angiogenic agents targeting the endothelium. Although not the first report of endothelial cells in peripheral blood,[46,47] the study from the Hebbel laboratory revealed human adult peripheral blood as a potential source of endothelial cells for building autologous vascular networks or endothelizing vascular grafts. An essential point from their study that was unfortunately overlooked for much of the past decade is that the highly proliferative endothelial cells (hereafter called EPCs) in adult peripheral blood require nearly 4 weeks to become predominant in the culture dish. Endothelial-like colonies appearing within days of plating are monocytes which mimic an endothelial phenotype by expressing low levels of CD31, VEGFRs, TIE2 and engulfing acetylated-LDL. These cells, lacking significant proliferative capacity, typically disappear from the culture by 2–3 weeks.

Subsequent to the Lin paper, many groups went on to isolate highly proliferative EPCs from adult peripheral blood and used the cells for *in vivo* and *in vitro* models of blood vessel assembly and function. Our group showed that EPCs expanded ex vivo in a culture dish could be used to create functional small diameter vascular grafts.[48] In this study, autologous EPCs from sheep were used to endothelialize 4 cm long,

4 mm diameter decellularized segments from a porcine iliac artery. The EPC-seeded constructs were patent and functional over 4 months when implanted as carotid interposition grafts. Ovine and human peripheral blood-derived EPCs were shown to maintain a stable endothelial phenotype when adherent on novel biomaterials such as poly-glycolic acid/poly-4-hydroxybutyrate fibers,[49] poly-glycolic acid/poly-L-lactic acid (PGA/PLLA)[32] and on fibroin silk fiber meshes,[50] providing an indication of the adaptability of the cells to potentially diverse matrices that might be used in tissue engineering.

The question of feasibility of obtaining EPCs from patients of advanced age, with cardiovascular disease, or with other pathological conditions has often arisen and suggested as a potential limitation on the use of patient-derived autologous EPCs for neovascularization of organs and tissues. A recent study by Stroncek and colleagues quantified EPCs in 50 mL peripheral blood samples from 13 patients with coronary artery disease and 13 normal controls.[51] EPCs were isolated from 7/13 patients and 9/13 healthy donors; there were no clinical characteristics that correlated with success or failure to obtain EPCs. The mean age was 62 for patients and 26 for healthy donors. EPCs from both groups were highly proliferative, growing to 10^7 cells by 36 days in culture, similar to what we reported for adult EPCs.[1] Consistent with Lin and the Hebbell laboratory,[43] EPC colonies did not begin to appear until at least 21 days in culture. Although EPCs were not obtained from 100% of the blood samples tested in this study, it is reasonable to propose success would approach 100% if larger starting blood samples were used or improved techniques were developed. Recently, there have been considerable advances made in isolating EPCs from adult peripheral blood. Reinisch and colleagues have shown remarkable improvements by omitting the density gradient sedimentation step and instead plating cells from whole blood onto non-coated culture dishes.[52] In addition, fetal bovine serum was replaced with a human platelet lysate, which resulted in a procedure free of animal-derived products. These improvements lead to successful isolation of EPCs from peripheral blood samples of approximately 25 mL from several healthy volunteers and several volunteers with cardiovascular disease. It was calculated that an average of 4 EPC colonies per milliliter of whole blood could be obtained; the cells were easily propagated and shown to undergo at least 30 population doublings. These improvements, which not only simplify the laboratory steps needed, but greatly increase their yield and reproducibility, represent a tremendous advance toward using human EPCs for cell-based therapies.

2.2.1.2 Umbilical Cord Blood

Human umbilical cord blood is an excellent source of EPCs as their concentration is estimated to be 20-fold higher than in adult peripheral blood.[53] Thus, cord blood has been the preferred source of EPCs for comparison with mature endothelial cells from the vessel wall,[54,55] *in vitro* studies of vessel assembly,[32,56,57] assembly of vessels in engineered skin substitutes,[3] and in ischemic myocardium.[55] Cord blood EPCs provide a potentially new source of postnatal cells for use in tissue-engineering and regenerative medicine applications. In the future, cord blood EPCs could be typed for histocompatibility antigens and used for cell-based therapies in appropriately-matched donors. It is conceivable that future cord blood banks would be of sufficient depth to offer appropriately matched cells for much of the population. For children with congenital defects diagnosed before birth, autologous cord blood EPCs could be harvested at birth, or even in utero, to provide cells for cardiovascular repair via tissue-engineering approaches. Ingram's group showed that levels of EPCs in the cord blood increase with gestational age, by full term, obtaining approximately 8 EPC colonies per 100 million mononuclear cells plated.[58] In summary, cord blood is a readily available source of non-embryonic cells with tremendous potential for tissue engineering and regeneration.

2.2.1.3   Origin of Human EPCs

The ability to obtain proliferative and functional EPCs from human peripheral blood or umbilical cord blood is clearly established, yet the precise origin of these cells is under debate. Initially, experiments pointed to a bone marrow origin.[59] This study described a population of multipotent adult progenitor cells (MAPCs), isolated from human bone marrow samples obtained from 55 healthy volunteer donors, that could differentiate into mature endothelial cells. However, there is a paucity of reports that show isolation of EPCs, as defined here—cells with clonogenic growth potential combined with demonstrable endothelial function—from bone marrow. Evidence for EPCs in residence in the vessel wall has been shown. Vessel wall-derived EPCs from HUVECs and human aortic endothelial cells (HAECs) show comparable proliferative and clonogenic potential to human EPCs from cord blood to.[60] These studies bring up the possibility that circulating EPCs originate from EPCs residing in the vessel wall. Two reports describing vascular wall resident progenitor cells in adult arteries support the concept of the vessel wall as a reservoir of vasculogenic cells.[61,62] These multipotent cells are suggested to give rise to resident EPCs, hematopoietic progenitor cells, and MSCs.[62] Indeed, vascular wall progenitor cells have been proposed as an additional source of adult multipotent stem cells that could be used to bioengineer new blood vessels for tissue-engineered organs.[63] Determining the ontogeny of these cells will require cell fate–mapping tools yet to be developed. However, the practical application of vessel wall–derived EPCs to tissue engineering and regenerative medicine can advance as the developmental origins of EPCs are delineated.

2.2.2   Sources of Pericyte/Smooth Muscle Cell Progenitor Cells

2.2.2.1   Bone Marrow

In contrast to the relative wealth of data on human EPCs generated from several laboratories, the origins and phenotypes of pericytes and smooth muscle cells has been less studied. However, the potential to isolate human cells with smooth muscle progenitor-like features from bone marrow was reported in 1993.[64] In this study, human bone marrow cultures were shown to contain stromal cells whose phenotype was reminiscent of subendothelial intimal smooth muscle cells; the specific markers expressed were α-SMA, calponin, h-caldesmon, and smooth muscle myosin heavy chain (SM-MHC).[64] Subsequently, bone marrow–derived precursors were shown to home to the intimal layer of aortic allografts in mice and to express α-SMA.[65] The

Verfaillie laboratory was able to induce bone marrow–derived MAPCs to differentiate into smooth muscle cells with functional L-type calcium channel activity.[65]

These bone marrow–derived cells with smooth muscle characteristics are thought
to be closely related to mesenchymal precursor/progenitor cells (MPCs), typically
called MSCs, which have been extensively described in the literature.[66] MSC and
MPC are terms applied broadly to cells with the following properties: ability to readily adhere to bacteriologic dishes or non-coated tissue culture dishes, positive for
expression of CD105 (endoglin, SH2), CD73 (5'nucleotidase, SH3), and CD90 (Thy-1
antigen), negative for expression of hematopoietic or endothelial markers and capable
of multi-lineage mesenchymal differentiation (adipogenesis, chondrogenesis, osteogenesis, and myogenesis). In this chapter, the term MPCs will be used to refer to
cells fitting this description. Additional cell surface markers, which may delineate a
subset of MPCs, are STRO-1 antigen and VCAM1/CD106.[67] Injection of these MPCs
into glioblastoma tumor tissue or into ischemic myocardium in nude rats resulted in
increased arteriogenesis, with the human MPCs located adjacent to vascular structures. Human bone marrow–derived MPCs have also been directly tested for ability
to function as smooth muscle cells in engineered small-diameter blood vessels.[18] In
this study, Niklason's group sought an alternative to vessel-derived smooth muscle
cells as human smooth muscle cells from clinically accessible venous segments do
not routinely proliferate adequately to create autologous bioengineered vascular
grafts. The authors devised a two-phase bioreactor regime which first stimulated
cellular proliferation, then differentiation. The resulting human MPC-derived engineered vessels were comparable to human umbilical arterial segments. From these
studies, it is clear that human bone marrow is a valuable source of MSCs or MPCs
with ability to take on the role of smooth muscle cells in the vascular wall.

2.2.2.2 Peripheral and Cord Blood

Other sources of human smooth muscle progenitor cells that have been reported
include peripheral blood,[68]cord blood,[58,69] and adipose tissue.[70] Simper and colleagues
reported a thorough phenotypic characterization of smooth muscle outgrowth cells
from peripheral blood samples from healthy volunteers. The mononuclear cell fraction was plated on type I collagen-coated dishes; smooth muscle differentiation was
stimulated by addition of PDGF-BB. The resulting smooth muscle cells expressed
α-SMA, calponin, and SM MHC,[68] but no functional assays were included in the
study. Subsequently, blood-derived smooth muscle outgrowth cells were shown to
have high proliferative potential.[71] Cord blood smooth muscle/MPCs have been isolated and shown to be essential for engineered vascular networks *in vivo* in Matrigel
implants.[4] Intriguingly, levels of human MSCs in cord blood from preterm infants at
24–31 weeks gestation were shown to be much higher than at full term.[58]

2.2.2.3 Adipose Tissue

Human subcutaneous adipose tissue has also been shown to be a robust source of
cells with features of MSCs and pericytes.[70] The adipose stromal cells (ASCs) are
CD34+ cells but do not express endothelial markers such as CD31/PECAM-1 or
CD144/VE-cadherin, nor express the hematopoietic marker CD45. The ASCs preferentially adhere to non-coated tissue culture dishes, which results in a significant

enrichment of these cells. The authors showed that the CD34+ cells in adipose tissue are positioned along the abluminal surface of CD31+ endothelium, and further that the ASC express pericyte markers NG-2 and PDGFR-α and -β. *In vitro*, the ASC were shown to have multi-lineage differentiation capabilities. Hence, the ASC can be considered perivascular/mural progenitor cells. In addition, ASCs were shown to secrete angiogenic factors, indicating these cells are well suited for building vascular networks *in vivo*.[70]

2.2.2.4   Vessel Wall

Earlier studies have shown the arterial vascular wall contains multipotent MPCs. A subpopulation of vascular cells predisposed to calcification, termed calcifying vascular cells (CVCs), was isolated from preparations of bovine aortic smooth muscle cells by single cell plating.[72] CVCs were found to be similar to MPCs but with weak or absent adipogenic differentiation potential. Studies in murine models have identified Sca-1+ progenitor cells in the adventia of vessels from a variety of tissues that were shown to differentiate into SMCs in response to PDGF-BB.[73] These adventitial progenitor cells were shown to migrate into atherosclerotic lesions *in vivo*, and thus proposed to contribute to atherosclerosis, but the authors also noted such cells might be used for tissue-engineering approaches to treat vascular disease. The following year, Howson and colleagues in the Nicosia laboratory reported on the presence of pericyte progenitor cells in the rat aorta.[74] Sca-1+/Hoechst^{low} "side population" or SP cells were also isolated from adult murine aortas and shown to differentiate *in vitro* toward an endothelial phenotype in response to VEGF-A or toward a smooth muscle phenotype in response to TGF-β1 and PDGF-BB.[61] It was not determined by clonal analysis whether these results were due to a bipotential cell type or due to two separate subpopulations within the Sca-1+ population. Nevertheless, the study provides additional evidence for vascular progenitor cells in the vessel wall.

2.3   ISOLATION STRATEGIES

2.3.1   Human EPCs from Blood

The procedures, techniques, and reagents needed to isolate human EPCs from cord blood or adult peripheral blood have been thoroughly described by our group[75] and by Ingram, Yoder, and colleagues.[76] In addition, a simplified and robust procedure has been reported by Strunk and colleagues.[52] Here, we will emphasize some key considerations but refer the reader to these sources for methodology.

The most efficient approach to obtain human functional EPCs is to rely on the robust proliferative capacity of the cells when plated in an endothelial growth medium. Partially enriched preparations of mononuclear cells are obtained by centrifugation of whole blood layered onto Ficoll-Paque Plus in Accuspin tubes. The "buffy coat" is retrieved, washed, and plated on fibronectin- or collagen-coated dishes in endothelial growth medium. Cells are allowed to adhere (48 h for cord blood and 96 h for adult peripheral blood) before aspirating and adding fresh media. Colonies with the distinctive cobblestone morphology begin to appear at 7–10 days for cord blood and up to 21 days for adult peripheral blood. Cultures are fed with

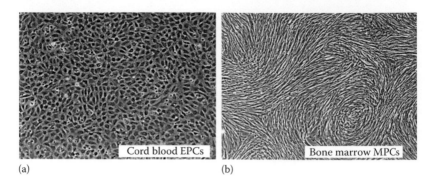

(a)   (b)

FIGURE 2.2   Low-passage cord blood–derived EPCs in two-dimensional culture: (a) shows cobblestone morphology; (b) shows, highly confluent bone marrow–derived MPCs, with spindle morphology.

fresh medium every 2–3 days until the colonies have expanded. Cord blood–derived colonies will expand and merge together to form a nearly confluent monolayer covering the culture dish. At this point, cells are removed from the dish with trypsin and dissociated to achieve a single cell suspension. Endothelial cells are selected using antihuman CD31-conjugated magnetic beads, and more than 90% of the total cells bind to the antibody-coated beads, indicating that EPCs outgrow other adherent cells from blood. The phase contrast image of CD31-selected cord blood EPCs shown in Figure 2.2 depicts the cobblestone morphology that is typical for endothelial cells. Alternatives to using antihuman CD31 are Ulex europeus-I-conjugated magnetic beads[77] and recently available anti-VE-cadherin-coated magnetic beads from R&D systems. EPC colonies from adult peripheral blood may be isolated more efficiently by retrieving individual colonies from the dish using cloning rings. Cells from different colonies can be kept separate if desired or pooled. Adult EPCs can then be dispersed onto a new culture dish for further expansion before anti-CD31-magnetic bead selection. As reported, although adult EPCs require more time for expansion, they exhibit equivalent microvessel-forming ability, i.e., vasculogenic activity, when combined with human MPCs at a ratio of 40:60 and implanted *in vivo*.[4] In summary, the most efficient strategy for purifying EPCs is to rely not on antibody selection or sorting techniques but instead on the robust proliferative capacity of human EPCs. Antibody selection is used as a clean-up step prior to testing the cell preparation for endothelial functions.

2.3.2   HUMAN MPCS FROM BONE MARROW OR CORD BLOOD

A variety of methods using positive and negative antibody selection techniques have been employed to enrich for mesenchymal stem/progenitor cells. For example, positive selection using anti-Stro-1 monoclonal antibodies and negative selection with anti-hematopoeitic cell markers have been used to enrich for mesenchymal stem/progenitor cells from bone marrow. However, to date, there is no definitive cell surface marker that allows one to isolate MPCs using antibody-coated magnetic beads. Currently, the oft-used isolation protocol to obtain MPCs from bone marrow is similar to the original method described by Friedenstein.[78] To start, bone marrow is subjected to density-gradient fractionation on Ficoll-Paque Plus to separate mononucleated cells from plasma and red blood cells. Mononuclear cells are recovered from the "buffy coat" layer, washed, and then plated on non-coated cell culture dishes or bacteriologic culture dishes to take advantage of the superior adhesion of MPCs. Cells are expanded in Dulbecco's Modified Eagle's Medium with 10% fetal bovine serum or in a modified endothelial growth medium from which VEGF, basic FGF (also known as FGF-2), and heparin have been removed.[4] The phase contrast image in Figure 2.2 shows MPCs from adult bone marrow. Optimal plating density and influence of basic FGF on human MSC has been investigated.[79] Similar to the approach implemented for EPCs, the Strunk laboratory has also reported rapid expansion of adult human bone marrow MPCs by omitting the Ficoll-Paque density-gradient fractionation step and replacing fetal bovine serum with human platelet lysate.[80] Minimal essential medium with L-glutamine, heparin, and 10% human platelet lysate was sufficient to obtain nearly 10^9 cells within 16 days. Once expanded sufficiently, MPCs are characterized by flow cytometry using antibodies directed against cell-surface proteins that MPCs are known to express: CD105 (also known as endoglin or SH2), the β1 integrin CD29, CD44 (hyaluronate receptor), and CD90 (Thy-1 antigen). MPC do not express hematopoietic or endothelial markers; antibodies against such markers are used as negative controls or to assess the phenotypic heterogeneity in the cell population. The definitive test is to show the MPCs can be induced to differentiate into multiple mesenchymal lineages— adipocytes, chondrocytes, osteocytes, and myocytes—using well-described differentiation media.[81]

2.4 BUILDING BLOOD VESSELS FROM HUMAN EPCs AND MPCs

2.4.1 Vascular Networks Delivered by Implantation

Early attempts to create cell-mediated microvascular networks for tissue engineering focused on the culture of mature endothelial cells, without perivascular cells, in three-dimensional extracellular matrix gels. In a pioneering study, Schechner and colleagues suspended HUVEC transduced with the *BCL-2* anti-apoptotic gene in gels composed of type I collagen and fibronectin.[82,83] The HUVEC/type I collagen/fibronectin suspensions were allowed to polymerize and incubated *in vitro* overnight to allow formation of multicellular, branching structures believed to be precursors of blood vessels. Each gel was then implanted into a bluntly dissected subcutaneous pouch on an immune-deficient mouse. Thirty to sixty days later, constructs were removed for analysis. The plant lectin Ulex europeus I was used to detect human endothelial cells, while anti-murine CD31 antibody was used to detect murine vessels that grew into the construct by angiogenesis. Lumenal structures composed of human endothelial cells were evident at 30 days and exhibited increased vascular complexity and coverage by murine perivascular cells by 60 days. Despite the limitations of using *BCL-2*-transduced HUVECs, this was a remarkable achievement in that it provided the first indication that *in vitro* assembled human endothelial cords could survive and form functional microvascular networks *in vivo*. The model was extended by implanting the cell-seeded type I collagen/fibronectin gels into the gastrocnemius muscle and showing vascularization within the implant.[83] Subsequently, the same group tested cord blood– and adult blood–derived EPCs in comparison to *BCL-2*-HUVECs in this model. Both sources of human EPCs formed vessels that were invested with murine perivascular cells after 21 days *in vivo*,[3] lending support to the concept of using autologous EPCs as a means to build vascular networks within a tissue implant. Importantly, this model has been applied successfully, as well as modified, for a number of experimental approaches which have centered on creating bioengineered vascular networks from adult progenitor cells.[17,84–86]

A 2004 study by our laboratory demonstrated the ability of human cord blood–derived EPCs to form microvessel-like structures when co-seeded with mature human smooth muscle cells on a polyglycolic acid-poly-L-lactic acid (PGA/PLLA) nonwoven mesh. EPCs seeded alone were viable but did not form microvessels in this setting, which underscored the importance of a collaborative interaction between endothelial and perivascular cells in building de novo vascular networks.[32] The potential of EPCs and perivascular cells to form microvessels was corroborated by the Jain group.[2] In this study, cord blood EPCs or adult peripheral blood EPCs were tested by implanting the cells, with or without a murine embryonic cell line, called 10T1/2, into a murine cranial window. The authors demonstrated the formation of perfused, functional, and long-lasting vessels using cord blood–derived EPCs and 10T1/2 cells. However, they found that vessels formed with adult EPCs and 10T1/2 cells were only detected for a short period of 11–21 days after implantation. The transient nature of the adult blood EPC-engineered vessels may in part be attributed to use of a murine cell line as the perivascular support cell. A follow-up paper by the same group showed that human bone marrow–derived MSCs combined with HUVECs were stable for over 4 months *in vivo*.[17] This underscores the importance of the perivascular support cell in the collaborative interaction with the endothelium.

Similarly, Keith March's group used EPCs in combination with stromal cells derived from human subcutaneous adipose tissue.[84] A suspension of these two cell types in collagen and fibronectin, polymerized *in vitro* and implanted subcutaneously into immune-compromised mice, formed a perfused network in as little as 4 days. In addition to this rapid network formation, it was shown that EPCs and ASCs could form a perfused vasculature in the presence of mature human adipocytes or porcine pancreatic islet cells, which stained for insulin *in vivo* after 2 weeks.

The laboratories of Hughes and George used a different strategy to form a vascular template *in vitro* and showed that the prevascularization step significantly increased the onset of anastomosis formation and construct perfusion *in vivo*.[87] These investigators co-seeded HUVECs and human lung fibroblasts in fibrin gels and allowed the cells to form capillary-like networks over the course of 7 days *in vitro* prior to being implanted subcutaneously into immunodeficient mice. *In vitro*, the cellular networks consisted of multicellular lumens showing spatial association of endothelial cells with fibroblasts. Importantly, the prevascularized constructs connected with host vessels within 4–5 days, whereas constructs allowed only one day of pre-assembly required 14 days to connect with the host vasculature. Importantly, this study provided the first evidence that the cellular assembly and cord formation that is observed *in vitro* corresponds to more rapid establishment of perfusion *in vivo*. This finding bodes well for the strategy of pre-constructing vascular networks in tissue-engineered organs prior to implantation. When constructs were created using HUVECs alone, the tissue never became perfused, again highlighting the importance of the paracrine interaction between endothelial cells and a supporting cell, in this case, fibroblasts. The investigators went on to compare human cord blood–derived EPCs to HUVECs in their model and, in addition, increased the ratio of fibroblasts to endothelial cells from 1:5 to 2:1. Strikingly, EPCs combined with fibroblasts at a ratio of 2:1 (fibroblasts:EPCs) formed vascular connections *in vivo* in 1–3 days.[88] The increased number of fibroblasts accelerated anastomoses formation *in vivo* for both EPCs and HUVECs by about 3 days, but EPC constructs achieved perfusion more rapidly overall, suggesting that cord blood–derived EPCs have greater potential for connecting with angiogenic murine vessels. The increased number of fibroblasts relative to endothelial cells may have contributed to vascular connections by providing increased amounts of soluble angiogenic factors, matrix remodeling enzymes and by stabilizing the nascent vessels.

2.4.2   VASCULAR NETWORKS FORMED IN SITU

Based on our initial observations showing that EPCs formed vessel-like structures on biopolymeric scaffolds only when co-seeded with human SMCs,[32] we set out to test the vasculogenic potential of human EPCs and SMCs suspended as single cells in liquid Matrigel at 4°C and implanted immediately by subcutaneous injection into immunodeficient mice, wherein the cell/Matrigel suspension forms a gel at 37°C. Within 5–7 days, the Matrigel implants were filled with human EPC-lined blood vessels surrounded by smooth muscle cells. That the human vessels had connected with the host circulation was inferred from the presence of red and white blood cells within the lumens. Microvessel density was shown to correlate with number of EPC/SMCs implanted, suggesting that vascular density can be dialed up or down by adjusting the input cell number. In addition, Melero-Martin showed that EPCs from adult peripheral blood were vasculogenic in this model.

The most recent work by Melero-Martin of the Bischoff group demonstrates the successful creation of complex, cell-mediated microvascular networks on a short timescale using cells sourced from adult blood and bone marrow.[4] In this work, MPCs isolated from adult bone marrow were used instead of the mature HSVSMC which filled the perivascular role in the lab's prior studies. MPCs were combined with EPCs expanded from umbilical cord blood or adult peripheral blood. After suspension of these cells in Matrigel and subcutaneous injection, a vascular network formed that became anastomosed with the host circulation in 4–7 days. The bioengineered vascular network was found to maintain a steady blood vessel density for up to 4 weeks *in vivo*. Spatial association of EPCs and MPCs was observed using GFP-tagged populations, and the persistence of the perfused vascular networks was observed using bioluminescence live imaging, in which luciferase-tagged EPCs luminesce upon delivery of the substrate luciferin via the bloodstream.[4] This work proved that a robust, fully injectible construct can be created using accessible adult cells. A similar injectible two-cell system was also developed by the Langer group, who found that embryonic stem cells differentiated into endothelial-like and smooth muscle–like cells formed perfused blood vessel networks when injected in Matrigel subcutaneously,[89] but to date there is little indication that vascular networks built using embryonic stem cells are superior to those built from adult progenitor cells.

The Augustin laboratory has developed a cell spheroid-based method of forming vascular networks.[90] In this model, HUVECs are cultured into aggregate spheroids by the hanging droplet method. Cells are next suspended in a Matrigel/fibrin blend supplemented with vascular endothelial growth factor (VEGF) and basic fibroblast growth factor (bFGF), then implanted subcutaneously into immunodeficient mice. Between 8 and 20 days, increasing numbers of HUVEC-lined lumens formed: some were perfused and some had recruited pericyte coverage from the host. To demonstrate higher-order functionality of these vessels, Alajati and colleagues showed that spheroid-derived vessels reacted to injection of tumor necrosis factor-α by expressing leukocyte adhesion proteins and increasing monocyte recruitment and extravasation. In summary, the study showed success in forming a vessel network with fully differentiated human ECs, and as such, provides another model system to study the cellular, molecular, and biochemical requirements for human endothelial cells to form de novo blood vessels.

In summary, recent work in the field of cell-mediated vascular network formation suggests that the basic two-cell blood vessel concept is valid using multiple endothelial and mesenchymal cell sources, different anatomical locations, and various extracellular matrix compositions. This model has been adapted for two basic delivery methods: *in vitro* polymerization and incubation of a gel followed by surgical implantation, as well as for direct injection of the vessel-building constituents (see Figure 2.3). Furthermore, the studies indicate that blood vessel networks persist for up to 4 months *in vivo*[2] and suggests great potential for supporting a population of parenchymal cells for future therapies.[84]

2.4.3 Systemic Delivery of Vascular Progenitors

There has been interest in the ability of vascular progenitors to home to sites of tumors and injuries, and how such homing ability could be used for therapeutic purposes. In 2008, Gérard Tobelem's lab demonstrated the homing capability of cord blood–derived EPCs and MPCs to sites of injury[91] and tumor vasculature.[92] Tobelem's group isolated populations of EPCs and MPCs from umbilical cord blood, then showed that upon co-injecting these cells into the mouse systemic circulation during a hindlimb ischemia experiment, the EPCs homed to the injured region and improved hindlimb perfusion. This group found that expression of angiopoeitin-1 by MPCs and Tie-2 by EPCs was a key signaling element for vasculogenesis *in vitro* and *in vivo*.[91] This study demonstrates that EPCs participate in dynamic tissue repair and also provides a ligand/receptor basis for the EPC/pericyte association observed in the labs of Bischoff, Jain, March, Augustin, and others.

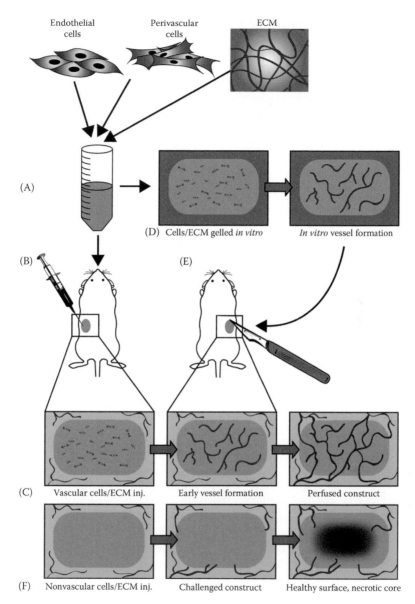

FIGURE 2.3 (See color insert.)   Two emerging paradigms for creation of cell-mediated vascular networks. (A) Endothelial-lineage cells, perivascular cells, and a solubilized extracellular matrix (ECM) are the essential constituents of engineered blood vessel networks. These two cell types are suspended in the ECM and, in one method, (B) injected directly into the animal, where (C) the individual cells form a provisional network and anastomose with the systemic circulation, resulting in a perfused construct. (D) In a second method, the cell/ECM suspension is polymerized and incubated *in vitro*, allowing initial network formation to occur. Upon surgical implantation (E), this network anastomoses with the host circulation to achieve perfusion. (F) It is hypothesized that implantation of a metabolically demanding cell population (cells not depicted) in the absence of vascular cells will result in necrosis within the construct.

2.4.4 Vascular Networks and Parenchymal Cells

A number of studies have been published demonstrating the ability of endothelial-lineage and perivascular cells to form blood vessels in order to support populations of parenchymal cells. Ingram and March showed that EPCs and ASCs could vascularize implanted adipose and pancreatic tissue constructs.[84] As described above, D'Amore and Langer have shown that both HUVECs and human embryonic stem cell–derived endothelial cells can form vessels in the presence of mouse skeletal myoblasts in 2–4 weeks, and addition of murine embryonic fibroblasts increased blood vessel formation.[93] In this study, a bioresorbable polymer sponge soaked with Matrigel served as the extracellular matrix in *in vivo* experiments that introduced this construct to the abdominal and quadriceps muscles of immunodeficient mice. *In vivo* for 2 weeks, myoblasts formed elongated, multinucleated fibrils staining for skeletal muscle markers desmin and myogenin. Blood vessels expressing CD31 and vWF penetrated the engineered tissues and were shown to be perfused based on binding of tomato lectin (helix pomatia) that had been injected intravenously. Although scaffolds containing no cells, and containing myoblasts alone showed some blood vessel ingrowth, constructs containing endothelial cells and fibroblasts showed a moderate but significant increase in vascularity. Unlike many studies, in this model, HUVECs and embryonic-derived endothelial cells showed equivalent vascular density,[93] perhaps due to the intrinsic angiogenic nature of the scaffold.

Several studies of cardiac tissue engineering were published in 2009, and these reinforced the importance of vascular cells for future therapeutic applications. The Murry group has used human embryonic stem cell–derived cardiomyocytes to create engineered cardiovascular grafts.[94] These constructs were created by culturing cells on a rotating orbital shaker where they coalesced into tissues. Upon transplanting constructs consisting only of embryonic cardiomyocytes onto rat hearts, most cells died, but upon adding HUVECs and mouse embryonic fibroblasts to the graft, myocyte survival and differentiation increased, as determined by β-myosin heavy chain staining. These grafts contained CD31-expressing lumens perfused with red blood cells. Furthermore, the grafts contracted in response to exogenous electrical stimulation.[94] Levenberg and Gepstein published a similar study using equivalent cell types seeded onto a biopolymer and Matrigel sponge, as described above, documenting vascularization and differentiation of the grafts, though not electrically-stimulated contraction.[95] These two studies show the importance of a vascular supply for successful myocardial tissue-engineering models.

2.5 FUTURE DIRECTIONS

Over a half dozen laboratories have shown the feasibility of building vascular networks *in vivo* using adult human endothelial and MPCs. The various experimental approaches in these studies underscore the versatility and power of these human cells. This success brings a full slate of questions and challenges that must be addressed. One important question is to determine how the vasculogenic potential of the cells is influenced by the three-dimensional extracellular matrix into which the cells are suspended, seeded, or delivered. The studies to date have used type I collagen/fibronectin gels and fibrin gels for pre-assembly *in vitro* prior to implantation, while Matrigel and Matrigel/fibrin have been used for injecting cells into subcutaneous sites. The matrices have not been compared head-to-head in the same model system to determine if a given extracellular matrix preparation correlates with superior microvessel density. Beyond microvessel number, how the ECM might influence vessel maturation to a microvascular, arterial, or venous architecture and to further differentiate to organ-specific vascular beds will be essential to building functional vascular networks.

In this regard, the critical question is whether or not the bioengineered vessels can contribute to organ or tissue function. For example, can direct delivery of EPCs and MPCs to ischemic tissue accelerate the return of blood flow and in turn contribute to tissue repair and regeneration? Or would intercalation of a preformed vascular network into muscle, as done by Enis and colleagues, contribute to muscle function? The studies by Foubert and colleagues and Stevens and colleagues are encouraging in this regard.[91,94] Many more studies are needed to understand the extent to which bioengineered vascular networks can be integrated with the host circulation and contribute to organ health and function. The potential applications of EPC- and MPC-mediated tissue vascularization are numerous, but a few examples would be in wound healing and skin grafts, to increase survival of pancreatic islet transplants, and to revascularize limbs to alleviate peripheral vascular disease.

The presence of EPCs and MPCs in cord blood, adult peripheral blood, and bone marrow raises the question of the physiologic role of these cells. Several studies have shown that circulating EPC concentration (defined as CD34+/VEGFR-2+/CD45[low]) correlates with cardiovascular health,[96,97] decreased risk of death from cardiovascular events,[98] and increased physical exercise in school age children.[99] However, the precise mechanism by which EPCs affect the vasculature is unknown. The interpretation of these studies is also limited by the fact that there is no specific marker for the identification of EPCs. The cell surface markers CD34 and VEGFR-2 are expressed by EPCs but also subsets of hematopoietic progenitor cells. Therefore, one can only speculate as to whether the beneficial effects correlated with increased CD34+/VEGFR2+ cells can be attributed precisely to the EPCs we have defined here as cells with robust proliferative and vasculogenic activity.

Another potential application of vascular progenitor cells is to use them as a platform for the delivery of therapeutic proteins to the circulation. Hebbel and colleagues have engineered human EPCs to provide continuous, systemic delivery of Factor VIII, the coagulation factor deficient in hemophilia A patients. Such patients are currently treated with repeated intravenous injections of either recombinant or donor-derived Factor VIII. The first study in 2002 showed that human EPCs (referred to as BOECs), which were genetically engineered to express Factor VIII and delivered by intravenous injection, homed to the bone marrow and produced therapeutic levels of Factor VIII in the plasma for 156 days.[8] A subsequent study showed similar results when BOEC-expressing Factor VIII were implanted subcutaneously in Matrigel.[100] These studies provide an exciting paradigm for using autologous EPCs for systemic protein delivery which could have broad applicability. For example, EPCs could be engineered to deliver anti-angiogenic proteins such as endostatin or to deliver therapeutic antibodies that directly target the tumor endothelium. This would potentially provide steady doses of the desired protein over months to years.

REFERENCES

1. Melero-Martin, J.; Khan, Z. A.; Picard, A.; Wu, X.; Paruchuri, S.; Bischoff, J., In vivo vasculogenic potential of human blood-derived endothelial progenitor cells. *Blood* 2007, 109(11), 4761–4768.
2. Au, P.; Daheron, L. M.; Duda, D. G.; Cohen, K. S.; Tyrrell, J. A.; Lanning, R. M.; Fukumura, D.; Scadden, D. T.; Jain, R. K., Differential in vivo potential of endothelial progenitor cells from human umbilical cord blood and adult peripheral blood to form functional long-lasting vessels. *Blood* 2008, 111(3), 1302–1305.
3. Shepherd, B. R.; Enis, D. R.; Wang, F.; Suarez, Y.; Pober, J. S.; Schechner, J. S., Vascularization and engraftment of a human skin substitute using circulating progenitor cell-derived endothelial cells. *Faseb J* 2006, 20(10), 1739–1741.
4. Melero-Martin, J. M.; De Obaldia, M. E.; Kang, S. Y.; Khan, Z. A.; Yuan, L.; Oettgen, P.; Bischoff, J., Engineering robust and functional vascular networks in vivo with human adult and cord blood-derived progenitor cells. *Circ Res* 2008, 103(2), 194–202.
5. Ingram, D. A.; Caplice, N. M.; Yoder, M. C., Unresolved questions, changing definitions, and novel paradigms for defining endothelial progenitor cells. *Blood* 2005, 106(5), 1525–1531.
6. Yoder, M. C.; Mead, L. E.; Prater, D.; Krier, T. R.; Mroueh, K. N.; Li, F.; Krasich, R.; Temm, C. J.; Prchal, J. T.; Ingram, D. A., Redefining endothelial progenitor cells via clonal analysis and hematopoietic stem/progenitor cell principals. *Blood* 2007, 109(5), 1801–1809.
7. Timmermans, F.; Plum, J.; Yoder, M. C.; Ingram, D. A.; Vandekerckhove, B.; Case, J., Endothelial progenitor cells: Identity defined? *J Cell Mol Med* 2009, 13(1), 87–102.
8. Lin, Y.; Chang, L.; Solovey, A.; Healey, J. F.; Lollar, P.; Hebbel, R. P., Use of blood outgrowth endothelial cells for gene therapy for hemophilia A. *Blood* 2002, 99(2), 457–462.
9. Urbich, C.; Dimmeler, S., Endothelial progenitor cells: Characterization and role in vascular biology. *Circ Res* 2004, 95(4), 343–353.
10. Horrevoets, A. J., Angiogenic monocytes: Another colorful blow to endothelial progenitors. *Am J Pathol* 2009, 174(5), 1594–1596.
11. Sieveking, D. P.; Buckle, A.; Celermajer, D. S.; Ng, M. K., Strikingly different angiogenic properties of endothelial progenitor cell subpopulations: Insights from a novel human angiogenesis assay. *J Am Coll Cardiol* 2008, 51(6), 660–668.
12. De Palma, M.; Naldini, L., Tie2-expressing monocytes (TEMs): Novel targets and vehicles of anticancer therapy? *Biochim Biophys Acta* 2009, 1796(1), 5–10.
13. Somani, A.; Nguyen, J.; Milbauer, L. C.; Solovey, A.; Sajja, S.; Hebbel, R. P., The establishment of murine blood outgrowth endothelial cells and observations relevant to gene therapy. *Transl Res* 2007, 150(1), 30–39.
14. Majesky, M. W., Developmental basis of vascular smooth muscle diversity. *Arterioscler Thromb Vasc Biol* 2007, 27(6), 1248–1258.
15. Crisan, M.; Chen, C. W.; Corselli, M.; Andriolo, G.; Lazzari, L.; Peault, B., Perivascular multipotent progenitor cells in human organs. *Ann NY Acad Sci* 2009, 1176, 118–123.
16. Schor, A. M.; Canfield, A. E.; Sutton, A. B.; Arciniegas, E.; Allen, T. D., Pericyte differentiation. *Clin Orthop Relat Res* 1995, (313), 81–91.
17. Au, P.; Tam, J.; Fukumura, D.; Jain, R. K., Bone marrow-derived mesenchymal stem cells facilitate engineering of long-lasting functional vasculature. *Blood* 2008, 111(9), 4551–4558.
18. Gong, Z.; Niklason, L. E., Small-diameter human vessel wall engineered from bone marrow-derived mesenchymal stem cells (hMSCs). *Faseb J* 2008, 22(6), 1635–1648.
19. DeLeve, L. D.; Wang, X.; McCuskey, M. K.; McCuskey, R. S., Rat liver endothelial cells isolated by anti-CD31 immunomagnetic separation lack fenestrae and sieve plates. *Am J Physiol Gastrointest Liver Physiol* 2006, 291(6), G1187–G1189.

20. Alva, J. A.; Zovein, A. C.; Monvoisin, A.; Murphy, T.; Salazar, A.; Harvey, N. L.; Carmeliet, P.; Iruela-Arispe, M. L., VE-Cadherin-Cre-recombinase transgenic mouse: A tool for lineage analysis and gene deletion in endothelial cells. *Dev Dyn* 2006, 235(3), 759–767.

21. Fraser, S. T.; Ogawa, M.; Yokomizo, T.; Ito, Y.; Nishikawa, S.; Nishikawa, S., Putative intermediate precursor between hematogenic endothelial cells and blood cells in the developing embryo. *Dev Growth Differ* 2003, 45(1), 63–75.

22. Wu, X.; Lensch, M. W.; Wylie-Sears, J.; Daley, G. Q.; Bischoff, J., Hemogenic endothelial progenitor cells isolated from human umbilical cord blood. *Stem Cells* 2007, 25(11), 2770–2776.

23. Weibel, E. R.; Palade, G. E., New cytoplasmic components in arterial endothelia. *J Cell Biol* 1964, 23, 101–112.

24. Michaux, G.; Cutler, D. F., How to roll an endothelial cigar: The biogenesis of Weibel-Palade bodies. *Traffic* 2004, 5(2), 69–78.

25. Rafii, S., Circulating endothelial precursors: Mystery, reality, and promise. *J Clin Invest* 2000, 105(1), 17–19.

26. Rohde, E.; Malischnik, C.; Thaler, D.; Maierhofer, T.; Linkesch, W.; Lanzer, G.; Guelly, C.; Strunk, D., Blood monocytes mimic endothelial progenitor cells. *Stem Cells* 2006, 24(2), 357–367.

27. Solovey, A.; Lin, Y.; Browne, P.; Choong, S.; Wayner, E.; Hebbel, R. P., Circulating activated endothelial cells in sickle cell anemia. *N Engl J Med* 1997, 337(22), 1584–1590.

28. Crisan, M.; Yap, S.; Casteilla, L.; Chen, C. W.; Corselli, M.; Park, T. S.; Andriolo, G.; Sun, B.; Zheng, B.; Zhang, L.; Norotte, C.; Teng, P. N.; Traas, J.; Schugar, R.; Deasy, B. M.; Badylak, S.; Buhring, H. J.; Giacobino, J. P.; Lazzari, L.; Huard, J.; Peault, B., A perivascular origin for mesenchymal stem cells in multiple human organs. *Cell Stem Cell* 2008, 3(3), 301–313.

29. Nakatsu, M. N.; Hughes, C. C., An optimized three-dimensional in vitro model for the analysis of angiogenesis. *Methods Enzymol* 2008, 443, 65–82.

30. Stratman, A. N.; Saunders, W. B.; Sacharidou, A.; Koh, W.; Fisher, K. E.; Zawieja, D. C.; Davis, M. J.; Davis, G. E., Endothelial cell lumen and vascular guidance tunnel formation requires MT1-MMP-dependent proteolysis in 3-dimensional collagen matrices. *Blood* 2009, 114(2), 237–247.

31. Dvorin, E. L.; Jacobson, J.; Roth, S. J.; Bischoff, J., Human pulmonary valve endothelial cells express functional adhesion molecules for leukocytes. *J Heart Valve Dis* 2003, 12(5), 617–624.

32. Wu, X.; Rabkin-Aikawa, E.; Guleserian, K. J.; Perry, T. E.; Masuda, Y.; Sutherland, F. W.; Schoen, F. J.; Mayer, J. E., Jr.; Bischoff, J., Tissue-engineered microvessels on three-dimensional biodegradable scaffolds using human endothelial progenitor cells. *Am J Physiol Heart Circ Physiol* 2004, 287(2), H480–H487.

33. Chen, C. W.; Montelatici, E.; Crisan, M.; Corselli, M.; Huard, J.; Lazzari, L.; Peault, B., Perivascular multi-lineage progenitor cells in human organs: Regenerative units, cytokine sources or both? *Cytokine Growth Factor Rev* 2009, 20(5–6), 429–434.

34. Paquet-Fifield, S.; Schluter, H.; Li, A.; Aitken, T.; Gangatirkar, P.; Blashki, D.; Koelmeyer, R.; Pouliot, N.; Palatsides, M.; Ellis, S.; Brouard, N.; Zannettino, A.; Saunders, N.; Thompson, N.; Li, J.; Kaur, P., A role for pericytes as microenvironmental regulators of human skin tissue regeneration. *J Clin Invest* 2009, 119(9), 2795–2806.

35. Bondjers, C.; Kalen, M.; Hellstrom, M.; Scheidl, S. J.; Abramsson, A.; Renner, O.; Lindahl, P.; Cho, H.; Kehrl, J.; Betsholtz, C., Transcription profiling of platelet-derived growth factor-B-deficient mouse embryos identifies RGS5 as a novel marker for pericytes and vascular smooth muscle cells. *Am J Pathol* 2003, 162(3), 721–729.

36. Cho, H.; Kozasa, T.; Bondjers, C.; Betsholtz, C.; Kehrl, J. H., Pericyte-specific expression of Rgs5: Implications for PDGF and EDG receptor signaling during vascular maturation. *Faseb J* 2003, 17(3), 440–442.

37. Brachvogel, B.; Moch, H.; Pausch, F.; Schlotzer-Schrehardt, U.; Hofmann, C.; Hallmann, R.; von der Mark, K.; Winkler, T.; Poschl, E., Perivascular cells expressing annexin A5 define a novel mesenchymal stem cell-like population with the capacity to differentiate into multiple mesenchymal lineages. *Development* 2005, 132(11), 2657–2668.

38. Brachvogel, B.; Pausch, F.; Farlie, P.; Gaipl, U.; Etich, J.; Zhou, Z.; Cameron, T.; von der Mark, K.; Bateman, J. F.; Poschl, E., Isolated Anxa5+/Sca-1+ perivascular cells from mouse meningeal vasculature retain their perivascular phenotype in vitro and in vivo. *Exp Cell Res* 2007, 313(12), 2730–2743.

39. Stratman, A. N.; Malotte, K. M.; Mahan, R. D.; Davis, M. J.; Davis, G. E., Pericyte recruitment during vasculogenic tube assembly stimulates endothelial basement membrane matrix formation. *Blood* 2009, 114(24), 5091–5101.

40. Darland, D. C.; D'Amore, P. A., TGF beta is required for the formation of capillary-like structures in three-dimensional cocultures of 10T1/2 and endothelial cells. *Angiogenesis* 2001, 4(1), 11–20.

41. Yoshida, T.; Owens, G. K., Molecular determinants of vascular smooth muscle cell diversity. *Circ Res* 2005, 96(3), 280–291.

42. Korff, T.; Kimmina, S.; Martiny-Baron, G.; Augustin, H. G., Blood vessel maturation in a 3-dimensional spheroidal coculture model: Direct contact with smooth muscle cells regulates endothelial cell quiescence and abrogates VEGF responsiveness. *Faseb J* 2001, 15(2), 447–457.

43. Lin, Y.; Weisdorf, D. J.; Solovey, A.; Hebbel, R. P., Origins of circulating endothelial cells and endothelial outgrowth from blood. *J Clin Invest* 2000, 105(1), 71–77.

44. Jacques, N.; Vimond, N.; Conforti, R.; Griscelli, F.; Lecluse, Y.; Laplanche, A.; Malka, D.; Vielh, P.; Farace, F., Quantification of circulating mature endothelial cells using a whole blood four-color flow cytometric assay. *J Immunol Methods* 2008, 337(2), 132–143.

45. Bertolini, F.; Shaked, Y.; Mancuso, P.; Kerbel, R. S., The multifaceted circulating endothelial cell in cancer: Towards marker and target identification. *Nat Rev Cancer* 2006, 6(11), 835–845.

46. George, F.; Poncelet, P.; Laurent, J. C.; Massot, O.; Arnoux, D.; Lequeux, N.; Ambrosi, P.; Chicheportiche, C.; Sampol, J., Cytofluorometric detection of human endothelial cells in whole blood using S-Endo 1 monoclonal antibody. *J Immunol Methods* 1991, 139(1), 65–75.

47. Sbarbati, R.; de Boer, M.; Marzilli, M.; Scarlattini, M.; Rossi, G.; van Mourik, J. A., Immunologic detection of endothelial cells in human whole blood. *Blood* 1991, 77(4), 764–769.

48. Kaushal, S.; Amiel, G. E.; Guleserian, K. J.; Shapira, O. M.; Perry, T.; Sutherland, F. W.; Rabkin, E.; Moran, A. M.; Schoen, F. J.; Atala, A.; Soker, S.; Bischoff, J.; Mayer, J. E., Jr., Functional small-diameter neovessels created using endothelial progenitor cells expanded ex vivo. *Nat Med* 2001, 7(9), 1035–1040.

49. Dvorin, E. L.; Wylie-Sears, J.; Kaushal, S.; Martin, D. P.; Bischoff, J., Quantitative evaluation of endothelial progenitors and cardiac valve endothelial cells: Proliferation and differentiation on poly-glycolic acid/poly-4-hydroxybutyrate scaffold in response to vascular endothelial growth factor and transforming growth factor beta1. *Tissue Eng* 2003, 9(3), 487–493.

50. Fuchs, S.; Motta, A.; Migliaresi, C.; Kirkpatrick, C. J., Outgrowth endothelial cells isolated and expanded from human peripheral blood progenitor cells as a potential source of autologous cells for endothelialization of silk fibroin biomaterials. *Biomaterials* 2006, 27(31), 5399–5408.

51. Stroncek, J. D.; Grant, B. S.; Brown, M. A.; Povsic, T. J.; Truskey, G. A.; Reichert, W. M., Comparison of endothelial cell phenotypic markers of late-outgrowth endothelial progenitor cells isolated from patients with coronary artery disease and healthy volunteers. *Tissue Eng Part A* 2009, 15(11), 3473–3486.

52. Reinisch, A.; Hofmann, N. A.; Obenauf, A. C.; Kashofer, K.; Rohde, E.; Schallmoser, K.; Flicker, K.; Lanzer, G.; Linkesch, W.; Speicher, M. R.; Strunk, D., Humanized large-scale expanded endothelial colony-forming cells function in vitro and in vivo. *Blood* 2009, 113(26), 6716–6725.

53. Ingram, D. A.; Mead, L. E.; Tanaka, H.; Meade, V.; Fenoglio, A.; Mortell, K.; Pollok, K.; Ferkowicz, M. J.; Gilley, D.; Yoder, M. C., Identification of a novel hierarchy of endothelial progenitor cells using human peripheral and umbilical cord blood. *Blood* 2004, 104(9), 2752–2760.

54. Bompais, H.; Chagraoui, J.; Canron, X.; Crisan, M.; Liu, X. H.; Anjo, A.; Tolla-Le Port, C.; Leboeuf, M.; Charbord, P.; Bikfalvi, A.; Uzan, G., Human endothelial cells derived from circulating progenitors display specific functional properties compared with mature vessel wall endothelial cells. *Blood* 2004, 103(7), 2577–2584.

55. Ott, I.; Keller, U.; Knoedler, M.; Gotze, K. S.; Doss, K.; Fischer, P.; Urlbauer, K.; Debus, G.; von Bubnoff, N.; Rudelius, M.; Schomig, A.; Peschel, C.; Oostendorp, R. A., Endothelial-like cells expanded from CD34+ blood cells improve left ventricular function after experimental myocardial infarction. *Faseb J* 2005, 19(8), 992–994.

56. Schmidt, D.; Breymann, C.; Weber, A.; Guenter, C. I.; Neuenschwander, S.; Zund, G.; Turina, M.; Hoerstrup, S. P., Umbilical cord blood derived endothelial progenitor cells for tissue engineering of vascular grafts. *Ann Thorac Surg* 2004, 78(6), 2094–2098.

57. Schmidt, D.; Asmis, L. M.; Odermatt, B.; Kelm, J.; Breymann, C.; Gossi, M.; Genoni, M.; Zund, G.; Hoerstrup, S. P., Engineered living blood vessels: Functional endothelia generated from human umbilical cord-derived progenitors. *Ann Thorac Surg* 2006, 82(4), 1465–1471; discussion 1471.

58. Javed, M. J.; Mead, L. E.; Prater, D.; Bessler, W. K.; Foster, D.; Case, J.; Goebel, W. S.; Yoder, M. C.; Haneline, L. S.; Ingram, D. A., Endothelial colony forming cells and mesenchymal stem cells are enriched at different gestational ages in human umbilical cord blood. *Pediatr Res* 2008, 64(1), 68–73.

59. Reyes, M.; Dudek, A.; Jahagirdar, B.; Koodie, L.; Marker, P. H.; Verfaillie, C. M., Origin of endothelial progenitors in human postnatal bone marrow. *J Clin Invest* 2002, 109(3), 337–346.

60. Ingram, D. A.; Mead, L. E.; Moore, D. B.; Woodard, W.; Fenoglio, A.; Yoder, M. C., Vessel wall-derived endothelial cells rapidly proliferate because they contain a complete hierarchy of endothelial progenitor cells. *Blood* 2005, 105(7), 2783–2786.

61. Sainz, J.; Al Haj Zen, A.; Caligiuri, G.; Demerens, C.; Urbain, D.; Lemitre, M.; Lafont, A., Isolation of "side population" progenitor cells from healthy arteries of adult mice. *Arterioscler Thromb Vasc Biol* 2006, 26(2), 281–286.

62. Zengin, E.; Chalajour, F.; Gehling, U. M.; Ito, W. D.; Treede, H.; Lauke, H.; Weil, J.; Reichenspurner, H.; Kilic, N.; Ergun, S., Vascular wall resident progenitor cells: A source for postnatal vasculogenesis. *Development* 2006, 133(8), 1543–1551.

63. Tilki, D.; Hohn, H. P.; Ergun, B.; Rafii, S.; Ergun, S., Emerging biology of vascular wall progenitor cells in health and disease. *Trends Mol Med* 2009, 15(11), 501–509.

64. Galmiche, M. C.; Kotelianski, V. E.; Briere, J.; Herve, P.; Charbord, P., Stromal cells from human long-term marrow cultures are mesenchymal cells that differentiate following a vascular smooth muscle differentiation pathway. *Blood* 1993, 82(1), 66–76.

65. Shimizu, K.; Sugiyama, S.; Aikawa, M.; Fukumoto, Y.; Rabkin, E.; Libby, P.; Mitchell, R. N., Host bone-marrow cells are a source of donor intimal smooth- muscle-like cells in murine aortic transplant arteriopathy. *Nat Med* 2001, 7(6), 738–741.

66. Wagner, W.; Ho, A. D., Mesenchymal stem cell preparations—Comparing apples and oranges. *Stem Cell Rev* 2007, 3(4), 239–248.
67. Martens, T. P.; See, F.; Schuster, M. D.; Sondermeijer, H. P.; Hefti, M. M.; Zannettino, A.; Gronthos, S.; Seki, T.; Itescu, S., Mesenchymal lineage precursor cells induce vascular network formation in ischemic myocardium. *Nat Clin Pract Cardiovasc Med* 2006, 3(1), S18–S22.
68. Simper, D.; Stalboerger, P. G.; Panetta, C. J.; Wang, S.; Caplice, N. M., Smooth muscle progenitor cells in human blood. *Circulation* 2002, 106(10), 1199–1204.
69. Kogler, G.; Sensken, S.; Wernet, P., Comparative generation and characterization of pluripotent unrestricted somatic stem cells with mesenchymal stem cells from human cord blood. *Exp Hematol* 2006, 34(11), 1589–1595.
70. Traktuev, D. O.; Merfeld-Clauss, S.; Li, J.; Kolonin, M.; Arap, W.; Pasqualini, R.; Johnstone, B. H.; March, K. L., A population of multipotent CD34-positive adipose stromal cells share pericyte and mesenchymal surface markers, reside in a periendothelial location, and stabilize endothelial networks. *Circ Res* 2008, 102(1), 77–85.
71. Metharom, P.; Liu, C.; Wang, S.; Stalboerger, P.; Chen, G.; Doyle, B.; Ikeda, Y.; Caplice, N. M., Myeloid lineage of high proliferative potential human smooth muscle outgrowth cells circulating in blood and vasculogenic smooth muscle-like cells in vivo. *Atherosclerosis* 2008, 198(1), 29–38.
72. Tintut, Y.; Alfonso, Z.; Saini, T.; Radcliff, K.; Watson, K.; Bostrom, K.; Demer, L. L., Multilineage potential of cells from the artery wall. *Circulation* 2003, 108(20), 2505–2510.
73. Hu, Y.; Zhang, Z.; Torsney, E.; Afzal, A. R.; Davison, F.; Metzler, B.; Xu, Q., Abundant progenitor cells in the adventitia contribute to atherosclerosis of vein grafts in ApoE-deficient mice. *J Clin Invest* 2004, 113(9), 1258–1265.
74. Howson, K. M.; Aplin, A. C.; Gelati, M.; Alessandri, G.; Parati, E. A.; Nicosia, R. F., The postnatal rat aorta contains pericyte progenitor cells that form spheroidal colonies in suspension culture. *Am J Physiol Cell Physiol* 2005, 289(6), C1396–C1407.
75. Melero-Martin, J. M.; Bischoff, J., Chapter 13. An in vivo experimental model for postnatal vasculogenesis. *Methods Enzymol* 2008, 445, 303–329.
76. Mead, L. E.; Prater, D.; Yoder, M. C.; Ingram, D. A., Isolation and characterization of endothelial progenitor cells from human blood. *Curr Protoc Stem Cell Biol* 2008, Chapter 2, Unit 2C 1.
77. Jackson, C. J.; Garbett, P. K.; Nissen, B.; Schrieber, L., Binding of human endothelium to Ulex europaeus I-coated Dynabeads: Application to the isolation of microvascular endothelium. *J Cell Sci* 1990, 96(Pt 2), 257–262.
78. Friedenstein, A. J., Marrow stromal fibroblasts. *Calcif Tissue Int* 1995, 56(1), S17.
79. Sotiropoulou, P. A.; Perez, S. A.; Salagianni, M.; Baxevanis, C. N.; Papamichail, M., Characterization of the optimal culture conditions for clinical scale production of human mesenchymal stem cells. *Stem Cells* 2006, 24(2), 462–471.
80. Schallmoser, K.; Rohde, E.; Reinisch, A.; Bartmann, C.; Thaler, D.; Drexler, C.; Obenauf, A. C.; Lanzer, G.; Linkesch, W.; Strunk, D., Rapid large-scale expansion of functional mesenchymal stem cells from unmanipulated bone marrow without animal serum. *Tissue Eng Part C Methods* 2008, 14(3), 185–196.
81. Pittenger, M. F.; Mackay, A. M.; Beck, S. C.; Jaiswal, R. K.; Douglas, R.; Mosca, J. D.; Moorman, M. A.; Simonetti, D. W.; Craig, S.; Marshak, D. R., Multilineage potential of adult human mesenchymal stem cells. *Science* 1999, 284(5411), 143–147.
82. Schechner, J. S.; Nath, A. K.; Zheng, L.; Kluger, M. S.; Hughes, C. C.; Sierra-Honigmann, M. R.; Lorber, M. I.; Tellides, G.; Kashgarian, M.; Bothwell, A. L.; Pober, J. S., In vivo formation of complex microvessels lined by human endothelial cells in an immunodeficient mouse. *Proc Natl Acad Sci USA* 2000, 97(16), 9191–9196.

83. Enis, D. R.; Shepherd, B. R.; Wang, Y.; Qasim, A.; Shanahan, C. M.; Weissberg, P. L.; Kashgarian, M.; Pober, J. S.; Schechner, J. S., Induction, differentiation, and remodeling of blood vessels after transplantation of Bcl-2-transduced endothelial cells. *Proc Natl Acad Sci USA* 2005, 102(2), 425–430.

84. Traktuev, D. O.; Prater, D. N.; Merfeld-Clauss, S.; Sanjeevaiah, A. R.; Saadatzadeh, M. R.; Murphy, M.; Johnstone, B. H.; Ingram, D. A.; March, K. L., Robust functional vascular network formation in vivo by cooperation of adipose progenitor and endothelial cells. *Circ Res* 2009, 104(12), 1410–1420.

85. Koike, N.; Fukumura, D.; Gralla, O.; Au, P.; Schechner, J. S.; Jain, R. K., Tissue engineering: Creation of long-lasting blood vessels. *Nature* 2004, 428(6979), 138–139.

86. Polverini, P. J.; Nor, J. E.; Peters, M. C.; Mooney, D. J., Growth of human blood vessels in severe combined immunodeficient mice. A new in vivo model system of angiogenesis. *Methods Mol Med* 2003, 78, 161–177.

87. Chen, X.; Aledia, A. S.; Ghajar, C. M.; Griffith, C. K.; Putnam, A. J.; Hughes, C. C.; George, S. C., Prevascularization of a fibrin-based tissue construct accelerates the formation of functional anastomosis with host vasculature. *Tissue Eng Part A* 2009, 15(6), 1363–1371.

88. Chen, X.; Aledia, A. S.; Popson, S. A.; Him, L. K.; Hughes, C. C.; George, S., Rapid anastomosis of endothelial precursor cell-derived vessels with host vasculature is promoted by a high density of co-transplanted fibroblasts. *Tissue Eng Part A* 2010, 16(2), 585–594.

89. Ferreira, L. S.; Gerecht, S.; Shieh, H. F.; Watson, N.; Rupnick, M. A.; Dallabrida, S. M.; Vunjak-Novakovic, G.; Langer, R., Vascular progenitor cells isolated from human embryonic stem cells give rise to endothelial and smooth muscle like cells and form vascular networks in vivo. *Circ Res* 2007, 101(3), 286–294.

90. Alajati, A.; Laib, A. M.; Weber, H.; Boos, A. M.; Bartol, A.; Ikenberg, K.; Korff, T.; Zentgraf, H.; Obodozie, C.; Graeser, R.; Christian, S.; Finkenzeller, G.; Stark, G. B.; Heroult, M.; Augustin, H. G., Spheroid-based engineering of a human vasculature in mice. *Nat Methods* 2008, 5(5), 439–445.

91. Foubert, P.; Matrone, G.; Souttou, B.; Lere-Dean, C.; Barateau, V.; Plouet, J.; Le Ricousse-Roussanne, S.; Levy, B. I.; Silvestre, J. S.; Tobelem, G., Coadministration of endothelial and smooth muscle progenitor cells enhances the efficiency of proangiogenic cell-based therapy. *Circ Res* 2008, 103(7), 751–760.

92. Le Ricousse-Roussanne, S.; Barateau, V.; Contreres, J. O.; Boval, B.; Kraus-Berthier, L.; Tobelem, G., Ex vivo differentiated endothelial and smooth muscle cells from human cord blood progenitors home to the angiogenic tumor vasculature. *Cardiovasc Res* 2004, 62(1), 176–184.

93. Levenberg, S.; Rouwkema, J.; Macdonald, M.; Garfein, E. S.; Kohane, D. S.; Darland, D. C.; Marini, R.; van Blitterswijk, C. A.; Mulligan, R. C.; D'Amore, P. A.; Langer, R., Engineering vascularized skeletal muscle tissue. *Nat Biotechnol* 2005, 23(7), 879–884.

94. Stevens, K. R.; Kreutziger, K. L.; Dupras, S. K.; Korte, F. S.; Regnier, M.; Muskheli, V.; Nourse, M. B.; Bendixen, K.; Reinecke, H.; Murry, C. E., Physiological function and transplantation of scaffold-free and vascularized human cardiac muscle tissue. *Proc Natl Acad Sci USA* 2009, 106(39), 16568–16573.

95. Lesman, A.; Habib, M.; Caspi, O.; Gepstein, A.; Arbel, G.; Levenberg, S.; Gepstein, L., Transplantation of a tissue-engineered human vascularized cardiac muscle. *Tissue Eng Part A* 2010, 16(1), 115–125.

96. Antonio, N.; Fernandes, R.; Rodriguez-Losada, N.; Jimenez-Navarro, M. F.; Paiva, A.; de Teresa Galvan, E.; Goncalves, L.; Ribeiro, C. F.; Providencia, L. A., Stimulation of endothelial progenitor cells: A new putative effect of several cardiovascular drugs. *Eur J Clin Pharmacol* 2010, 66(3), 219–230.

97. Rossig, L.; Urbich, C.; Dimmeler, S., Endothelial progenitor cells at work: Not mature yet, but already stress-resistant. *Arterioscler Thromb Vasc Biol* 2004, 24(11), 1977–1979.

98. Werner, N.; Kosiol, S.; Schiegl, T.; Ahlers, P.; Walenta, K.; Link, A.; Bohm, M.; Nickenig, G., Circulating endothelial progenitor cells and cardiovascular outcomes. *N Engl J Med* 2005, 353(10), 999–1007.

99. Walther, C.; Gaede, L.; Adams, V.; Gelbrich, G.; Leichtle, A.; Erbs, S.; Sonnabend, M.; Fikenzer, K.; Korner, A.; Kiess, W.; Bruegel, M.; Thiery, J.; Schuler, G., Effect of increased exercise in school children on physical fitness and endothelial progenitor cells: A prospective randomized trial. *Circulation* 2009, 120(22), 2251–2259.

100. Matsui, H.; Shibata, M.; Brown, B.; Labelle, A.; Hegadorn, C.; Andrews, C.; Hebbel, R. P.; Galipeau, J.; Hough, C.; Lillicrap, D., Ex vivo gene therapy for hemophilia A that enhances safe delivery and sustained in vivo factor VIII expression from lentivirally engineered endothelial progenitors. *Stem Cells* 2007, 25(10), 2660–2669.

3 Induced Pluripotent Stem Cells

*Ji Woong Han, Rebecca Diane
Levit, and Young-sup Yoon*

CONTENTS

3.1 INTRODUCTION

In 1493, Spanish explorer Juan Ponce de León left Spain with Christopher Columbus on an epic quest for the mythical and biblical Fountain of Youth. He believed that drinking this legendary water would give him eternal life. To a biologist, a cellular "fountain of youth" would allow a cell or cell line to not only survive indefinitely but also retain the plasticity of youth, the ability of a cell to differentiate into any cell type in the body. Scientists have been studying embryonic cells' ability to do this for more than 50 years; however, what gives these cells this ability and how to create it in other cells have been elusive. While Ponce de León's quest was ultimately fruitless, recent advancements in the field of stem cell biology have made the molecular quest for longevity and plasticity within reach.

In the process of embryonic development, cells differentiate into specialized cell types and become restricted in their developmental potential (Table 3.1). Although totipotent and pluripotent cells exist only in the early embryo, adult organisms still have many different types of multipotent (e.g., hematopoietic stem cells) and unipotent stem cells to replace tissues damaged by injury or normal aging (Figure 3.1).[1] For more than 50 years, scientists have been experimenting with mature cells to determine what factors contribute to their plasticity and differentiation ability. The first nuclear-transfer experiments were performed more than 50 years ago by Briggs and King, and showed that the injection of the nucleus from differentiated blastula cells into an enucleated frog oocyte could give rise to tadpoles.[2] Later, Gurdon showed that the nuclei from even more differentiated frog intestinal cells could give rise to adult animals, although that efficiency was quite low. These experiments show that the nuclei of differentiated adult cells are equivalent to those of the early embryo and retain nuclear plasticity if placed in the appropriate environment.[3,4] In 1996, Wilmut and colleagues reported the first cloned animal, "Dolly" the sheep, by the nuclear transfer (NT) from

---

TABLE 3.1
Definition of Different Potency

Potency	Definition	Cells
Totipotent	Ability of a single cell to divide and form all the types of differentiated cells in the organism	Zygote, first cleavage blastomeres in mammals
Pluripotent	Ability of the cell to differentiate into all lineages of body, except for extra embryonic tissue	ES cells, iPS cells
Multipotent	Ability of the cell to give rise to a limited number of lineages	Hematopoietic stem cells
Unipotent	Ability of the cells to develop into only one type of cell type	SSCs, hepatocytes, epidermal stem cells.

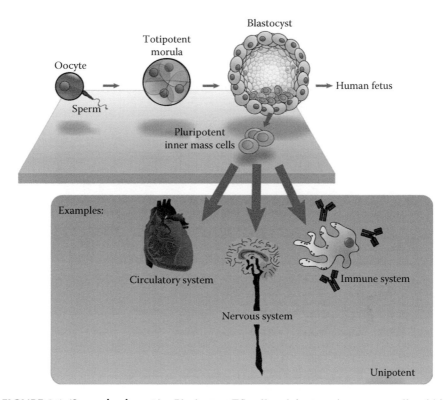

FIGURE 3.1 (See color insert.) Pluripotent ES cells originate as inner mass cells within a blastocyst. The stem cells can become any tissue in the body, excluding a placenta. Only the morula's cells are totipotent, able to become all tissues and a placenta. (Image retrieved from Wikipedia. Jones, M., http://commons.wikimedia.org/wiki/File:Stem_cells_diagram. png. With permission.)

adult cells into denucleated eggs.[5,6] These achievements support earlier findings that the epigenetic state of nuclei of differentiated cells is changeable and retains the ability to be reprogrammed by factors that are in oocytes and embryonic stem (ES) cells.

In 2006, Takahashi and Yamanaka made a revolutionary discovery; they reported that mouse fibroblasts could be reprogrammed to a pluripotent state by retroviral transduction of small number of defined factors but did not require NT.[7] The creation of these special cells, termed induced pluripotent stem (iPS) cells, is a major breakthrough in stem cell research and has shed light on the genetic mechanisms that are necessary to establish and maintain pluripotency. iPS cells from mouse and human have now been generated by various methods (Table 3.2). Mouse and human iPS cells are similar to ES in morphological, molecular, and developmental features. In contrast to somatic cell nuclear-transfer (SCNT) methods and derivation of ES cells from the inner cell mass (ICM) of the blastocyst, direct reprogramming provides a realistic and reliable way of generating sufficient numbers of patient-specific pluripotent stem cells without any sophisticated and expensive equipment.[8] The iPS cell technology facilitates the generation of disease- and patient-specific cell lines[9,10]

TABLE 3.2
Methods of iPS Cell Generation by Ectopic TF Expression

Species	Expression Methods	TFs	Cell Types	Other Factors	References
Mouse	Retroviral vectors	OSKM	Fibroblasts		[7]
		OSKM	Hepatocytes, gastric epithelial cells		[76]
		OSKM	Bone marrow mononuclear cells		[260]
		OSKM	Amniotic and yolk sac cells		[261]
		OSKM/OSK	Fibroblasts	5′-azaC, VPA	[93]
		OSKM/OSK	Fibroblasts, keratinocytes, *Ink4/Arf-/-, p53-/-, p21-/-*	Ink4a/Arf shRNA	[118]
		OSKM/OSK	Fibroblasts/T-lymphocytes, p53-/-		[116]
		OSK	Fibroblasts		[71]
		OSK	Fibroblasts	miR-291-3p, miR-294, miR-295	[114]
		OSK	Fibroblasts, *p53-/-, Terc-/-*		[119]
		OSK/OS	Fibroblasts, *p53-/-*	p53 shRNA, p21 shRNA, Ink4a/Arf shRNA	[117]
		OKM/OK	Fibroblasts	RepSox	[108]
		SKM/OK	NPCs	BIX01294	[97]
		OK/OM	NSCs		[171]
		OK	NSCs	PD0325901, CHIR99021	[106]
		OK	Fibroblasts	BIX01294, BayK8644	[96]
		OK	Fibroblasts	CHIR99021	[262]
		O	NSCs		[263]
	Lentiviral vectors	OSKM	Pancreatic β cells		[264]
	Inducible lentiviral vectors (tet)	OSKM	Fibroblasts		[73]
		OSKM	Fibroblasts, B-lymphocytes	5′-azaC	[113]
		OSKM	B-lymphocytes		[265]

Species	Method	Factors	Cell type	Small molecules	Reference
		OSKM/OKM/OSK	Fibroblasts	Alk5 inhibitor	[266]
		OSKM/OKM/OSK	Melanocytes		[267]
	Lentiviral vectors, single polycistronic	OSK	Fibroblasts	Wnt3a	[105]
		OSM	Fibroblasts	Kenpaullone	[112]
		OSKM	Fibroblasts		[268]
		OSK	Fibroblasts		[269]
	Inducible lentiviral vectors (tet), single polycistronic	OSKM	Fibroblasts		[81,82]
	Adenoviral vectors	OSKM	Fibroblasts, p53-/-, Ink4/Arf-/-		[120]
		OSKM	Fibroblasts, hepatocytes		[74]
	Retroviral vectors + adenoviral vectors	OSK	Hepatocytes		[75]
	PB transposon vectors	OSKM	Fibroblasts	PD173074	[87]
		OSKM	Fibroblasts		[86,88]
	Plasmid vectors	OSKM	Fibroblasts		[75]
	Plasmid vectors, single polycistronic	OSKM	Fibroblasts		[270]
	Recombinant proteins	OSKM	Fibroblasts	VPA	[90]
Rat	Retroviral vectors	OSKM	Fibroblasts		[271]
	Retroviral vectors	OSK	Liver epithelial cells	PD0325901, CHIR99021, A-83-01	[262]
Dog	Retroviral vectors	OSKM	Fibroblasts	VPA, PD0325901, CHIR99021, A-83-01	[272]
Pig	Retroviral vectors	OSKM	Fibroblasts		[273]
	Inducible lentiviral vectors (tet)	OSKMNL/OSKM	Fibroblasts		[274]
Monkey	Retroviral vectors	OSKM	Fibroblasts		[69]

(continued)

TABLE 3.2 (continued)
Methods of iPS Cell Generation by Ectopic TF Expression

Species	Expression Methods	TFs	Cell Types	Other Factors	References
Human	Retroviral vectors	OSKM	Fibroblasts		[109]
		OSNL	Fibroblasts		[70]
		OSKML	Fibroblasts		[110]
		OSKM	Fibroblasts	hTERT, LargeT, ROCKi	[103]
		OSKM	Fibroblasts	p53 shRNA, p53DD	[116]
		OSKM	Fibroblasts from patients		[9,10,275,276]
		OSKM	Blood cell, CD34+		[277,278]
		OSKM	Amniotic fluid-derived cells		[279]
		OSKM	Amniotic and yolk sac cells		[261]
		OSKM/OSK	Fibroblasts	p53 shRNA, p53DD	[117]
		OSKM/OSK	Fibroblasts	INK4a/ARF shRNA	[118]
		OSK	Fibroblasts		[71]
		OSK	Fibroblasts	p53 shRNA	[119]
		OSK	Fibroblasts from patients		[280]
		OS	Fibroblasts	VPA	[95]
		OS	Cord blood cells, CD133+		[281]
		OK	Fibroblasts	CHIR99021, Parnate, PD0325901, SB431542	[262]
		OK	NSCs		[282]
		OK/O	NSCs		[283]
	Lentiviral vectors	OSKM/OSNL	Fibroblasts	Large T	[284]
		OSNL	Fibroblasts from patients		[285]
		OSNL	Cord blood cells		[286]

Inducible lentiviral vectors (tet)	OSKM/OKM/OSK	Melanocytes	[267]
Inducible lentiviral vectors (tet), single polycistronic	OSKM	Keratinocytes	[81]
Lentiviral vectors, cre-recombinase excisable	OSKM/OSK	Fibroblasts from patients	[134]
Adenoviral vectors	OSKM	Fibroblasts	[287]
Sendaiviral vectors	OSKM	Fibroblasts	[288]
PB vectors	OSKM	Fibroblasts	[86]
Plasmid vectors	OSKMNL	Fibroblasts	[289]
Recombinant proteins	OSKM	Fibroblasts	[89]

O, Oct4; S, Sox2; K, Klf4; M, c-Myc; N, Nanog; L, Lin28.

that can be used as disease models in high-throughput screening and mechanistic studies[11] as well as for regenerative and therapeutic purposes, as demonstrated in rodent models.[12,13]

Pluripotency is the ability of the cell to differentiate into all cell types of an adult organism.[14] Pluripotency occurs naturally only in early embryos and may be maintained *in vitro* in cultured ES cells harvested from ICM of blastocysts. Isolated ES cells can maintain their population by proliferating indefinitely and self-renewing and have the potential to differentiate into every type of lineage in the body.[15,16] Self-renewal allows ES cells in culture to go through numerous cell cycles including cell division without losing the pluripotency under specific conditions.[15,16] Mouse ES cells require coculture with a feeder layer of cells that provide unknown but essential factors. The culture media must also contain leukemia inhibitory factor (LIF) or cytokines to prevent mouse ES cells from spontaneous differentiation[15,16] or bone morphogenetic proteins (BMPs) and fibroblast growth factors (FGFs) instead of LIF to prevent differentiation of human ES cells.[8] Without feeders or cytokines, ES cells undergo spontaneous differentiation and lose their pluripotency. During differentiation, stem cells commit to one cell lineage while losing the plasticity to change into other types of the cell lineages. Reprogramming, the process used to make iPS cells, is the reverse of differentiation: differentiated cells convert back into pluripotent cells.[1,17] Many of the advancements in the field of stem cell research have been achieved using mice; however, the ultimate goal of this work is the production of patient-specific human pluripotent cells to use in the treatments of human disease. Both human ES cell and SCNT have technical and ethical problems that may limit or prevent their therapeutic use in humans. iPS cell technology circumvents many of these problems and is regarded as the best method for producing patient-specific pluripotent stem cells for application in regenerative medicine. While not as simplistic as drinking water from a "fountain of youth," these cells have enormous therapeutic potential to cure disease, regenerate damaged limbs, and replace tissue lost in heart attack and stroke.

3.2 GENERATION OF PLURIPOTENT STEM CELLS FROM SOMATIC CELLS

Generation of pluripotent stem cells from adult cells is an artificial manipulation that may not produce cells identical to naturally occurring pluripotent stem cells. However, some aspects of iPS cell generation may parallel the innate genetic processes that occur during embryonic development, including the reprogramming of the gamete pronuclei at fertilization under the influence of factors in oocytes. Genetically matched patient-specific pluripotent cells may be a limitless source of stem cell–based therapy for degenerative diseases without immune rejection.[18] Although tremendous effort has been put into generating immune compatible patient-specific stem cells, promising methods were not successful until iPS cell technology with defined transcription factors was developed (Figure 3.2).[7]

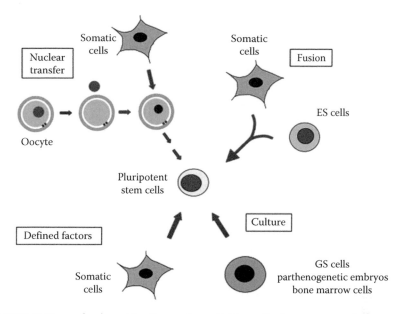

FIGURE 3.2 (See color insert.) Currently available methods to generate pluripotent stem cells from adult somatic or germ cells. In mouse models, three methods have been reported to generate pluripotent stem cells from somatic cells: NT, fusion, and forced expression of defined factors. Also reported is the generation of chimera-competent pluripotent stem cells after the long-term culture of bone marrow cells. In addition, pluripotent stem cells can be established from mouse adult germ cells: multipotent GS cells and parthenogenetic ES cells. (From *Cell Stem Cell*, 1(1), Yamanaka, S., Strategies and new developments in the generation of patient-specific pluripotent stem cells, 39–49. Copyright 2007. With permission from Elsevier.)

3.2.1 REPROGRAMMING METHODS

3.2.1.1 Somatic Cell Nuclear Transfer

The first successful NT was performed with Northern leopard frog (*Rana pipiens*) by Briggs and King.[2] They showed that transfer of nuclei from blastula stage embryos into enucleated eggs resulted in normal tadpoles. Gurdon and colleagues then succeeded in producing fertile adult African clawed frogs (*Xenopus laevis*) by transferring nuclei from more differentiated tadpole intestinal cell into enucleated eggs in 1996.[19] When they used the nuclei from more differentiated adult cells, the egg could not develop to the adult stages. In 1975, Bromhall reported that nuclei transfer of rabbit morula cell into enucleated oocytes give rise to the morula stage embryo albeit with low efficiency.[20] The first cloned adult mammal by SCNT was claimed by Illmensee and Hoppe in 1981 with mice inner cell nuclei.[21] However, these experiments could not be duplicated by other groups.[22] In 1983, McGrath and Solter[23] obtained 17 live mouse progeny when they transferred pronuclei from one zygote to another zygote. However, when they used nuclei from four-cell stage, eight-cell stage, and ICM, embryos did not develop into blastocysts. In investigating the reasons for this, genomic imprinting was discovered. During the gametogenesis, different sets of genes are selectively repressed so that the pronuclei of sperm and egg contribute unequally to normal development.[24] In 1986, Willadsen obtained fully viable sheep by combining a blastomere from 8- or 16-cell embryos with half of the cytoplasm of an unfertilized egg after fusion of embryos by Sendai virus or electrofusion apparatus.[25] Subsequently, other cloned animals were obtain from rabbits,[26] pigs,[27] mice,[28] cows,[29] and monkeys[30] by SCNT with undifferentiated nuclei. However, it remained difficult to generate cloned animals with the nuclei from more differentiated adult cells.

In 1996 Wilmut and colleagues showed an adult sheep, famously known as "Dolly," could be produced with more differentiated nuclei from follicle cells.[5,6] Subsequently, many other mammals including mice,[31] cows,[32] goats,[33] pigs,[34] rabbits,[35] and a cat[36] have been produced with adult cell nuclei. One interesting mature cell type used was the mature B- and T-lymphocytes, which have a genomic rearrangement in immunoglobulin genes. Adult mice generated from these cells gave rise to adult "monoclonal" mice by injecting ES cells from cloned blastocysts into tetraploid blastocysts.[37] The same group also cloned mice by SCNT from neural stem cells with the highest success rates using NT-derived ES cells.[38] In contrast to the difficulties of obtaining cloned animals from low efficiency, it is relatively easy to obtain ES cells from cloned blastocysts after NT with comparable efficiency.[39] A significant ethical issue of generating patient-specific human ES cell lines is the availability of human oocytes. However, Egli and colleagues have shown the possibility that discarded human *in vitro* fertilization (IVF) embryos could potentially be used as recipients for human NT ES cell derivation instead of oocytes by generating pluripotent cells by NT of adult somatic cells to zygotes treated with drug for arresting mitosis.[40]

3.2.1.2 Cell Fusion

In 1976, Miller and Ruddle produced somatic cell hybrids by fusing pluripotent embryonic carcinoma (EC) cells with thymocytes from young adult mice.[41] Similar experiments were performed by *in vitro* hybridization with ES cells by electric fusion, resulting in ES–thymocyte hybrid cells able to contribute to all three primary germ layers of chimeric embryos.[42] In 2002, two groups showed that mouse bone marrow cells and brain cells could regain their pluripotency by fusing with ES cells.[43,44] Recently, reprogramming of various human somatic cells, foreskin fibroblasts,[45] and $CD45^+CD33^+$myeloperoxidase$^+$ myeloid precursors[46] by fusion with human ES cells was reported. The factors responsible for reprogramming by cell fusion with ES cells may reside in the nucleus[47] or in cytoplasm.[48] Smith and colleagues observed a marked increase in reprogrammed cell colonies when Nanog was overexpressed in ES cell fusion experiments.[49] Overexpression of Nanog in mouse and human ES cells enables them to grow without normally required growth factors such as LIF[50,51] and feeder cells.[52] Nanog-deficient ES cells lost pluripotency and spontaneously differentiated into extraembryonic endoderm lineages even in culture medium including LIF.[51] Another group reported that somatic cell–derived Nanog was reactivated through cell fusion with ES cells as well as SCNT.[53] Smith and colleagues showed that the forced overexpression of Nanog in ES cells promoted pluripotent gene activation when they fused neural stem cells, thymocytes, and fibroblasts to Nanog overexpressing ES cells.[49] However, the tetraploidy of fused pluripotent cells would prevent this strategy from ever being used for cell-based therapy. Although selective elimination of some ES cell–derived chromosomes is possible,[54] it may be impossible to generate diploid cells without large-scale genomic instability. In another approach, ES cell–derived cell extract was used to reprogram permeabilized somatic cells by short-term incubation,[55] but convincing evidence for reprogramming of the somatic cell genome is still lacking.

3.2.1.3   Culture-Induced Reprogramming

Physiologically, ES cells need to be transformed and reprogrammed by the long-term culture of ICM. Primordial germ cells (PGCs) or spermatogonial stem cells (SSCs) from gonads are unipotent *in vivo*; however, these cells could be transformed to pluripotent ES-like cells[56] or multipotent adult germline stem cells (maGSCs)[57] after prolonged *in vitro* culture. Recently, multipotent adult spermatogonial stem cells (MASCs) were derived from adult spermatogonial progenitor cells (SPCs). This required long-term cultivation on mitotically inactivated testicular feeders containing CD34[+] stromal cells.[58] The resulting cells had a different gene expression profile from that of ES cells[58] but similar to epiblast stem cells (EpiSCs), which were derived from the epiblast of postimplantation mouse embryos.[59,60] Also both MASCs and EpiSCs were unable to form chimeras in contrast to other types of pluripotent cells such as ES cells. Most recently, pluripotent-related gene expression could be induced from human fibroblasts by changing oxygen and FGF2 concentration during cell culture.[61] Whether truly pluripotent cells can be derived from somatic tissues from the postnatal animal by long-term culture is unknown.

SSCs are not available for reprogramming in female organisms; however, oocytes can be coaxed to undergo parthenogenesis, or asexual reproduction, under specific culture conditions. This process requires that meiosis is arrested in metaphase in the presence of cytochalasin to prevent extrusion of the second polar body forming a pseudozygote.[62,63] The resulting parthenogenetic embryonic stem cells (PGESs) show a loss of heterozygosity in the major histocompatibility complex (MHC) and thus may be rejected after transplantation due to the lack of paternal set of histocompatibility antigens. Daley and colleagues developed methods to maintain both of the maternal MHC loci in mouse parthenogenetic ES cells.[64] While there has been some success using cell culture–based manipulations of gonadal and rapidly proliferating cells to create pluripotency, these techniques are both cumbersome, have low efficiency, and result in cells with genetic aberrations and epigenetic abnormalities. There is likely a limit to how much external media conditions can fundamentally alter a cells regenerative capacity.

3.2.1.4   Ectopic Overexpression of Transcription
Factors; Induced Pluripotent Stem Cell

Based on the success and limitations of previous NT and cell culture experiments, scientist began to experiment with more direct manipulation of cells' genetic information to create developmental plasticity in mature, differentiated cells. The first successful generation of iPS cells from somatic cells was performed by ectopic overexpression of pluripotent specific transcription factors.[7] They introduced a mini-library of 24 candidate reprogramming factors, known as pluripotency-associated genes, which were known to be expressed in ES cells. The genes were introduced into mouse embryonic fibroblasts (MEFs) *in vitro* and carried a fusion of the β-galactosidase and neomycin-resistance genes expressed from the *Fbx15* locus.[65] In this system, reactivation of the *Fbx15* gene, which resulted in resistance to the antibiotic G418, suggested that the cell was reprogrammed for pluripotency. When they infected MEFs with all 24 genes and cultured them on feeder cells in ES medium in the presence of G418, drug-resistant colonies emerged that had ES-like proliferation, gene expression, and morphology. The cells were round and had large nucleoli and scant cytoplasm. To narrow down the essential factors, combinations of the 24 factors were attempted until four transcription factors, *Oct3/4*, *Sox2*, *Klf4*, and *c-Myc*, were identified as essential factors. The resultant cells, which had many pluripotent features, were termed as iPS cells.[7] The four essential factors are often referred to as Yamanaka factors in recognition of Shinya Yamanaka, the stem cell biologist who designed and directed these experiments. Since 2006, several groups have improved Yamanaka's reprogramming technique and generated iPS cells that are epigenetically and developmentally indistinguishable from ES cells.[66–68] Direct reprogramming has been replicated by others using the same factors in mouse, rhesus macaque, and human cells.[10,66–69] One group combined *Oct3/4* and *Sox2* with *Lin28* and *Nanog* to derive human iPS cells.[70] Remarkably, the same four factors identified in the murine system were able to confer pluripotency in primate cells, indicating that the fundamental transcriptional network governing pluripotency is conserved across these species (Table 3.2). Several groups have shown that the *c-Myc* gene is dispensable for reprogramming,[10,71] although with lower efficiency. *c-Myc* reactivation can trigger tumorigenicity of iPS cell derivatives, and the elimination of this property will be important to translate iPS cells to the clinic.

Expression of the reprogramming factors in differentiated cells appears to initiate a sequence of stochastic events that eventually lead to a small fraction of cells becoming iPS cells. The enforced expression of the four genes by retroviral vectors is required only temporarily at the initiation stage of reprogramming and silenced followed by the replacement of endogenous genes.[72] Thus, ectopic expression of *Oct4*, *Sox2*, c-*Myc*, and *Klf4* may trigger a sequence of epigenetic events such as chromatin modifications or changes in DNA methylation that eventually results in the pluripotent state. Many unanswered questions remain concerning these important cells, including how to increase their generation efficiency for clinical application, by what mechanism these genes create pluripotency, and if these cells are truly the same in terms of regenerative capacity compared to each other and compared to pluripotent ES cells.

3.2.1.4.1 Integrating Viral Vectors

Several types of viral vectors, including retroviral[7,10,67] and lentiviral[13,73,74] vectors, have been used in iPS cell generation. Retroviral vectors were the first type used to create iPS cells and the site of viral integration has been closely studied. It is reported that viral integration into specific loci does not contribute to direct reprogramming.[75] However, it is generally accepted that the presence of multiple retroviral integration sites (RISs) in individual iPS clones prohibits their clinical use, although the use of any "gene therapy" in humans is fraught with safety and ethical issues. Takahashi and Yamanaka noted approximately 20 RISs per iPS clone in their initial report.[7] In a recent report comparing reprogramming of MEFs to that of murine hepatocytes or gastric epithelial cells, Aoi and colleagues examined the number of RIS for each of the four retroviruses by Southern blot. They detected 1–9 RIS in MEF-derived iPS clones and 1–4 RIS in gastric epithelium or hepatocytes-derived clones, suggesting that tissues of epithelial origin may be more readily reprogrammed. The integration sites were random and did not show common viral integration sites or any difference in functional characterization.[76] Several RIS found by Aoi had previously been identified by retroviral-mediated tumor induction in mice, so the safety of these cells is still not established.[77,78] Bioinformatics analysis revealed no enrichment of a specific gene function, gene network, or canonical pathway by retroviral insertions.[79] Retroviruses were first selected for iPS generation because of their more robust integration behavior. They have a propensity to integrate near transcription start sites and may cause more malignant transformations. To date, no analysis has shown that retroviruses used to make iPS cells congregate in dangerous areas of the genome or show any preference for any specific area. This danger still exists, which will likely limit the clinical application of this technique.

Lentiviral vectors have a hypothetical safety advantage over retroviral vectors, namely they lack the propensity to integrate near transcription start sites.[80] To date, no insertion site analysis has been conducted, and thus the biological relevance of vector differences remains theoretical. An additional advantage of lentivirus is its ability to transcribe large genetic packages, and two recent publications detail the use of polycistronic lentiviral vectors that delivered the four reprogramming factors in a single construct. Previously four separate vectors were used, each carrying one gene. Both papers demonstrate the derivation of iPS clones from a single vector integration, which may minimize mutagenesis caused by viral insertion and increase efficiency.[81,82]

3.2.1.4.2 Nonintegrating Viral Vectors

There are many human diseases where important tissues die and need to be replaced that may be treatable with pluripotent cells. Previously, no source of patient specific stem cells were available to regenerate needed tissue, whether it be heart muscle after heart attack, brain after stroke, kidney after chronic kidney disease, etc. The ability to differentiate cells with pluripotency may allow for regeneration of needed tissue from patient specific iPS cells, bypassing the issue of immune rejection. However, concerns about oncogenicity, insertional mutagenesis, and vector safety are prominent. In the first report on germline competent iPS cells, ~20% of chimeric mice had tumors caused by the reactivation of the integrated *c-Myc* proviral transgene into host genome.[67] Another studied showed cancer-related mortality in 18 of 36 (50%) iPS chimeric mice.[13] Chimeric mice derived from ES cells usually did not show any malignancies, suggesting the tumorigenicity of iPS cells is robust. When *c-Myc* was omitted during reprogramming to reduce tumorigenicity, the efficiency of iPS cell generation decreased by one order of magnitude and the timing of colony formation was significantly delayed.[71] Insertional mutagenesis due to the integration of viral vectors into critical sites of the host genome leading to malignant transformation has been observed in preclinical and clinical gene therapy trials.[83-85] Because of the these limitations and safety concerns, alternative methods of iPS cell generation have been sought and have focused on completely eliminating integration of retroviral vectors from the reprogramming procedure. This is generally considered an essential condition for clinical application of iPS cells.

The fact that the four Yamanaka factors only need to be expressed transiently was utilized by Stadtfeld and colleagues who used transiently active and nonintegrating adenoviral vectors to give rise to iPS cell lines.[74] These cells could contribute to the formation of teratomas and chimeric mice, but were unable to pass through the germline.[74] By contrast, Okita and colleagues were unable to obtain murine hepatocyte iPS clones when the four reprogramming factors were introduced by adenovirus alone and required additional transfections of *Oct3/4* and *Klf4* or *Oct3/4* and *Sox2* by retrovirus.[75] For unknown reasons, some cell types may be amenable to the safer adenoviral vector transduction, while other cell types cannot be made into iPS cells using this technique. The biologic basis of this cell specific plasticity is a fundamental but unanswered question in stem cell biology.

3.2.1.4.3   Nonviral Reprogramming

While the partial success with adenovirus in generating iPS cells suggests that this safer vector may be useful, the widespread use of any viral vector technology in humans is likely impossible. In the 1990s, several gene therapy phase 1 and 2 clinical trials were halted due to excessive morbidity and mortality that has made this technology unethical. Because of the great potential of iPS cells, investigators have turned to nonviral vector systems for the generation of iPS cells. Yamanaka and colleagues used a polycistronic expression plasmid containing the *Oct3/4*, *Sox2*, and *Klf4* cDNAs linked by the foot-and-mouth disease virus 2A self-cleaving peptide.[75] When this construct, which lacks viral genetic material, was repeatedly transfected into MEFs together with a separate *c-Myc* cDNA expression vector over a 1 week time period (on days 1, 3, 5, and 7), 1–29 iPS colonies emerged in 7 out of 10 independent experiments. In 6 of 10 experiments, no evidence of plasmid integration into the host genome was detected by PCR or Southern blot analysis.[75] While the efficiency of iPS generation by plasmid transfection was greatly decreased, and the potentially oncogenic *c-Myc* viral vector transduction was needed, this report provides proof-of-concept for the generation of iPS cells without transgene integration of viral vector.

Several recent studies have reported the use of transient viral vector approach to iPS generation that would increase efficiency and safety.[86-88] This method begins with the incorporation of all four Yamanaka genes into a single piggyBac (PB) vector. Each gene was separated with a viral 2A oligopeptide, which allows synthesis of the four factors as a single transcript followed by posttranslational cleavage of the proteins at the appropriate location. The most important and unique feature of this approach is that the reprogramming genes are removed from the genome by transient transfection of PB transposase.[88] Thus, permanent alteration of the genome could be avoided and many safety concerns circumvented. This technique maintains a relatively robust reprogramming efficiency. Using a similar idea, loxP-CRE recombinase system was used to splice out the vector-integrated DNA after gene transcription was no longer needed.[87] However, this system is less precise and may leave some residual elements outside the loxP sites including the transposon repeats that may be mutagenic. These studies confirm that transient expression of the reprogramming factors is sufficient for reprogramming of somatic cells to acquire pluripotency. They also confirm that it is possible to remove integrated vector material when it is no longer needed and thus minimize the risk of late cancer development. This paves the way toward the use of iPS cells for patient- or disease-specific regenerative medicine.

3.2.1.4.4   Recombinant Proteins of Transcription Factors

The integration of viral vectors into target cell DNA causes the cells intrinsic machinery to transcribe and translate the protein products, ultimately resulting in cell reprogramming. As an alternate approach, the protein products of the reprogramming genes have been directly delivered to cells, without viral DNA integration.[89,90] Two separate groups made use of a short peptide sequence derived from human immunodeficiency virus protein that is permeable to cell membranes.[89,90] When this peptide is linked to each factor, the resulting fusion proteins can enter cells and cause reprogramming, albeit at a low efficiency. Although promising, this strategy currently requires multiple rounds of factor delivery and it has a low success rate.

3.2.1.4.5   Other Factors

Recent data suggest that specific epigenetic information is needed to maintain the ES cell state and that this information, in the form of histone acetylation and DNA methylation, may be important during the generation of iPS cells. The chromatin-remodeling factor *Chd1* maintains the pluripotency of ES cells by keeping the chromatin structure open and able to replicate. When present during the induction of iPS cells, it increases the efficiency of induction of pluripotency.[91] Previous work in SCNT showed a modest (two to fivefold) beneficial effect of histone deacetylase (HDAC) inhibitors on nuclear reprogramming efficiency.[38,92]

Subsequently, studies have reported the effects of epigenetic modification on iPS cell generation. Huangfu and colleagues tested the effects of small-molecule chemicals involved in chromatin modification on reprogramming using an *Oct4-GFP* reporter mouse cells.[93,94] Treatment of the four factor infected MEFs with the DNA methyltransferase inhibitor 5′-azacytidine increased the reprogramming efficiency, and three known HDAC inhibitors—suberoylanilide hydroxamic acid, trichostatin A (TSA), and valproic acid (VPA)—also greatly increased the efficiency of reprogramming with VPA being the most effective compound, leading to a >100-fold increase.[93] This demonstration of enhanced reprogramming efficiency by HDAC inhibitors suggests that chromatin modification is a key step in defining a cells' pluripotent state. When VPA was used in the generation of human iPS cells from primary human fibroblasts, just two transcription factors (*Oct4* and *Sox2*) were needed.[95] On average, between one and five iPS lines were isolated from 1×10^5 primary human fibroblasts under these conditions, an efficiency similar to a three factors (*Oct4*, *Sox2*, and *Klf4*) system in the absence of VPA. This suggests that VPA treatment can effectively replace *Klf4*.[95]

Chemical screens have identified two other compounds, BIX01294 and BayK8644, which, in combination with two factors (*Oct4* and *Klf4*), enhanced the reprogramming efficiency of mouse neural progenitors[96] and MEFs.[97] BIX01294 is inhibitor of the G9a histone methyltransferases (HMTs), which methylate histone H3 at the position of lysine 9 (H3K9).[98] G9a HMT is reported to silence *Oct4* expression during early embryogenesis by subsequent de novo DNA methylation at the promoter region, thereby preventing reprogramming.[99–102] Several other small molecules have been found that may indirectly influence a cell's epigenetic state. BayK8644 is an L-channel calcium agonist[96] with an interesting mode of action: it might exert its effect through upstream signaling pathways rather than direct epigenetic remodeling. Another small-molecule involved in signaling pathways also increases the efficiency of reprogramming. Rho-associate kinase (ROCK) inhibitor, Y-27632, could increase the human iPS cell induction by enhancing cell survival.[103,104] Other inhibitors for Wnt signaling,[105–107] MEK,[106,107] FGF,[87] and TGFβ receptor[107,108] also have been shown to exert effects on the generation and maintenance of pluripotency of iPS cells. Taken together, these studies suggest that chromatin remodeling and signaling pathways that modify the epigenetic state are important for creating and maintaining pluripotency. It was also shown that if the appropriate molecules are present, *Klf4* is dispensable for reprogramming, similar to *c-Myc*[93]. In fact, *Oct4* and *Sox2* are the only two factors thus far consistently found by all investigators necessary to derive iPS cell lines from human somatic cells.[10,70,109,110] By acting as master regulators of pluripotency, *Oct4* and *Sox2* appear to be critical to direct reprogramming.[111] Once small molecules were found that could replace *Klf4* in the generation of iPS cells, investigators began working for similar substitutes for the other "required" factors,[108,112] including DNA demethylating agents[113] and agents that modify histone acetylating and methylation.[96,97] Due to the limitations of using viral vectors, or partial viral vector systems, there is great interest in developing chemical means to produce iPS cells.

While some researchers are investigating novel chemicals, others are investigating the potential of microRNA (miRNA) to produce iPS cells. miR-291-3p, miR-294, and miR-295, which are expressed specifically in mouse ES cells, enhance the efficiency of pluripotency induction from MEFs.[114] Although it remains unclear whether these miRNAs have overlapping targets during reprogramming, studying them in the iPS cells will allow further understanding of the biology of miRNAs associated with pluripotency. Also, it will be interesting to test whether inhibition of miRNAs expressed in differentiated cells but not in stem cells, such as let-7 and miR-21, facilitate reprogramming.

Cell cycle regulatory proteins can also change the efficiency of iPS cell generation in ways that are just beginning to be understood. Inhibition of the tumor suppressor protein *p53*-related signal pathway facilitates the generation of iPS cells, suggesting that it represses dedifferentiation.[115–120] iPS cell technology can be used to study how *p53* modulates the stability of the differentiated state. Inhibition of *p53* directly by *Mdm2* and indirectly by downregulation of *Arf*[118,120] enhances the progression of normal fibroblasts into iPS cells through its direct target, *p21*, which promotes cellular senescence.[116,117] Deficiency of *p53* improves the efficiency and kinetics of iPS cell generation with two factors, *Oct4* and *Sox2*.[117] Other tumor suppressors, such as *Rb* or *Pten*, are also candidate repressors of dedifferentiation that can be investigated using the iPS cell technology.[121] Lineage-specific transcription factors and epigenetic modifications such as DNA methylation or lack of H3K4 methylation at certain loci also may repress dedifferentiation.[113] Screening for repressors of dedifferentiation, using, for example, large-scale RNAi screens, for inducing pluripotency in different adult somatic cells may reveal some genes that could be used as reprogramming factors.

3.2.2 EFFICIENCY AND KINETICS OF IPS CELL REPROGRAMMING

Four strategies for induction of pluripotency from differentiated cells by reprogramming were described: SCNT, cell fusion, culture-induced reprogramming, and iPS cell generation by ectopic expression of pluripotency-related transcription factors. However, culture-induced reprogramming of embryonic germ cells and SSCs does not constitute an example of radical reprogramming. Of all of the successful methods, cell fusion seems the least likely to have clinical and research relevance. This technique results in the creation of tetraploid cells instead of diploid, and a high risk for genetic aberrancy.[45] By contrast, both SCNT and iPS cell formation with defined factors have succeeded in terms of reprogramming and stable diploid status, although efficiency of generation is still low. The best recorded efficiencies using mouse fibroblasts are 3.4% for SCNT[39] and 1%–3% for iPS cell generation.[122–124] The timeline of iPS cell generation is usually protracted; inappropriate changes might be reversed over time. In SCNT, the time required for reprogramming before initiation of development is relatively short. *Oct4* was detectable in mouse SCNT blastocysts within 12–24 h, although the reactivation of embryonic genes was variable.[125] Rapid *Oct4* and *Ssea4* expression is also seen in the somatic nucleus of 13%–16% of mouse ES cell–somatic cell heterokaryons.[126] In both processes, significant and rapid nuclear swelling, which is thought to indicate chromatin decondensation, precedes reprogramming. By contrast, during iPS cell formation, *Oct4* becomes detectable only after roughly 2 weeks[73,74] and only in a small proportion of treated cells. These data indicate that the processes share a stochastic nature but that, in SCNT and cell fusion, reprogramming occurs much faster.

Depending on the starting cell type, vector, reprogramming mix, and infection methods, reprogramming efficiencies range from 0.001% to 11%.[123,127] Mouse fibroblasts dedifferentiated *in vitro* from animals made from iPS cells that were established by doxycycline-inducible vectors showed that the efficiency of reprogramming is ~2%–4%, compared with ~0.05% in direct reprogramming from fibroblasts.[68,124] One group reported that treatment with VPA increased the reprogramming efficiency up to 11% for mouse fibroblasts.[93] However, a different group observed reduced reprogramming when they used the PB transposon with VPA treatment.[88] The lack of standardization of characterizing iPS cell formation makes comparing reprogramming efficiencies between different laboratories difficult.[123] Specific types of somatic cells such as keratinocytes, stomach cells, and liver cells are more easily converted than fibroblasts.[76,128] These differences might be caused by variable delivery efficiency of factors or by the cell status, which might be amenable to change. The system of doxycycline-inducible iPS cell generation is useful for measuring the efficiency and kinetics of reprogramming in murine fibroblasts.[73,74,124] By withdrawal of doxycycline after various periods, it was shown that transgenes expression was essential for a minimum of 10–12 days to initiate cellular reprogramming. Longer exposure to doxycycline resulted in an increased number of reprogrammed cells.[73,74] This system revealed the kinetics of pluripotency marker appearance during reprogramming.[73,74] The expression of alkaline phosphatase (AP), a key marker for pluripotency, is followed by *Oct4* expression. The stochastic nature of reprogramming could be suggested by the activation of endogenous *Oct4*-driven or *Nanog*-driven reporter genes at different times.[72]

A recent experiment used secondary' iPS cells from mice derived from "primary" iPS cells with drug-inducible provirus. The segregation of each factor by germline transmission was identifiable and showed that a single copy of each of the four factors is sufficient to allow optimal reprogramming frequency and kinetics.[129] Furthermore, by using introduction of complementary factors into cell lines carrying single copies of each factor, it has been shown that *Klf4* and *c-Myc* act earlier during reprogramming, conceivably by inducing epigenetic alterations that assist the binding of *Oct4* and *Sox2* to their target genes.[129] These conclusions were confirmed and extended by Plath and colleagues showing that the ectopic expression of *c-Myc* is necessary only during the initial stages of reprogramming and can be substituted by VPA.[130] *c-Myc* or VPA might enhance the interaction of *Oct4, Sox2*, and *Klf4* with their target genes to repress somatic cell–specific gene expression and initiate the pluripotency-related gene.[71,130,131]

Two recent studies demonstrated that molecular markers are sequentially expressed during reprogramming of mouse fibroblasts.[73,74] In the first 3–5 days after viral transduction, the fibroblast-specific marker THY-1 was downregulated,[74] and ES cell–specific marker AP was upregulated[73] in a large proportion of fibroblasts. In subsequent days, a population of SSEA-1-positive cells emerged within the previously THY-1-negative or AP-positive cells. Around 10–14 days after the initial viral transduction, the endogenous *Oct4* or *Nanog* locus was reactivated in a small percentage of cells within the SSEA-1-positive population. Fully reprogrammed iPS clones that are independent from ectopically expressed factors can only be isolated at this stage.

3.3   CHARACTERIZATION OF iPS CELLS AND ES CELLS

IPS cells express ES cell–specific genes that maintain the developmental potential to differentiate into all three primary germ layers, ectoderm, endoderm, and mesoderm. Several functional tests, including *in vitro* differentiation, DNA methylation analysis, *in vivo* teratoma formation, chimera formation, germ-line transmission, and tetraploid aggregation, have been used to define and compare pluripotency in iPS and ES cells. Generally, ES cells and iPS cells seem to have similar functional properties; both cells can contribute to chimera formation, germ-line transmission, and even the most rigorous test, tetraploid complementation assay, by which live animals are born.[132,133]

3.3.1 GENOME

When iPS cells are created, maintaining the genomic integrity is of crucial importance because alterations can cause neoplastic disease and limit therapeutic application. Several groups have investigated the karyotype of mouse[68] and human iPS cell lines[70,109] to determine how much genetic alteration is present. Nevertheless, Aasen and colleagues showed that continuous passaging of human iPS cells resulted in frequent chromosomal abnormalities starting as early as passage 13.[128] This finding suggests that the long-term culture of human iPS cells, similar to the situation for human ES cells, has to be monitored carefully for culture-induced genetic abnormalities.

The use of retroviral and lentiviral vectors for expressing the reprogramming transcription factors has the inherent risk of insertional mutagenesis, which would be a problem if the cells were used in regenerative medicine. However, Aoi and colleagues reported no common insertion sites in hepatocyte- and stomach cell–derived iPS cells.[76] In addition, recent adenoviral and plasmid-based strategies have a much lower risk of insertional mutagenesis because the genomic DNA is not broken by the virus.[74,75] However, even without viral integration, genetic changes might occur as part of the reprogramming process. Another component to DNA integrity, telomere length, is not altered in mouse[74] and human hTERT-[10,109] derived iPS cells.[109] Studies of iPS cells have suggested that DNA integrity is maintained throughout the generation process; however, only long-term studies may show if these cells are truly free of malignant potential *in vivo*. In addition, this risk will have to be weighed against their vast therapeutic potential.

3.3.2 TRANSCRIPTOME

Comparative global gene expression analyses of the ES cell and iPS cell transcriptomes using microarrays have been performed for human and mouse lines.[7,66,68] Mikkelsen and colleagues reported that whole genome expression profiles of iPS cells and ES cells of the same species are no more different than those of individual ES cell lines.[113] Nonetheless, most other groups noted that iPS cells are not identical to ES cells. Takahashi and colleagues compared the global gene-expression profile of human iPS and human ES cells for 32,266 transcripts.[109] Notably, 1267 (~4%) of the genes were detected with >5-fold difference in up- or downregulation between iPS cells and human ES cells. Soldner and colleagues compared the transcriptional profiles of human iPS cell lines, in which the Cre-recombinase excisable exogenous viral sequences had been removed (factor-free human iPS cells), with those of human iPS cells before transgene excision.[134] The transcriptomes of factor-free human iPS cells more closely resemble those of human ES cells than the parental human iPS cells with integration by viral sequence. This could be caused by the loss of any downstream gene activation by the residual expression of the exogenous transcription factors or the loss of the epigenetic memory of the somatic state after initial reprogramming event. However, it remains difficult to compare these differences because most groups used genetically unrelated cell lines. More experiments are needed to clarify these discrepancies such as comparing the factor-containing and factor-free iPS cells from differentiated cells from human ES cell line.

A well-characterized gene expression pattern occurs after ectopic expression of four factors in MEFs, including an initial downregulation of cell-type-specific transcription factors[74,113] and upregulation of genes involved in proliferation, DNA replication, and cell cycle progression.[113] During the reprogramming process, many self-renewal-related genes are reactivated, including FGF4 as well as polycomb genes.[7,113] However, a large fraction of pluripotency-related genes are only upregulated during the late stages of reprogramming.[73,74,113] In a different study, the expression of key pluripotency-related genes, such as *Oct4*, *Sox2*, and *Rex1*, was approximately twofold lower in the iPS cells compared with two human ES cell lines, HSF1 and H9.[110] Pluripotent cells are highly sensitive to the levels of these transcription factors,[135] and there is a notable amount of normal transcriptional heterogeneity in human ES cell cultures.[136] Therefore, the observed variation could reflect differences in the culture conditions rather than incomplete reprogramming. More work on human ES cells is thus required to better understand the extent of normal transcriptional variation within human and also mouse ES cells and to fully understand how iPS cells compare.

3.3.3 EPIGENOME

Epigenetic modifications have an essential role in controlling access to genes and regulatory elements in the genome by regulating chromatin structure.[73] The differences in epigenetic status between a somatic cell and pluripotent stem cell are huge and dedifferentiation requires global epigenetic reprogramming. Analysis of the epigenetic state allows determination of the degree of reprogramming in iPS cells. For instance, pluripotent stem cells contain bivalent domains, which are characteristic chromatin signatures.[137,138] These are regions enriched for repressive histone H3 lysine 27 trimethylation (H3K27me3) and simultaneously for histone H3 lysine 4 trimethylation (H3K4me3) as an activating signal.[139] It was assumed initially that bivalent domains might be ES cell specific because they were first identified using chromatin immunoprecipitation (ChIP) followed by hybridization to microarrays (ChIP-Chip) that featured key developmental regulators. All of these resolved either to a univalent (H3K4me3 only or H3K27me3 only) state or lost both marks in differentiated cells.[137] Using ChIP-seq (ChIP followed by high-throughput sequencing) technology, Mikkelsen and colleagues later showed that bivalent domains are more generally indicative of genes that remain in a poised state. Consequently, pluripotent cells were found to contain large numbers of bivalent domains (~2500) compared with multipotent neural progenitor cells (NPCs) (~200) that still retain multilineage potential but are more restricted than ES cells.[140]

Several studies of the murine iPS cell have identified a small number of representative loci that have consistent chromatin and DNA methylation patterns.[7,66,68] Maherali and colleagues used ChIP-Chip to investigate the presence of H3K4me3 and H3K27me3 in the promoter regions of 16,500 genes and results showed that iPS cells were highly similar to ES cells in epigenetic state.[66] In addition, the H3K4me3 pattern was similar across all samples, indicating that reprogramming was largely associated with changes in H3K27me3 rather than H3K4me3.[66] Mikkelsen and colleagues have used the more comprehensive ChIP-Seq technique to determine genome-wide chromatin maps in several iPS lines, which are derived with different methods: drug selection using an *Oct4*–neomycin-resistance gene,[68] drug selection using a *Nanog*–neomycin-resistance gene,[68] and by morphological appearance.[72] Overall global levels of repressive H3K27me3 and the characteristic bivalent chromatin structure are retained in the different iPS cell lines. The restoration of repressive chromatin marks appears crucial to stably silence lineage-specific genes that are active in somatic cells and inactive in undifferentiated pluripotent cells. Failure to establish the repressive marks results in incompletely reprogrammed cells. Activating H3K4me3 patterns are also crucial for complete reprogramming and have been observed to be restored genome-wide, in particular around the promoters of pluripotency- associated genes, such as *Oct4* and *Nanog*, in the fully reprogrammed iPS lines in this study.[113]

A second component of the epigenetic machinery is DNA methylation, which is a stable and heritable mark that is involved in gene silencing including genomic imprinting and X-chromosome inactivation. DNA methylation patterns are dynamic during early embryonic development and are essential for normal postimplantation development.[141] Overall DNA methylation levels remain stable during ES cell differentiation, although they are not static for any given individual gene.[140] The 5′-promoter regions of many transcriptional units contain clusters of the dinucleotide CpG, which are methylated at transcriptionally silent genes and demethylated upon activation. In differentiated cells, the *Oct4*, *Nanog*, and *Sox2* promoter regions are highly methylated as an inactivated status, whereas in ES cells, these promoters are unmethylated to be activated. During reprogramming, almost complete demethylation of these promoters has been observed.[66–68,113] Therefore, the loss of DNA methylation at the promoters of pluripotent-related genes is essential for achieving complete reprogramming. Takahashi and Yamanaka found that iPS cells selected with *Fbx15* did not reactivate endogenous *Oct4* and *Nanog* genes and the respective promoters remained methylated.[7] Interestingly, loss of DNA methylation at this class of genes seems to be a rather late event in the reprogramming process because cells that have already acquired self-renewing properties still showed high levels of DNA methylation.[113]

3.3.4 Developmental Potential: Pluripotency

Research on transcriptional and epigenetic state of iPS cells is highly informative, and it might ultimately be possible to characterize newly derived iPS cell lines based on their genomic profiling alone. Before selecting the most informative markers, it is important to use *in vivo* assays to analyzing the interplay between transcriptome, epigenome, and developmental potential. Recently, Jaenisch and Young provided a detailed comparison of the different strategies for assessing developmental potential and their stringency.[1] *In vitro* differentiation is the least stringent assay and tetraploid-embryo complementation is the most stringent assay for testing developmental potential.[1,142] These strategies could be used to determine the pluripotency of mouse iPS cells, but only *in vitro* differentiation and teratoma formation could be applied to test those of human iPS cells; experiments such as generation of chimera are ethically inappropriate using human cells. From this point of view, epigenetic and transcriptional analyses are more important to assess the developmental potential of human iPS cells. Mouse iPS cells appear to have a developmental potential similar to that of ES cells by histological analysis of teratoma formation and by high contribution to chimera formation with germline transmission, as shown by many different groups.[7,66,68] To show the final step of developmental potential of iPS cells as equivalent to that of ES cell, three separate groups injected mouse iPS cells (2N) into tetraploid blastocysts (4N), which are capable of producing placental and other extraembryonic tissues but not the embryo itself, and have created live mice from iPS cells.[132,133,143] The procedure, called "tetraploid complementation," is the most stringent test for pluripotency. If the stem cells that are injected into the tetraploid blastocyst differentiate into embryonic tissues that produce a mouse, the stem cells are considered truly pluripotent.

3.4 MECHANISM OF REPROGRAMMING

The most widely used set of reprogramming factors—*Oct4, Sox2, Klf4*, and *c-Myc*—is sufficient for mouse iPS cell and human iPS cell generation (Table 3.2). Other groups later identified a partially overlapping combination of factors—*Oct4, Sox2, Nanog*, and *Lin28*.[70] Therefore, *Oct4* and *Sox2* are key transcription factors in iPS cell generation, and the complexity of downstream events maintains pluripotency and blocks differentiation.[135] There are several hundreds of downstream genes of *Oct4* and *Sox2* with complex interactions and signaling networks.[111,144,145] The other two factors, *Klf4* and *c-Myc*, are known to promote cellular proliferation, chromatin remodeling, and the prevention of cell death.[146] Studies from Yamanaka and Daley employed different combinations of three to six genes to reprogram cells,[10,109] adding SV40 large T antigen and telomerase[10] or withholding either *Klf4* or *c-Myc*.[10] Plath and colleagues added a fifth gene, the pluripotency-promoting transcription factor *Nanog*, to the mix, although they also showed that it was not essential.[110] The research reported from Thomson and colleagues used *Oct4, Sox2*, and *Nanog* but not *c-Myc* or *Klf4*.[70] Instead, a completely different type of gene known as *Lin28* conferred potent reprogramming activity when added to the cocktail.[70] *Lin28* is a regulator of miRNAs previously shown to be downregulated during the differentiation of ES cells.[147] *Lin28* appears to function by inhibiting let-7 miRNAs that in turn inhibit the growth-promoting oncogenes leading to cellular proliferation.[148] Surprisingly, many roads appear capable of leading to the induction of pluripotency. The type of gene may be of greater importance than the individual identity of the gene used.

3.4.1 REPROGRAMMING FACTORS

3.4.1.1 POU Domain, Class 5, Transcription Factor 1 (Pou5f1, Oct4)

Oct4 (octamer-binding transcription factor 4, also known as *Oct 3* and *Pou5f1*) was first described as a protein present in unfertilized oocytes, ES cells, and PGCs.[149] Its expression is essential for the development of the ICM *in vivo*, the derivation of ES cells, and the maintenance of a pluripotent state.[150] The precise levels of *Oct4* in ES cells cause differentiation into different fates of cells; primitive endoderm and mesoderm can be differentiated by increase in expression, whereas repression of

Oct4 induces loss of pluripotency and differentiation into trophectoderm.[135] This suggests that ES cells possess a regulatory network to keep *Oct4* expression at the optimal level to ensure pluripotency.[151] Biochemically, *Oct4* has been shown to be a DNA-binding protein with a bipartite POU/homeodomain.[152] *Oct4* relies on two transactivation domains flanking the DNA-binding domain to exert its transcription activities.[153] The nuclear localization signal of *Oct4* is required for its transcription activity, and its ablation leads to the generation of a dominant-negative form of *Oct4*, which is capable of inducing ES cell differentiation by interfering with wild-type *Oct4* activity.[151]

3.4.1.2   SRY-Box Containing Gene 2 (Sox2)

Sex-determining region Y (SRY)-box 2, known as *Sox2*, is a transcription factor involved in the self-renewal of ES cells. It has an important role in maintaining ES cell pluripotency and heterodimerizes in a complex with *Oct4*.[154] Like *Oct4*, *Sox2* has also been implicated in the regulation of FGF4 expression.[155] *Sox2* is required for epiblast and extraembryonic ectoderm formation, suggesting that *Sox2* and *Oct4* cooperatively specify the fate of pluripotent stem cells at implantation.[156] Recent studies demonstrated that *Sox2* is necessary for regulating multiple transcription factors that affect *Oct4* expression, thus stabilizing ES cells in a pluripotent state by maintaining the requisite level of *Oct4* expression.[157,158] In addition to ES cells, *Sox2* is also expressed in the extra-embryonic ectoderm, trophoblast stem (TS) cells, and the developing central nervous system (neural stem cells).[156,158] In these cell lineages, *Sox2* expression is restricted to cells with stem cell characteristics, supporting their self-renewal capability, and is lost in cells with a more restricted developmental potential.[156] Interestingly, forced expression of *Oct4* can compensate for loss of *Sox2* in ES cells[158] and, in direct reprogramming, *Sox2* can be replaced by *Sox1*, *Sox3*, and, to a lesser extent, *Sox15* or *Sox18*.[71]

3.4.1.3   Myelocytomatosis Oncogene (Myc, c-Myc)

c-Myc (Myelocytomatosis oncogene) is a pleiotropic transcription factor that has been linked to several cellular functions, including cell cycle regulation, cell proliferation, growth, differentiation, and metabolism.[159] This factor is highly expressed in rapidly proliferating cells and is generally low or absent during quiescence.[160] It has a multitude of target sites in the genome[161] and plays a role in regulation of both protein coding genes[162,163] as well as noncoding miRNA genes.[164,165] The induction of target genes involved in proliferation such as *p21*, *p15*, *Cyclin-D-CDK4*, and *E2F2* may enable cells to proliferate and might be an important function during the reprogramming. *c-Myc* also functions in self-renewal and differentiation of stem and progenitor cells, particularly in interactions between stem cells and the local microenvironment.[158] *c-Myc* appears to be involved in recruiting chromatin-remodeling activators to promoters[160,166,167] and may also regulate global chromatin structure.[168] A recent study showed that overexpression of *c-Myc* increased replication origin activity and that it affected the kinetics of reprogramming.[169] Although the exact role of *c-Myc* in reprogramming is not clear, it is dispensable for the generation of iPS cells in mouse and human,[13,71,95,170,171] but the efficiency of reprogramming decreases dramatically. Generating iPS cells in the absence of *c-Myc* is considered safer because the spontaneous reactivation of the proviral *c-Myc* causes a high incidence of tumors in chimeric mice generated from iPS cells.[67]

3.4.1.4   Kruppel-Like Factor 4 (Klf4)

Kruppel-like factor 4 (*Klf4*) is a transcription factor expressed in a variety of tissues, including the epithelium of the intestine, kidney, and the skin.[172] Depending on the target gene and interaction partner, *Klf4* can both activate and repress transcription,[173] and can function both as an oncoprotein and a tumor suppressor. Constitutive expression of *Klf4* suppresses cell proliferation by blocking G1–S progression of the cell cycle. As *p53* has been shown to suppress *Nanog* expression during differentiation,[174] the ability of *Klf4* to inhibit *p53* might be important in the reprogramming process by preventing cell cycle exit.[175] This may also be important for counteracting the apoptotic and differentiation effects induced by sustained activation of *c-Myc* in human ES cells.[176] *Klf4* has been shown to cooperate with *Oct4* and *Sox2* to activate the *Lefty1* core promoter,[177] suggesting that this activity may be important for maintenance of the pluripotency transcriptional program. It has been demonstrated that the forced overexpression of *Klf4* in ES cells inhibits differentiation in erythroid progenitors, suggesting a role for this factor in ES cell function.[178] Its exact role in the reprogramming process is not fully understood, and it can be replaced with other Klf family members (*Klf2* and *Klf5*)[71] or the unrelated factors *Nanog* and *Lin28*.[70] Clearly, *Klf4* transduction is not essential for the reprogramming process as it is dispensable for human reprogramming.

3.4.1.5   Nanog Homeobox (Nanog)

Nanog was first described as a factor that was involved in maintaining ES cell self-renewal and pluripotency.[50,51] Smith and colleagues termed the factor *Nanog*, after the mythological Celtic land of the ever-young, "Tir nan Og."[50] *Nanog* was discovered based on its ability to sustain stem cell self-renewal in the absence of LIF.[50,51] Although it was originally believed that *Nanog* prevents ES cells from differentiation in the absence of LIF by repressing the expression of differentiation genes, *Nanog* behaves as a strong activator of the *Oct4* promoter, thus participating in the regulation of *Oct4* expression in ES cells.[179,180] *Nanog* expression is found in the interior cells of the compacted morula and the ICM of the blastocyst. On implantation, *Nanog* expression is detected only in the epiblast and is eventually restricted to PGCs.[50] More recently, loss of *Nanog* predisposes ES cells to differentiation but does not mark commitment and is reversible.[181] Interestingly, *Nanog* is not an essential factor for iPS cell generation and does not appear to notably affect the efficiency.

3.4.1.6   Lin-28 Homolog (Lin28)

Lin28 is a conserved RNA-binding protein involved in developmental timing in *Caenorhabditis elegans*.[182] Its homologs in *Drosophila*, *Xenopus*, and mouse appear to be expressed and downregulated during development, consistent with a conserved role for this factor as a regulator of developmental timing.[183] *Lin28* appears to shuttle from the nucleus to the cytoplasm, where it localizes to processing bodies, or P-bodies, which are sites of mRNA degradation and miRNA regulation.[184] This suggests *Lin28* regulates mRNA translation or stability. Therefore, in the context of reprogramming,

Lin28 could act to stabilize newly synthesized transcripts to allow for adoption of an ES cell transcriptional network. In mammals, *Lin28* is expressed in ES cells and during early embryogenesis, but its expression becomes restricted to several tissues during late embryogenesis and adult life. It can increase the stability of specific mRNAs and contribute to identity establishment of the tissue in which it is expressed.[185] Gregory and colleagues showed that *Lin28* blocks processing of the *let-7* miRNA, thus suggesting a role in controlling miRNA-mediated differentiation in stem cells.[148]

3.4.2 Silencing of Integrated Retroviral Vectors in iPS Cells

Retroviral vectors used for iPS cell generation, such as pMXs and pLIB, are derived from Moloney murine leukemia virus (MoMLV), and transgene transcription is initiated at the long terminal repeat (LTR) promoter of these vectors. Historically, MoMLV has been extensively studied for retroviral silencing since Teich and colleagues found that MoMLV did not replicate in EC cells.[186] This finding lead to the identification of negative regulatory elements, called silencers, in MoMLV sequences and the known silencers have been mapped to LTR regions including the negative control region (NCR), direct repeat (DR) enhancer, CpG rich promoter, and primer binding site (PBS).[187] These silencers play an important role in transcriptional shut down of MoMLV LTR expression in ES cells. Other retroviral vectors designed to overcome stem cell silencing have deletions or mutations at some of the silencers. One example already used in iPS cell experiments is the murine stem cell virus (MSCV) derived vector (pMIG). In this case, MSCV transduced pluripotency transgenes are still silenced in human iPS cells,[103] which is an important feature of fully reprogrammed iPS cells that facilitates their subsequent differentiation into embryoid bodies or specific cell types.

Lentiviral vectors were originally reported to escape silencing in ES cells and transgenic mice,[188,189] and therefore may be less suitable for reprogramming experiments. More recent experiments found that lentiviral vectors are often completely silent in ES cells[190,191] and have been used for successful induction of mouse and human iPS cells.[70,192] Many studies have shown that multiple epigenetic pathways are involved in the silencing of retroviruses and lentiviruses.[191] Histone modification and DNA methylation have been evaluated by several investigators and play a central role in marking the silenced state. Methylation at H3K9 or 27 H3K27 represses that area of DNA, while H3K4 methylation activates transcription. Wolf and Goff[187] identified enriched H3K9 dimethylation and HP1γ recruited by Trim28 near the PBS silencer element as a silencing mechanism. Infection of P19 EC cells with lentivirus causes a loss of H3K4 dimethylation and increase of H3K9 dimethylation.[193] Currently, several HMTs have been identified that increase H3K9 methylation: Ehmt1 (GLP) and Ehmt2 (G9a) increase monomethylation, Setdb1 (ESET) increases mono- and dimethylation, and Suv39hi and Suv39h2 increase di- and trimethylation.[194] These H3K9 methylations are read by the HP1 family of proteins that may also contribute to the recruitment of DNA methyltransferases for DNA methylation. Finally, the YY-1 factor recruits polycomb group (PcG) complexes to methylate H3K27 and deacetylate histones.[195] It will be of great interest to determine the function of these factors on retroviral silencing in pluripotent stem cells.

Other chromatin factors also influence retroviral silencing. Brm, a catalytic subunit of the SWI/SNF complex that controls nucleosome positioning, is required for consistent viral expression in human tumor cell lines.[196] In addition, the linker histone H1 facilitates the folding of chromatin into condensed 30 nm fibers. Histone H1 is generally associated with transcriptional repression and is detected on silent retroviruses and lentiviruses in ES cells.[190] Notably, histone H1 has been shown to affect DNA methylation of imprinted genes in knockout ES cells,[197] suggesting a potential role for histone H1–mediated recruitment of DNA methylation. Finally, small RNAs in the form of rasiRNA (repeat associated siRNA) have been shown to repress endogenous retrotransposons in mouse oocytes[198] and could have effects on retroviral silencing in stem cells.

DNA methylation also influences retroviral silencing during iPS cell generation. For example, retroviral transcription is inversely correlated with the Dnmt3a2 (Dnmt3a isoform 2) expression levels.[67] Dnmt3a2 levels peak at day 3 of reprogramming, while Dnmt3b and Trim28 levels gradually increase with their highest expression at day 13 of iPS cell induction.[74] Finally, silent retrovirus in iPS cells can be reactivated by knockdown of Dnmt1, which is responsible for the maintenance of DNA methylation.[68] Once the pluripotent cell epigenetic and transcriptional networks are established, retroviral silencing also occurs.

Recent genome wide chromatin evaluations of mouse ES cells indicate that pluripotent stem cells have more open chromatin than mature differentiated cells.[199] This chromatin is hyperdynamic, dependent on the kinetics of HP1 and histone H1.[200] Moreover, ES cells have extensive regions of bivalent chromatin that are marked by both silent (H3K27me) and active (H3K4me) histone modifications.[138] These reports suggest that most RISs in ES cells are in open or bivalent chromatin, consistent with their known preference for integrating into promoter elements. Even lentivirus vectors prefer to integrate into the body of expressing genes marked by H3Ac, H4Ac, and H3K4me1, but disfavored to H3K27me3 and DNA methylation sites.[201] Since genes that are transcriptionally active tend to be located near transcription factors in more central neighborhoods of the nucleus, it is also possible that variegated retroviral vectors are directed to peripheral nuclear locations when silent and more central locations when active.[202] Finally, it has been reported that lentiviral LTRs loop to interact with each other when transcribed for proper RNA processing.[203] It is possible that variegated virus temporarily loops to interact with nearby endogenous regulatory elements that repress or activate the virus.

The significance of retroviral silencing in iPS cell biology is derived from the fact that the total amount of *Oct4* must be precisely regulated during the intermediate stages of reprogramming. If viral transgenes get silenced earlier than the establishment of the endogenous pluripotency transcriptional network, then partially reprogrammed iPS cells could easily revert back to the somatic state or progress down the trophectoderm lineage. At the same time, if viral transgenes do not get silenced after the establishment of pluripotency, then the cells could be forced to differentiate due to excess amounts of *Oct4* or may become stable, partially reprogrammed iPS cell lines. In fact, extended expression of the factors by a doxycycline-inducible promoter does stabilize them into partially reprogrammed iPS cells.[113]

Therefore, accurate switching between exogenous and endogenous gene expression while maintaining factor levels consistent with pluripotency contributes to the efficiency of iPS cell induction.

3.4.3 SIGNAL NETWORKS

3.4.3.1 Transcription Factor Networks

In mammals, it was thought that the differentiation process was irreversible until the successful cloning of Dolly by SCNT.[5] This cloning experiment demonstrated that somatic cells can be reprogrammed back to the totipotent zygotic state by cellular factors in unfertilized eggs. The complex signaling interactions took a considerable amount of time and effort to identify. *Oct4* represses itself when overexpressed while *Nanog*, *Sox2*, and *FoxD3* activate its expression.[135] Additional repressors may yet be unidentified that provide counterbalance for maintaining *Oct4* expression levels in ES cells.

Based on large-scale data sets, Young and colleagues[111] proposed that *Oct4*, *Sox2*, and *Nanog* collaborate to form regulatory circuitry consisting of autoregulatory and feed-forward loops that dictate pluripotency and self-renewal. This autoregulatory circuitry suggests that the three factors function collectively to stably maintain their own expression.[204] Autoregulatory loops appear to be a general feature of master regulators of cell state.[205] Functional studies have confirmed that *Oct4* and *Sox2* co-occupy and activate the *Oct4* and *Nanog* genes,[206,207] and experiments with an inducible *Sox2* null murine ES cell line have provided compelling evidence for this interconnected autoregulatory loop and its role in the maintenance of pluripotency.[158] The loop formed by *Oct4*, *Sox2*, and *Nanog* also suggests how the core regulatory circuitry of iPS cells might be triggered when *Oct4*, *Sox2*, and other transcription factors are overexpressed in fibroblasts.[7,66-68] When these factors are ectopically overexpressed, they may directly activate endogenous *Oct4*, *Sox2*, and *Nanog*, the products of which in turn contribute to the maintenance of their own gene expression. Similar approaches were attempted by other groups to identify these networks by high-throughput technologies, and more elaborate mechanisms were proposed.[144,208-210] *Oct4*, *Sox2*, and *Nanog* co-occupy several hundred genes, often at apparently overlapping genomic sites.[111,144] A large multi-protein complex containing *Oct4* and *Nanog* can be obtained by immunoprecipitation (IP) in ES cells.[208] Other pluripotency factors may function in complexes to control their target genes, and this phenomenon may explain why efficient iPS cell generation requires combined overexpression of multiple transcription factors. Not all components of this putative complex are required to initiate the process of reprogramming because exogenous *Nanog* is not necessary for iPS generation. It seems likely that exogenous *Oct4* and other factors induce expression of endogenous *Nanog* to levels sufficient to accomplish full reprogramming. The master regulators of pluripotency occupy the promoters of active genes encoding transcription factors, signal transduction components, and chromatin-modifying enzymes that promote ES cell self-renewal.[111,144] However, these transcriptionally active genes consist of only about half of the targets of *Oct4*, *Sox2*, and *Nanog* in ES cells. These master regulators also co-occupy the promoters of a large set of developmental specific transcription factors that are silent in ES cells, but whose expression is associated with lineage commitment and cellular differentiation.[111,144] Silencing of these developmental regulators is almost certainly a key feature of pluripotency, because expression of these developmental factors is associated with commitment to particular lineages. *MyoD*, for example, is a transcription factor capable of inducing a muscle gene expression program in a variety of cells.[211] Therefore, *Oct4*, *Sox2*, and *Nanog* likely help maintain the undifferentiated state of ES cells by contributing to repression of lineage specification factors.

In addition to *Oct4*, *Sox2*, and *Nanog*, many other factors required for pluripotency have been identified, including *Sall4, Dax1, Essrb, Tbx3, Tcl1, Rif1, Nac1*, and *Zfp281*.[144,208,212] These pluripotency factors regulate each other to form a complicated transcriptional regulatory network in ES cells.[210] For example, *Sall4*, a *spalt* family member, interacts with *Nanog* and co-occupies *Nanog* and *Sall4* enhancer regions. Additionally, *Sall4* also regulates *Oct4* expression by binding to the *Oct4* promoter.[213,214] *Essrb* and *Rif1* are primary targets of both *Oct4* and *Nanog*.[144] Beside their DNA-binding activities, these pluripotency-related proteins are extensively interconnected by protein–protein interaction. A protein interaction network in mouse ES cells has been constructed by tagging of Nanog and purification of Nanog-associated proteins.[208] This mini-interactome is highly enriched for proteins that are required for the survival or differentiation of the ICM and for early development. Many of the genes encoding proteins in the interaction network are targets of *Nanog* and/or *Oct4*, suggesting that the transcriptional network might have a feedback mechanism through the protein interaction network. The protein interaction network is linked to several cofactor pathways largely involved in transcriptional repression.[208] These data support a model wherein essential factors maintain the pluripotent state by simultaneously activating genes involved in pluripotency and repressing genes important for development.

3.4.3.2 Signal Transduction Pathway: Ground Level of Self-Renewal

ES and iPS cells require extrinsic growth factors for maintenance of pluripotency in culture, suggesting that pluripotency is an inherently unstable cell state and that ES cells are "primed" for rapid differentiation. Historically, ES cells were cultured in the presence of an underlying feeder cell layer of mitotically inactivated fetal fibroblast cells, which provides an environment capable of supporting pluripotency and blocking spontaneous differentiation. The necessary factor for self-renewal of mouse ES cells is LIF, a cytokine able to maintain mouse ES cells even in the absence of the fibroblast cell feeder layer.[215] LIF is not required for pluripotency of the ICM *in vivo*[216] and is unable to maintain pluripotency in human ES cells, suggesting that alternative mechanisms function in the maintenance of pluripotency within these contexts. Serum is also important for mouse ES cell maintenance, although bone morphogenic protein 4 (BMP4) is able to replace this requirement.[217,218] In addition, Wnt signaling has been found to act synergistically with LIF to maintain pluripotency in mouse ES cells and appears to have a role in human ES cells.[219,220] Autocrine loops of Activin/Nodal signaling have also been implicated in the maintenance of mouse ES cells.[220]

Mouse and human ES cells exhibit distinct growth habits: doubling rates for human ES cells are characteristically longer than those recorded for mouse ES cells, between 30 and 40 h.[221] Human ES cells require maintenance of cell–cell contacts for propagation, and spontaneous differentiation within human ES cell colonies is initiated from central cells,[222] whereas spontaneous differentiation of mouse ES cells occurs at the colony periphery. Human ES cells are routinely cultured on a fibroblast cell feeder layer, but their growth factor requirements also differ from those of mouse ES cells. The growth factors capable of promoting pluripotency in this system appear to be FGF2 produced by the feeder cell layer, and insulin-like growth factor (IGF) secreted by human ES cells, which set up interdependent paracrine loops.[223] Studies focusing on extrinsic signals required for maintaining human ES cells have found FGF2 is sufficient to support growth of these cells on Matrigel, a substrate made up predominantly of laminin and collagen but with additional unknown factors.[224] It is likely that extrinsic signals maintaining human ES cells will inhibit BMP signaling to sustain proliferation without differentiation. BMP4 has been shown to negatively regulate pluripotency and induce trophoblast-like cell formation from human ES cells.[225] Activin/Nodal and FGF2 are capable of maintaining human ES cells in the absence of feeder layers and other exogenous factors.[226] One role of Activin A in the maintenance of human ES cells has been proposed as inhibition of the BMP4 signaling pathway mediators Smads 1 and 5. Activin A and FGF2 have recently been shown to facilitate derivation and maintenance of pluripotent mouse epiblast-derived cell lines.[59,60]

Although culture conditions and extrinsic growth factors that can support pluripotent cell maintenance have been defined, it is poorly understood how the signaling pathways controlled by these factors maintain the transcription factor network required for pluripotency. In mouse ES cells, LIF activates JAK/STAT signaling and mitogen activated protein kinase (MAPK) pathways. The choice between pluripotency and differentiation is dependent on a balance between Stat3 and extracellular signal-regulated kinase (ERK) MAPK activity, respectively. In mouse ES cells, BMP4 prevents differentiation through the inhibition of ERK[218] and induction of other inhibitors of differentiation such as inhibitor of differentiation (Id) proteins.[217] Stat3 activates a number of genes that play important roles in pluripotency, including *c-Myc*,[227] *Nanog*,[50–52,228] *Eed*,[229] *Jmjd1a*,[230] and *GABPα*, which is required for the maintenance of *Oct4* expression.[231]

Neither STAT3 nor BMP4 activity are implicated in pluripotency of human ES cells,[225,232] while ERK activity is required for the maintenance of pluripotency.[233,234] Sustained activation of *c-Myc* in human ES cells induces differentiation and apoptosis.[176] Thus, consistent with the differing extrinsic requirements, the intracellular signals regulating pluripotency in mouse and human ES cells appear to be different, despite the conservation of the core transcription factor networks and functional similarity of the cells. Smith and colleagues reported that the self-renewal of ES cells is maintained by the inhibition of ERK pathway and glycogen synthase kinase 3 (GSK3) after the elimination of extrinsic stimuli, suggesting that ES cells have an innate program for self-replication.[235]

3.4.4   EPIGENETIC REGULATION OF CHROMATIN IN iPS CELLS

As the substrate of transcription, chromatin is subjected to various forms of epigenetic regulation including chromatin remodeling, histone modifications, histone variants, and DNA methylation. For example, trimethylation of lysine 9 and lysine 27 of histone 3 (H3K9 and H3K27) correlate with inactive regions of chromatin, whereas H3K4 trimethylation, and acetylation of H3 and H4 are associated with active transcription,[236] and DNA methylation generally represses gene expression.[237] Given that ES and somatic cells contain almost identical genomic DNA, epigenetic regulation is one of the major influences on their differentiation potential and pluripotency.

To maintain pluripotency in ES cells, differentiation triggering genes should be inactive. PcG proteins play important roles in silencing these developmental regulators. The PcG proteins function in two distinct Polycomb Repressive Complexes, PRC1 and PRC2. Genome-wide binding site analyses have been carried out for PRC1 and PRC2 in mouse ES cells and for PRC2 in human ES cells.[238,239] The genes regulated by the PcG proteins are co-occupied by nucleosomes with trimethylated H3K27. These genes are transcriptionally repressed in ES cells and are preferentially activated when differentiation is induced. Many of these genes encode transcription factors with important roles in development. Interestingly, the pluripotency factors *Oct4*, *Sox2,* and *Nanog* co-occupy a significant fraction of the PcG protein regulated genes when acting as transcription factors.[238,239] PcG proteins may facilitate pluripotency maintenance by suppressing developmental pathways. Most of the transcriptionally silent developmental regulators targeted by *Oct4*, *Sox2*, and *Nanog* are also occupied by the PcG,[138,238,239] which are epigenetic regulators that facilitate maintenance of cell state through gene silencing. The PcG form multiple polycomb PRCs, the components of which are conserved from *Drosophila* to humans.[240] PRC2 catalyzes H3K27 methylation, an enzymatic activity required for PRC2-mediated epigenetic gene silencing. H3K27 methylation is thought to provide a binding surface for PRC1, which facilitates oligomerization, condensation of chromatin structure, and inhibition of chromatin remodeling activity in order to maintain silencing. PRC1 also contains a histone ubiquitin ligase, *Ring1b*, whose activity appears likely to contribute to silencing in ES cells.[241] How the PcG are recruited to genes encoding developmental regulators in ES cells is not yet understood. Some of the most conserved vertebrate sequences are associated with genes encoding developmental regulators, and some of these may be sites for DNA-binding proteins that recruit PcG proteins. Recent studies revealed that the silent developmental genes that are occupied by *Oct4*, *Sox2*, *Nanog*, and PcG proteins experience an unusual form of transcriptional regulation.[242] These genes undergo transcription initiation but not productive transcript elongation in ES cells. The transcription initiation apparatus is recruited to developmental gene promoters, but RNA polymerase is incapable of fully transcribing these genes, presumably because of repression mediated by the PcG. This explains why the silent genes encoding developmental regulators are generally organized in "bivalent" domains that are occupied by nucleosomes with histone H3K4me3, which is associated with gene activity, and by nucleosomes with histone H3K27me3, which is associated with repression.[138,199,242]

The presence of inactive RNA polymerase at the promoters of genes encoding developmental regulators may explain why these genes are especially poised for transcription activation during differentiation.[238,239] PcG complexes and associated proteins may serve to pause RNA polymerase machinery at key regulators of development in pluripotent cells and in lineages where they are not expressed. When the cells are activated, PcGs and nucleosomes with H3K27 methylation are lost,[138,139,238,239] allowing the transcription apparatus to fully function and transcribe these genes. The mechanisms that lead to selective activation of genes encoding specific developmental regulators are not yet understood, but are under active investigation.[243] Beyond the specific regulation of development-related genes, ES cells maintain chromatin in a highly dynamic and transcriptionally permissive state. Fewer heterochromatin foci are detected in ES cell nuclei compared to differentiated cells. Fluorescence recovery after photobleaching and biochemical analyses reveal that ES cells, compared with differentiated cells, have an increased fraction of loosely bound or soluble architectural chromatin proteins, including core and linker histones and heterochromatin protein HP1. A hyperdynamic chromatin structure is functionally important for pluripotency.[200] The status of histone modifications also indicates that the chromatin in ES cells is more transcriptionally permissive than in differentiated cells. Consistent with the global dynamics of chromatin, ES cell differentiation is associated with a decrease in global levels of active histone marks, such as acetylated histone H3 and H4, and an increase in repressive histone marks, histone H3 lysine 9 methylation.[200,244] Taken together, these unique epigenetic characteristics of ES cells facilitate rapid but regulated transcription, allowing differentiation down different cell fate pathways as needed by the organism.

3.4.5 MicroRNAs and Pluripotency

Noncoding RNA comprises a large fraction of vertebrate transcriptomes. Although not all noncoding RNAs are functional, many play important regulatory roles. miRNA is small noncoding RNAs of ~22 nucleotide in length. They regulate gene expression via at least two distinct mechanisms: degradation of target mRNA transcripts and inhibition of mRNA translation.[245] ES cells express a unique set of miRNAs that are downregulated as ES cells differentiate into embryoid bodies. Some of these miRNAs are conserved between human and mouse and are clustered in the genome.[246,247] miRNAs might play a role in the maintenance of pluripotency in iPS cells as well as ES cells.[248] *Lin28*, one of the factors used to reprogram human fibroblasts,[70] has recently been shown to block processing of the *let-7* family miRNA in ES cells.[148,249] *Let-7* family members also have been implicated in the promotion of differentiation of cancer stem cells.[250,251] Thus, *Lin28* may facilitate reprogramming by repressing *let-7*-induced differentiation in fibroblasts. These data suggest that pluripotency may be maintained at the regulatory sites of pluripotency genes and at the sites of specific miRNA production.

3.4.6 Other Possible Mechanisms

Nuclear reprogramming is a complex process that is not well understood. There are many features that distinguish pluripotent ES cells and iPS cells from differentiated cells, including gene expression, miRNA expression, epigenetic modifications, cell cycle regulation, and telomerase activity. iPS cells have always been made from proliferating somatic cell populations, and reprogramming technology necessitates DNA replication and cell division. The duration of the cell cycle for ES and iPS cells is much shorter than for differentiated cells, mainly due to a shortened G1 phase. In MEFs, the G1 phase lasts 15–20 h and temporally accounts for more than 80% of the cell cycle. However, in both mouse and human ES cells, G1 lasts 2–4 h and temporally accounts for only 15%–20% of the cell cycle. This unique cell cycle pattern is characterized by hyperphosphorylated RB protein, constitutively high activity of cyclin E and A-associated kinases, and a lack of expression of major CDK inhibitors.[252] The role of a shortened G1 phase in maintaining pluripotency is not clear, though the exclusivity of it among cells that are pluripotent suggests it is important. Upon differentiation, the ES cell cycle pattern quickly switches to a MEF-like pattern.[253] Another difference between ES cells and somatic cells is the high-level telomerase activity in ES and many adult stem cells.

Similar to ES cells, iPS cells exhibit a cell cycle with a shortened G1 phase[66] and elevated telomerase activity.[7,70,109] During reprogramming, fibroblasts not only become pluripotent, they also become immortal. Fibroblasts proliferate a finite period of time before entering into senescence. In contrast, ES cells and iPS cells do not experience such a limitation. Immortalization requires that at least two barriers be overcome: cellular senescence and telomere shortening.[254,255] Rb and p53 are the key senescence-inducing factors. In ES cells, the Rb pathway is constitutively inactivated due to hyperphosphorylation,[253] while certain aspects of p53 function are compromised.[256] Inhibition of p53 function and Ink4/Arf promotes the reprogramming of somatic cells to pluripotent cells.[116–120]

Cells can enter a so-called replication crisis state where they undergo apoptosis if their telomere erodes below a critical length.[257,258] To avoid telomere shortening, the activity of telomerase must be upregulated. c-Myc directly upregulates the transcription of Tert, the gene encoding the enzymatic subunit of the telomerase.[259] It is unclear if elevated telomerase activity in iPS cells is due to ectopic expression of c-Myc and how much the resulting change of telomerase activity contributes to reprogramming. It is also unclear how the four factors find ways to inactivate Rb and p53 and to what extent this contributes to reprogramming. It is also worth stressing that immortalization by reprogramming is different from transformation. iPS cells are immortal in their undifferentiated state. Upon differentiation, both the Rb and p53 pathways become fully functional again through yet unknown mechanisms.

REFERENCES

1. Jaenisch, R.; Young, R., Stem cells, the molecular circuitry of pluripotency and nuclear reprogramming. *Cell* 2008, 132, (4), 567–582.
2. Briggs, R.; King, T. J., Transplantation of living nuclei from blastula cells into enucleated frogs' eggs. *Proc Natl Acad Sci USA* 1952, 38, (5), 455–463.
3. Gurdon, J. B., The developmental capacity of nuclei taken from intestinal epithelium cells of feeding tadpoles. *J Embryol Exp Morphol* 1962, 10, 622–640.
4. Gurdon, J. B.; Uehlinger, V., "Fertile" intestine nuclei. *Nature* 1966, 210, (5042), 1240–1241.

5. Campbell, K. H.; McWhir, J.; Ritchie, W. A.; Wilmut, I., Sheep cloned by nuclear transfer from a cultured cell line. *Nature* 1996, 380, (6569), 64–66.
6. Wilmut, I.; Schnieke, A. E.; McWhir, J.; Kind, A. J.; Campbell, K. H., Viable offspring derived from fetal and adult mammalian cells. *Nature* 1997, 385, (6619), 810–813.
7. Takahashi, K.; Yamanaka, S., Induction of pluripotent stem cells from mouse embryonic and adult fibroblast cultures by defined factors. *Cell* 2006, 126, (4), 663–676.
8. Thomson, J. A.; Itskovitz-Eldor, J.; Shapiro, S. S.; Waknitz, M. A.; Swiergiel, J. J.; Marshall, V. S.; Jones, J. M., Embryonic stem cell lines derived from human blastocysts. *Science* 1998, 282, (5391), 1145–1147.
9. Dimos, J. T.; Rodolfa, K. T.; Niakan, K. K.; Weisenthal, L. M.; Mitsumoto, H.; Chung, W.; Croft, G. F.; Saphier, G.; Leibel, R.; Goland, R.; Wichterle, H.; Henderson, C. E.; Eggan, K., Induced pluripotent stem cells generated from patients with ALS can be differentiated into motor neurons. *Science* 2008, 321, (5893), 1218–1221.
10. Park, I. H.; Arora, N.; Huo, H.; Maherali, N.; Ahfeldt, T.; Shimamura, A.; Lensch, M. W.; Cowan, C.; Hochedlinger, K.; Daley, G. Q., Disease-specific induced pluripotent stem cells. *Cell* 2008, 134, (5), 877–886.
11. Rubin, L. L., Stem cells and drug discovery: The beginning of a new era? *Cell* 2008, 132, (4), 549–552.
12. Hanna, J.; Wernig, M.; Markoulaki, S.; Sun, C. W.; Meissner, A.; Cassady, J. P.; Beard, C.; Brambrink, T.; Wu, L. C.; Townes, T. M.; Jaenisch, R., Treatment of sickle cell anemia mouse model with iPS cells generated from autologous skin. *Science* 2007, 318, (5858), 1920–1923.
13. Wernig, M.; Zhao, J. P.; Pruszak, J.; Hedlund, E.; Fu, D.; Soldner, F.; Broccoli, V.; Constantine-Paton, M.; Isacson, O.; Jaenisch, R., Neurons derived from reprogrammed fibroblasts functionally integrate into the fetal brain and improve symptoms of rats with Parkinson's disease. *Proc Natl Acad Sci USA* 2008, 105, (15), 5856–5861.
14. Niwa, H., How is pluripotency determined and maintained? *Development* 2007, 134, (4), 635–646.
15. Evans, M. J.; Kaufman, M. H., Establishment in culture of pluripotential cells from mouse embryos. *Nature* 1981, 292, (5819), 154–156.
16. Martin, G. R., Isolation of a pluripotent cell line from early mouse embryos cultured in medium conditioned by teratocarcinoma stem cells. *Proc Natl Acad Sci USA* 1981, 78, (12), 7634–7638.
17. Egli, D.; Birkhoff, G.; Eggan, K., Mediators of reprogramming: Transcription factors and transitions through mitosis. *Nat Rev Mol Cell Biol* 2008, 9, (7), 505–516.
18. Lerou, P. H.; Daley, G. Q., Therapeutic potential of embryonic stem cells. *Blood Rev* 2005, 19, (6), 321–331.
19. Gurdon, J. B.; Byrne, J. A., The first half-century of nuclear transplantation. *Proc Natl Acad Sci USA* 2003, 100, (14), 8048–8052.
20. Bromhall, J. D., Nuclear transplantation in the rabbit egg. *Nature* 1975, 258, (5537), 719–722.
21. Illmensee, K.; Hoppe, P. C., Nuclear transplantation in Mus musculus: Developmental potential of nuclei from preimplantation embryos. *Cell* 1981, 23, (1), 9–18.
22. McLaren, A., Mammalian development: Methods and success of nuclear transplantation in mammals. *Nature* 1984, 309, (5970), 671–672.
23. McGrath, J.; Solter, D., Nuclear transplantation in the mouse embryo by microsurgery and cell fusion. *Science* 1983, 220, (4603), 1300–1302.
24. Surani, M. A.; Barton, S. C.; Norris, M. L., Development of reconstituted mouse eggs suggests imprinting of the genome during gametogenesis. *Nature* 1984, 308, (5959), 548–550.
25. Willadsen, S. M., Nuclear transplantation in sheep embryos. *Nature* 1986, 320, (6057), 63–65.

26. Stice, S. L.; Robl, J. M., Nuclear reprogramming in nuclear transplant rabbit embryos. *Biol Reprod* 1988, 39, (3), 657–664.

27. Prather, R. S.; Sims, M. M.; First, N. L., Nuclear transplantation in early pig embryos. *Biol Reprod* 1989, 41, (3), 414–418.

28. Cheong, H. T.; Takahashi, Y.; Kanagawa, H., Birth of mice after transplantation of early cell-cycle-stage embryonic nuclei into enucleated oocytes. *Biol Reprod* 1993, 48, (5), 958–963.

29. Sims, M.; First, N. L., Production of calves by transfer of nuclei from cultured inner cell mass cells. *Proc Natl Acad Sci USA* 1994, 91, (13), 6143–6147.

30. Meng, L.; Ely, J. J.; Stouffer, R. L.; Wolf, D. P., Rhesus monkeys produced by nuclear transfer. *Biol Reprod* 1997, 57, (2), 454–459.

31. Wakayama, T.; Perry, A. C.; Zuccotti, M.; Johnson, K. R.; Yanagimachi, R., Full-term development of mice from enucleated oocytes injected with cumulus cell nuclei. *Nature* 1998, 394, (6691), 369–374.

32. Kato, Y.; Tani, T.; Sotomaru, Y.; Kurokawa, K.; Kato, J.; Doguchi, H.; Yasue, H.; Tsunoda, Y., Eight calves cloned from somatic cells of a single adult. *Science* 1998, 282, (5396), 2095–2098.

33. Baguisi, A.; Behboodi, E.; Melican, D. T.; Pollock, J. S.; Destrempes, M. M.; Cammuso, C.; Williams, J. L.; Nims, S. D.; Porter, C. A.; Midura, P.; Palacios, M. J.; Ayres, S. L.; Denniston, R. S.; Hayes, M. L.; Ziomek, C. A.; Meade, H. M.; Godke, R. A.; Gavin, W. G.; Overstrom, E. W.; Echelard, Y., Production of goats by somatic cell nuclear transfer. *Nat Biotechnol* 1999, 17, (5), 456–461.

34. Polejaeva, I. A.; Chen, S. H.; Vaught, T. D.; Page, R. L.; Mullins, J.; Ball, S.; Dai, Y.; Boone, J.; Walker, S.; Ayares, D. L.; Colman, A.; Campbell, K. H., Cloned pigs produced by nuclear transfer from adult somatic cells. *Nature* 2000, 407, (6800), 86–90.

35. Chesne, P.; Adenot, P. G.; Viglietta, C.; Baratte, M.; Boulanger, L.; Renard, J. P., Cloned rabbits produced by nuclear transfer from adult somatic cells. *Nat Biotechnol* 2002, 20, (4), 366–369.

36. Shin, T.; Kraemer, D.; Pryor, J.; Liu, L.; Rugila, J.; Howe, L.; Buck, S.; Murphy, K.; Lyons, L.; Westhusin, M., A cat cloned by nuclear transplantation. *Nature* 2002, 415, (6874), 859.

37. Hochedlinger, K.; Jaenisch, R., Monoclonal mice generated by nuclear transfer from mature B and T donor cells. *Nature* 2002, 415, (6875), 1035–1038.

38. Blelloch, R.; Wang, Z.; Meissner, A.; Pollard, S.; Smith, A.; Jaenisch, R., Reprogramming efficiency following somatic cell nuclear transfer is influenced by the differentiation and methylation state of the donor nucleus. *Stem Cells* 2006, 24, (9), 2007–2013.

39. Wakayama, T.; Tabar, V.; Rodriguez, I.; Perry, A. C.; Studer, L.; Mombaerts, P., Differentiation of embryonic stem cell lines generated from adult somatic cells by nuclear transfer. *Science* 2001, 292, (5517), 740–743.

40. Egli, D.; Rosains, J.; Birkhoff, G.; Eggan, K., Developmental reprogramming after chromosome transfer into mitotic mouse zygotes. *Nature* 2007, 447, (7145), 679–685.

41. Miller, R. A.; Ruddle, F. H., Pluripotent teratocarcinoma-thymus somatic cell hybrids. *Cell* 1976, 9, (1), 45–55.

42. Tada, M.; Takahama, Y.; Abe, K.; Nakatsuji, N.; Tada, T., Nuclear reprogramming of somatic cells by in vitro hybridization with ES cells. *Curr Biol* 2001, 11, (19), 1553–1558.

43. Terada, N.; Hamazaki, T.; Oka, M.; Hoki, M.; Mastalerz, D. M.; Nakano, Y.; Meyer, E. M.; Morel, L.; Petersen, B. E.; Scott, E. W., Bone marrow cells adopt the phenotype of other cells by spontaneous cell fusion. *Nature* 2002, 416, (6880), 542–545.

44. Ying, Q. L.; Nichols, J.; Evans, E. P.; Smith, A. G., Changing potency by spontaneous fusion. *Nature* 2002, 416, (6880), 545–548.

45. Cowan, C. A.; Atienza, J.; Melton, D. A.; Eggan, K., Nuclear reprogramming of somatic cells after fusion with human embryonic stem cells. *Science* 2005, 309, (5739), 1369–1373.

46. Yu, J.; Vodyanik, M. A.; He, P.; Slukvin, I. I. Thomson, J. A., Human embryonic stem cells reprogram myeloid precursors following cell-cell fusion. *Stem Cells* 2006, 24, (1), 168–176.

47. Do, J. T.; Scholer, H. R., Nuclei of embryonic stem cells reprogram somatic cells. *Stem Cells* 2004, 22, (6), 941–949.

48. Strelchenko, N.; Kukharenko, V.; Shkumatov, A.; Verlinsky, O.; Kuliev, A.; Verlinsky, Y., Reprogramming of human somatic cells by embryonic stem cell cytoplast. *Reprod Biomed Online* 2006, 12, (1), 107–111.

49. Silva, J.; Chambers, I.; Pollard, S.; Smith, A., Nanog promotes transfer of pluripotency after cell fusion. *Nature* 2006, 441, (7096), 997–1001.

50. Chambers, I.; Colby, D.; Robertson, M.; Nichols, J.; Lee, S.; Tweedie, S.; Smith, A., Functional expression cloning of Nanog, a pluripotency sustaining factor in embryonic stem cells. *Cell* 2003, 113, (5), 643–655.

51. Mitsui, K.; Tokuzawa, Y.; Itoh, H.; Segawa, K.; Murakami, M.; Takahashi, K.; Maruyama, M.; Maeda, M.; Yamanaka, S., The homeoprotein Nanog is required for maintenance of pluripotency in mouse epiblast and ES cells. *Cell* 2003, 113, (5), 631–642.

52. Darr, H.; Mayshar, Y.; Benvenisty, N., Overexpression of NANOG in human ES cells enables feeder-free growth while inducing primitive ectoderm features. *Development* 2006, 133, (6), 1193–1201.

53. Hatano, S. Y.; Tada, M.; Kimura, H.; Yamaguchi, S.; Kono, T.; Nakano, T.; Suemori, H.; Nakatsuji, N.; Tada, T., Pluripotential competence of cells associated with Nanog activity. *Mech Dev* 2005, 122, (1), 67–79.

54. Matsumura, H.; Tada, M.; Otsuji, T.; Yasuchika, K.; Nakatsuji, N.; Surani, A.; Tada, T., Targeted chromosome elimination from ES-somatic hybrid cells. *Nat Methods* 2007, 4, (1), 23–25.

55. Taranger, C. K.; Noer, A.; Sorensen, A. L.; Hakelien, A. M.; Boquest, A. C.; Collas, P., Induction of dedifferentiation, genomewide transcriptional programming, and epigenetic reprogramming by extracts of carcinoma and embryonic stem cells. *Mol Biol Cell* 2005, 16, (12), 5719–5735.

56. Kanatsu-Shinohara, M.; Inoue, K.; Lee, J.; Yoshimoto, M.; Ogonuki, N.; Miki, H.; Baba, S.; Kato, T.; Kazuki, Y.; Toyokuni, S.; Toyoshima, M.; Niwa, O.; Oshimura, M.; Heike, T.; Nakahata, T.; Ishino, F.; Ogura, A.; Shinohara, T., Generation of pluripotent stem cells from neonatal mouse testis. *Cell* 2004, 119, (7), 1001–1012.

57. Guan, K.; Nayernia, K.; Maier, L. S.; Wagner, S.; Dressel, R.; Lee, J. H.; Nolte, J.; Wolf, F.; Li, M.; Engel, W.; Hasenfuss, G., Pluripotency of spermatogonial stem cells from adult mouse testis. *Nature* 2006, 440, (7088), 1199–1203.

58. Seandel, M.; James, D.; Shmelkov, S. V.; Falciatori, I.; Kim, J.; Chavala, S.; Scherr, D. S.; Zhang, F.; Torres, R.; Gale, N. W.; Yancopoulos, G. D.; Murphy, A.; Valenzuela, D. M.; Hobbs, R. M.; Pandolfi, P. P.; Rafii, S., Generation of functional multipotent adult stem cells from GPR125+ germline progenitors. *Nature* 2007, 449, (7160), 346–350.

59. Brons, I. G.; Smithers, L. E.; Trotter, M. W.; Rugg-Gunn, P.; Sun, B.; Chuva de Sousa Lopes, S. M.; Howlett, S. K.; Clarkson, A.; Ahrlund-Richter, L.; Pedersen, R. A.; Vallier, L., Derivation of pluripotent epiblast stem cells from mammalian embryos. *Nature* 2007, 448, (7150), 191–195.

60. Tesar, P. J.; Chenoweth, J. G.; Brook, F. A.; Davies, T. J.; Evans, E. P.; Mack, D. L.; Gardner, R. L.; McKay, R. D., New cell lines from mouse epiblast share defining features with human embryonic stem cells. *Nature* 2007, 448, (7150), 196–199.

61. Page, R. L.; Ambady, S.; Holmes, W. F.; Vilner, L.; Kole, D.; Kashpur, O.; Huntress, V.; Vojtic, I.; Whitton, H.; Dominko, T., Induction of stem cell gene expression in adult human fibroblasts without transgenes. *Cloning Stem Cells* 2009, 11, (3), 417–426.

62. Allen, N. D.; Barton, S. C.; Hilton, K.; Norris, M. L.; Surani, M. A., A functional analysis of imprinting in parthenogenetic embryonic stem cells. *Development* 1994, 120, (6), 1473–1482.

63. Cibelli, J. B.; Grant, K. A.; Chapman, K. B.; Cunniff, K.; Worst, T.; Green, H. L.; Walker, S. J.; Gutin, P. H.; Vilner, L.; Tabar, V.; Dominko, T.; Kane, J.; Wettstein, P. J.; Lanza, R. P.; Studer, L.; Vrana, K. E.; West, M. D., Parthenogenetic stem cells in nonhuman primates. *Science* 2002, 295, (5556), 819.

64. Kim, K.; Lerou, P.; Yabuuchi, A.; Lengerke, C.; Ng, K.; West, J.; Kirby, A.; Daly, M. J.; Daley, G. Q., Histocompatible embryonic stem cells by parthenogenesis. *Science* 2007, 315, (5811), 482–486.

65. Tokuzawa, Y.; Kaiho, E.; Maruyama, M.; Takahashi, K.; Mitsui, K.; Maeda, M.; Niwa, H.; Yamanaka, S., Fbx15 is a novel target of Oct3/4 but is dispensable for embryonic stem cell self-renewal and mouse development. *Mol Cell Biol* 2003, 23, (8), 2699–2708.

66. Maherali, N.; Sridharan, R.; Xie, W.; Utikal, J.; Eminli, S.; Arnold, K.; Stadtfeld, M.; Yachechko, R.; Tchieu, J.; Jaenisch, R.; Plath, K.; Hochedlinger, K., Directly reprogrammed fibroblasts show global epigenetic remodeling and widespread tissue contribution. *Cell Stem Cell* 2007, 1, (1), 55–70.

67. Okita, K.; Ichisaka, T.; Yamanaka, S., Generation of germline-competent induced pluripotent stem cells. *Nature* 2007, 448, (7151), 313–317.

68. Wernig, M.; Meissner, A.; Foreman, R.; Brambrink, T.; Ku, M.; Hochedlinger, K.; Bernstein, B. E.; Jaenisch, R., In vitro reprogramming of fibroblasts into a pluripotent ES-cell-like state. *Nature* 2007, 448, (7151), 318–324.

69. Liu, H.; Zhu, F.; Yong, J.; Zhang, P.; Hou, P.; Li, H.; Jiang, W.; Cai, J.; Liu, M.; Cui, K.; Qu, X.; Xiang, T.; Lu, D.; Chi, X.; Gao, G.; Ji, W.; Ding, M.; Deng, H., Generation of induced pluripotent stem cells from adult rhesus monkey fibroblasts. *Cell Stem Cell* 2008, 3, (6), 587–590.

70. Yu, J.; Vodyanik, M. A.; Smuga-Otto, K.; Antosiewicz-Bourget, J.; Frane, J. L.; Tian, S.; Nie, J.; Jonsdottir, G. A.; Ruotti, V.; Stewart, R.; Slukvin, II; Thomson, J. A., Induced pluripotent stem cell lines derived from human somatic cells. *Science* 2007, 318, (5858), 1917–1920.

71. Nakagawa, M.; Koyanagi, M.; Tanabe, K.; Takahashi, K.; Ichisaka, T.; Aoi, T.; Okita, K.; Mochiduki, Y.; Takizawa, N.; Yamanaka, S., Generation of induced pluripotent stem cells without Myc from mouse and human fibroblasts. *Nat Biotechnol* 2008, 26, (1), 101–106.

72. Meissner, A.; Wernig, M.; Jaenisch, R., Direct reprogramming of genetically unmodified fibroblasts into pluripotent stem cells. *Nat Biotechnol* 2007, 25, (10), 1177–1181.

73. Brambrink, T.; Foreman, R.; Welstead, G. G.; Lengner, C. J.; Wernig, M.; Suh, H.; Jaenisch, R., Sequential expression of pluripotency markers during direct reprogramming of mouse somatic cells. *Cell Stem Cell* 2008, 2, (2), 151–159.

74. Stadtfeld, M.; Nagaya, M.; Utikal, J.; Weir, G.; Hochedlinger, K., Induced pluripotent stem cells generated without viral integration. *Science* 2008, 322, (5903), 945–949.

75. Okita, K.; Nakagawa, M.; Hyenjong, H.; Ichisaka, T.; Yamanaka, S., Generation of mouse induced pluripotent stem cells without viral vectors. *Science* 2008, 322, (5903), 949–953.

76. Aoi, T.; Yae, K.; Nakagawa, M.; Ichisaka, T.; Okita, K.; Takahashi, K.; Chiba, T.; Yamanaka, S., Generation of pluripotent stem cells from adult mouse liver and stomach cells. *Science* 2008, 321, (5889), 699–702.

77. Akagi, K.; Suzuki, T.; Stephens, R. M.; Jenkins, N. A.; Copeland, N. G., RTCGD: Retroviral tagged cancer gene database. *Nucleic Acids Res* 2004, 32, (Database issue), D523–D527.

78. Hawley, R. G., Does retroviral insertional mutagenesis play a role in the generation of induced pluripotent stem cells? *Mol Ther* 2008, 16, (8), 1354–1355.

79. Varas, F.; Stadtfeld, M.; de Andres-Aguayo, L.; Maherali, N.; di Tullio, A.; Pantano, L.; Notredame, C.; Hochedlinger, K.; Graf, T., Fibroblast-derived induced pluripotent stem cells show no common retroviral vector insertions. *Stem Cells* 2009, 27, (2), 300–306.

80. Wu, X.; Li, Y.; Crise, B.; Burgess, S. M., Transcription start regions in the human genome are favored targets for MLV integration. *Science* 2003, 300, (5626), 1749–1751.

81. Carey, B. W.; Markoulaki, S.; Hanna, J.; Saha, K.; Gao, Q.; Mitalipova, M.; Jaenisch, R., Reprogramming of murine and human somatic cells using a single polycistronic vector. *Proc Natl Acad Sci USA* 2009, 106, (1), 157–162.

82. Sommer, C. A.; Stadtfeld, M.; Murphy, G. J.; Hochedlinger, K.; Kotton, D. N.; Mostoslavsky, G., Induced pluripotent stem cell generation using a single lentiviral stem cell cassette. *Stem Cells* 2009, 27, (3), 543–549.

83. Li, Z.; Dullmann, J.; Schiedlmeier, B.; Schmidt, M.; von Kalle, C.; Meyer, J.; Forster, M.; Stocking, C.; Wahlers, A.; Frank, O.; Ostertag, W.; Kuhlcke, K.; Eckert, H. G.; Fehse, B.; Baum, C., Murine leukemia induced by retroviral gene marking. *Science* 2002, 296, (5567), 497.

84. Hacein-Bey-Abina, S.; Von Kalle, C.; Schmidt, M.; McCormack, M. P.; Wulffraat, N.; Leboulch, P.; Lim, A.; Osborne, C. S.; Pawliuk, R.; Morillon, E.; Sorensen, R.; Forster, A.; Fraser, P.; Cohen, J. I.; de Saint Basile, G.; Alexander, I.; Wintergerst, U.; Frebourg, T.; Aurias, A.; Stoppa-Lyonnet, D.; Romana, S.; Radford-Weiss, I.; Gross, F.; Valensi, F.; Delabesse, E.; Macintyre, E.; Sigaux, F.; Soulier, J.; Leiva, L. E.; Wissler, M.; Prinz, C.; Rabbitts, T. H.; Le Deist, F.; Fischer, A.; Cavazzana-Calvo, M., LMO2-associated clonal T cell proliferation in two patients after gene therapy for SCID-X1. *Science* 2003, 302, (5644), 415–419.

85. Howe, S. J.; Mansour, M. R.; Schwarzwaelder, K.; Bartholomae, C.; Hubank, M.; Kempski, H.; Brugman, M. H.; Pike-Overzet, K.; Chatters, S. J.; de Ridder, D.; Gilmour, K. C.; Adams, S.; Thornhill, S. I.; Parsley, K. L.; Staal, F. J.; Gale, R. E.; Linch, D. C.; Bayford, J.; Brown, L.; Quaye, M.; Kinnon, C.; Ancliff, P.; Webb, D. K.; Schmidt, M.; von Kalle, C.; Gaspar, H. B.; Thrasher, A. J., Insertional mutagenesis combined with acquired somatic mutations causes leukemogenesis following gene therapy of SCID-X1 patients. *J Clin Invest* 2008, 118, (9), 3143–3150.

86. Woltjen, K.; Michael, I. P.; Mohseni, P.; Desai, R.; Mileikovsky, M.; Hamalainen, R.; Cowling, R.; Wang, W.; Liu, P.; Gertsenstein, M.; Kaji, K.; Sung, H. K.; Nagy, A., piggyBac transposition reprograms fibroblasts to induced pluripotent stem cells. *Nature* 2009, 458, (7239), 766–770.

87. Kaji, K.; Norrby, K.; Paca, A.; Mileikovsky, M.; Mohseni, P.; Woltjen, K., Virus-free induction of pluripotency and subsequent excision of reprogramming factors. *Nature* 2009, 458, (7239), 771–775.

88. Yusa, K.; Rad, R.; Takeda, J.; Bradley, A., Generation of transgene-free induced pluripotent mouse stem cells by the piggyBac transposon. *Nat Methods* 2009, 6, (5), 363–369.

89. Kim, D.; Kim, C. H.; Moon, J. I.; Chung, Y. G.; Chang, M. Y.; Han, B. S.; Ko, S.; Yang, E.; Cha, K. Y.; Lanza, R.; Kim, K. S., Generation of human induced pluripotent stem cells by direct delivery of reprogramming proteins. *Cell Stem Cell* 2009, 4, (6), 472–476.

90. Zhou, H.; Wu, S.; Joo, J. Y.; Zhu, S.; Han, D. W.; Lin, T.; Trauger, S.; Bien, G.; Yao, S.; Zhu, Y.; Siuzdak, G.; Scholer, H. R.; Duan, L.; Ding, S., Generation of induced pluripotent stem cells using recombinant proteins. *Cell Stem Cell* 2009, 4, (5), 381–384.

91. Gaspar-Maia, A.; Alajem, A.; Polesso, F.; Sridharan, R.; Mason, M. J.; Heidersbach, A.; Ramalho-Santos, J.; McManus, M. T.; Plath, K.; Meshorer, E.; Ramalho-Santos, M., Chd1 regulates open chromatin and pluripotency of embryonic stem cells. *Nature* 2009, 460, (7257), 863–868.

92. Kishigami, S.; Mizutani, E.; Ohta, H.; Hikichi, T.; Thuan, N. V.; Wakayama, S.; Bui, H. T.; Wakayama, T., Significant improvement of mouse cloning technique by treatment with trichostatin A after somatic nuclear transfer. *Biochem Biophys Res Commun* 2006, 340, (1), 183–189.

93. Huangfu, D.; Maehr, R.; Guo, W.; Eijkelenboom, A.; Snitow, M.; Chen, A. E.; Melton, D. A., Induction of pluripotent stem cells by defined factors is greatly improved by small-molecule compounds. *Nat Biotechnol* 2008, 26, (7), 795–797.

94. Szabo, P. E.; Hubner, K.; Scholer, H.; Mann, J. R., Allele-specific expression of imprinted genes in mouse migratory primordial germ cells. *Mech Dev* 2002, 115, (1–2), 157–160.

95. Huangfu, D.; Osafune, K.; Maehr, R.; Guo, W.; Eijkelenboom, A.; Chen, S.; Muhlestein, W.; Melton, D. A., Induction of pluripotent stem cells from primary human fibroblasts with only Oct4 and Sox2. *Nat Biotechnol* 2008, 26, (11), 1269–1275.

96. Shi, Y.; Desponts, C.; Do, J. T.; Hahm, H. S.; Scholer, H. R.; Ding, S., Induction of pluripotent stem cells from mouse embryonic fibroblasts by Oct4 and Klf4 with small-molecule compounds. *Cell Stem Cell* 2008, 3, (5), 568–574.

97. Shi, Y.; Do, J. T.; Desponts, C.; Hahm, H. S.; Scholer, H. R.; Ding, S., A combined chemical and genetic approach for the generation of induced pluripotent stem cells. *Cell Stem Cell* 2008, 2, (6), 525–528.

98. Chang, Y.; Zhang, X.; Horton, J. R.; Upadhyay, A. K.; Spannhoff, A.; Liu, J.; Snyder, J. P.; Bedford, M. T.; Cheng, X., Structural basis for G9a-like protein lysine methyltransferase inhibition by BIX-01294. *Nat Struct Mol Biol* 2009, 16, (3), 312–317.

99. Feldman, N.; Gerson, A.; Fang, J.; Li, E.; Zhang, Y.; Shinkai, Y.; Cedar, H.; Bergman, Y., G9a-mediated irreversible epigenetic inactivation of Oct-3/4 during early embryogenesis. *Nat Cell Biol* 2006, 8, (2), 188–194.

100. Epsztejn-Litman, S.; Feldman, N.; Abu-Remaileh, M.; Shufaro, Y.; Gerson, A.; Ueda, J.; Deplus, R.; Fuks, F.; Shinkai, Y.; Cedar, H.; Bergman, Y., De novo DNA methylation promoted by G9a prevents reprogramming of embryonically silenced genes. *Nat Struct Mol Biol* 2008, 15, (11), 1176–1183.

101. Dong, K. B.; Maksakova, I. A.; Mohn, F.; Leung, D.; Appanah, R.; Lee, S.; Yang, H. W.; Lam, L. L.; Mager, D. L.; Schubeler, D.; Tachibana, M.; Shinkai, Y.; Lorincz, M. C., DNA methylation in ES cells requires the lysine methyltransferase G9a but not its catalytic activity. *EMBO J* 2008, 27, (20), 2691–2701.

102. Tachibana, M.; Matsumura, Y.; Fukuda, M.; Kimura, H.; Shinkai, Y., G9a/GLP complexes independently mediate H3K9 and DNA methylation to silence transcription. *EMBO J* 2008, 27, (20), 2681–2690.

103. Park, I. H.; Zhao, R.; West, J. A.; Yabuuchi, A.; Huo, H.; Ince, T. A.; Lerou, P. H.; Lensch, M. W.; Daley, G. Q., Reprogramming of human somatic cells to pluripotency with defined factors. *Nature* 2008, 451, (7175), 141–146.

104. Krawetz, R. J.; Li, X.; Rancourt, D. E., Human embryonic stem cells: Caught between a ROCK inhibitor and a hard place. *Bioessays* 2009, 31, (3), 336–343.

105. Marson, A.; Foreman, R.; Chevalier, B.; Bilodeau, S.; Kahn, M.; Young, R. A.; Jaenisch, R., Wnt signaling promotes reprogramming of somatic cells to pluripotency. *Cell Stem Cell* 2008, 3, (2), 132–135.

106. Silva, J.; Barrandon, O.; Nichols, J.; Kawaguchi, J.; Theunissen, T. W.; Smith, A., Promotion of reprogramming to ground state pluripotency by signal inhibition. *PLoS Biol* 2008, 6, (10), e253.

107. Li, W.; Wei, W.; Zhu, S.; Zhu, J.; Shi, Y.; Lin, T.; Hao, E.; Hayek, A.; Deng, H.; Ding, S., Generation of rat and human induced pluripotent stem cells by combining genetic reprogramming and chemical inhibitors. *Cell Stem Cell* 2009, 4, (1), 16–19.

108. Ichida, J. K.; Blanchard, J.; Lam, K.; Son, E. Y.; Chung, J. E.; Egli, D.; Loh, K. M.; Carter, A. C.; Di Giorgio, F. P.; Koszka, K.; Huangfu, D.; Akutsu, H.; Liu, D. R.; Rubin, L. L.; Eggan, K., A small-molecule inhibitor of Tgf-beta signaling replaces Sox2 in reprogramming by inducing Nanog. *Cell Stem Cell* 2009, 5, (5), 491–503

109. Takahashi, K.; Tanabe, K.; Ohnuki, M.; Narita, M.; Ichisaka, T.; Tomoda, K.; Yamanaka, S., Induction of pluripotent stem cells from adult human fibroblasts by defined factors. *Cell* 2007, 131, (5), 861–872.

110. Lowry, W. E.; Richter, L.; Yachechko, R.; Pyle, A. D.; Tchieu, J.; Sridharan, R.; Clark, A. T.; Plath, K., Generation of human induced pluripotent stem cells from dermal fibroblasts. *Proc Natl Acad Sci USA* 2008, 105, (8), 2883–2888.

111. Boyer, L. A.; Lee, T. I.; Cole, M. F.; Johnstone, S. E.; Levine, S. S.; Zucker, J. P.; Guenther, M. G.; Kumar, R. M.; Murray, H. L.; Jenner, R. G.; Gifford, D. K.; Melton, D. A.; Jaenisch, R.; Young, R. A., Core transcriptional regulatory circuitry in human embryonic stem cells. *Cell* 2005, 122, (6), 947–956.

112. Lyssiotis, C. A.; Foreman, R. K.; Staerk, J.; Garcia, M.; Mathur, D.; Markoulaki, S.; Hanna, J.; Lairson, L. L.; Charette, B. D.; Bouchez, L. C.; Bollong, M.; Kunick, C.; Brinker, A.; Cho, C. Y.; Schultz, P. G.; Jaenisch, R., Reprogramming of murine fibroblasts to induced pluripotent stem cells with chemical complementation of Klf4. *Proc Natl Acad Sci USA* 2009, 106, (22), 8912–8917.

113. Mikkelsen, T. S.; Hanna, J.; Zhang, X.; Ku, M.; Wernig, M.; Schorderet, P.; Bernstein, B. E.; Jaenisch, R.; Lander, E. S.; Meissner, A., Dissecting direct reprogramming through integrative genomic analysis. *Nature* 2008, 454, (7200), 49–55.

114. Judson, R. L.; Babiarz, J. E.; Venere, M.; Blelloch, R., Embryonic stem cell-specific microRNAs promote induced pluripotency. *Nat Biotechnol* 2009, 27, (5), 459–461.

115. Zhao, Y.; Yin, X.; Qin, H.; Zhu, F.; Liu, H.; Yang, W.; Zhang, Q.; Xiang, C.; Hou, P.; Song, Z.; Liu, Y.; Yong, J.; Zhang, P.; Cai, J.; Liu, M.; Li, H.; Li, Y.; Qu, X.; Cui, K.; Zhang, W.; Xiang, T.; Wu, Y.; Liu, C.; Yu, C.; Yuan, K.; Lou, J.; Ding, M.; Deng, H., Two supporting factors greatly improve the efficiency of human iPSC generation. *Cell Stem Cell* 2008, 3, (5), 475–479.

116. Hong, H.; Takahashi, K.; Ichisaka, T.; Aoi, T.; Kanagawa, O.; Nakagawa, M.; Okita, K.; Yamanaka, S., Suppression of induced pluripotent stem cell generation by the p53-p21 pathway. *Nature* 2009, 460, (7259), 1132–1135.

117. Kawamura, T.; Suzuki, J.; Wang, Y. V.; Menendez, S.; Morera, L. B.; Raya, A.; Wahl, G. M.; Belmonte, J. C., Linking the p53 tumour suppressor pathway to somatic cell reprogramming. *Nature* 2009, 460, (7259), 1140–1144.

118. Li, H.; Collado, M.; Villasante, A.; Strati, K.; Ortega, S.; Canamero, M.; Blasco, M. A.; Serrano, M., The Ink4/Arf locus is a barrier for iPS cell reprogramming. *Nature* 2009, 460, (7259), 1136–1139.

119. Marion, R. M.; Strati, K.; Li, H.; Murga, M.; Blanco, R.; Ortega, S.; Fernandez-Capetillo, O.; Serrano, M.; Blasco, M. A., A p53-mediated DNA damage response limits reprogramming to ensure iPS cell genomic integrity. *Nature* 2009, 460, (7259), 1149–1153.

120. Utikal, J.; Polo, J. M.; Stadtfeld, M.; Maherali, N.; Kulalert, W.; Walsh, R. M.; Khalil, A.; Rheinwald, J. G.; Hochedlinger, K., Immortalization eliminates a roadblock during cellular reprogramming into iPS cells. *Nature* 2009, 460, (7259), 1145–1148.

121. Zheng, H.; Ying, H.; Yan, H.; Kimmelman, A. C.; Hiller, D. J.; Chen, A. J.; Perry, S. R.; Tonon, G.; Chu, G. C.; Ding, Z.; Stommel, J. M.; Dunn, K. L.; Wiedemeyer, R.; You, M. J.; Brennan, C.; Wang, Y. A.; Ligon, K. L.; Wong, W. H.; Chin, L.; DePinho, R. A., p53 and Pten control neural and glioma stem/progenitor cell renewal and differentiation. *Nature* 2008, 455, (7216), 1129–1133.

122. Hockemeyer, D.; Soldner, F.; Cook, E. G.; Gao, Q.; Mitalipova, M.; Jaenisch, R., A drug-inducible system for direct reprogramming of human somatic cells to pluripotency. *Cell Stem Cell* 2008, 3, (3), 346–353.

123. Maherali, N.; Ahfeldt, T.; Rigamonti, A.; Utikal, J.; Cowan, C.; Hochedlinger, K., A high-efficiency system for the generation and study of human induced pluripotent stem cells. *Cell Stem Cell* 2008, 3, (3), 340–345.

124. Wernig, M.; Lengner, C. J.; Hanna, J.; Lodato, M. A.; Steine, E.; Foreman, R.; Staerk, J.; Markoulaki, S.; Jaenisch, R., A drug-inducible transgenic system for direct reprogramming of multiple somatic cell types. *Nat Biotechnol* 2008, 26, (8), 916–924.

125. Boiani, M.; Eckardt, S.; Scholer, H. R.; McLaughlin, K. J., Oct4 distribution and level in mouse clones: Consequences for pluripotency. *Genes Dev* 2002, 16, (10), 1209–1219.

126. Pereira, C. F.; Terranova, R.; Ryan, N. K.; Santos, J.; Morris, K. J.; Cui, W.; Merkenschlager, M.; Fisher, A. G., Heterokaryon-based reprogramming of human B lymphocytes for pluripotency requires Oct4 but not Sox2. *PLoS Genet* 2008, 4, (9), e1000170.

127. Amabile, G.; Meissner, A., Induced pluripotent stem cells: Current progress and potential for regenerative medicine. *Trends Mol Med* 2009, 15, (2), 59–68.

128. Aasen, T.; Raya, A.; Barrero, M. J.; Garreta, E.; Consiglio, A.; Gonzalez, F.; Vassena, R.; Bilic, J.; Pekarik, V.; Tiscornia, G.; Edel, M.; Boue, S.; Izpisua Belmonte, J. C., Efficient and rapid generation of induced pluripotent stem cells from human keratinocytes. *Nat Biotechnol* 2008, 26, (11), 1276–1284.

129. Markoulaki, S.; Hanna, J.; Beard, C.; Carey, B. W.; Cheng, A. W.; Lengner, C. J.; Dausman, J. A.; Fu, D.; Gao, Q.; Wu, S.; Cassady, J. P.; Jaenisch, R., Transgenic mice with defined combinations of drug-inducible reprogramming factors. *Nat Biotechnol* 2009, 27, (2), 169–171.

130. Sridharan, R.; Tchieu, J.; Mason, M. J.; Yachechko, R.; Kuoy, E.; Horvath, S.; Zhou, Q.; Plath, K., Role of the murine reprogramming factors in the induction of pluripotency. *Cell* 2009, 136, (2), 364–377.

131. Wernig, M.; Meissner, A.; Cassady, J. P.; Jaenisch, R., c-Myc is dispensable for direct reprogramming of mouse fibroblasts. *Cell Stem Cell* 2008, 2, (1), 10–12.

132. Kang, L.; Wang, J.; Zhang, Y.; Kou, Z.; Gao, S., iPS cells can support full-term development of tetraploid blastocyst-complemented embryos. *Cell Stem Cell* 2009, 5, (2), 135–138.

133. Zhao, X. Y.; Li, W.; Lv, Z.; Liu, L.; Tong, M.; Hai, T.; Hao, J.; Guo, C. L.; Ma, Q. W.; Wang, L.; Zeng, F.; Zhou, Q., iPS cells produce viable mice through tetraploid complementation. *Nature* 2009, 461, (7260), 86–90.

134. Soldner, F.; Hockemeyer, D.; Beard, C.; Gao, Q.; Bell, G. W.; Cook, E. G.; Hargus, G.; Blak, A.; Cooper, O.; Mitalipova, M.; Isacson, O.; Jaenisch, R., Parkinson's disease patient-derived induced pluripotent stem cells free of viral reprogramming factors. *Cell* 2009, 136, (5), 964–977.

135. Niwa, H.; Miyazaki, J.; Smith, A. G., Quantitative expression of Oct-3/4 defines differentiation, dedifferentiation or self-renewal of ES cells. *Nat Genet* 2000, 24, (4), 372–376.

136. Osafune, K.; Caron, L.; Borowiak, M.; Martinez, R. J.; Fitz-Gerald, C. S.; Sato, Y.; Cowan, C. A.; Chien, K. R.; Melton, D. A., Marked differences in differentiation propensity among human embryonic stem cell lines. *Nat Biotechnol* 2008, 26, (3), 313–315.

137. Bernstein, B. E.; Meissner, A.; Lander, E. S., The mammalian epigenome. *Cell* 2007, 128, (4), 669–681.

138. Bernstein, B. E.; Mikkelsen, T. S.; Xie, X.; Kamal, M.; Huebert, D. J.; Cuff, J.; Fry, B.; Meissner, A.; Wernig, M.; Plath, K.; Jaenisch, R.; Wagschal, A.; Feil, R.; Schreiber, S. L.; Lander, E. S., A bivalent chromatin structure marks key developmental genes in embryonic stem cells. *Cell* 2006, 125, (2), 315–326.

139. Mikkelsen, T. S.; Ku, M.; Jaffe, D. B.; Issac, B.; Lieberman, E.; Giannoukos, G.; Alvarez, P.; Brockman, W.; Kim, T. K.; Koche, R. P.; Lee, W.; Mendenhall, E.; O'Donovan, A.; Presser, A.; Russ, C.; Xie, X.; Meissner, A.; Wernig, M.; Jaenisch, R.; Nusbaum, C.; Lander, E. S.; Bernstein, B. E., Genome-wide maps of chromatin state in pluripotent and lineage-committed cells. *Nature* 2007, 448, (7153), 553–560.

140. Meissner, A.; Mikkelsen, T. S.; Gu, H.; Wernig, M.; Hanna, J.; Sivachenko, A.; Zhang, X.; Bernstein, B. E.; Nusbaum, C.; Jaffe, D. B.; Gnirke, A.; Jaenisch, R.; Lander, E. S., Genome-scale DNA methylation maps of pluripotent and differentiated cells. *Nature* 2008, 454, (7205), 766–770.

141. Rougier, N.; Bourc'his, D.; Gomes, D. M.; Niveleau, A.; Plachot, M.; Paldi, A.; Viegas-Pequignot, E., Chromosome methylation patterns during mammalian preimplantation development. *Genes Dev* 1998, 12, (14), 2108–2113.

142. Eggan, K.; Akutsu, H.; Loring, J.; Jackson-Grusby, L.; Klemm, M.; Rideout, W. M., 3rd; Yanagimachi, R.; Jaenisch, R., Hybrid vigor, fetal overgrowth, and viability of mice derived by nuclear cloning and tetraploid embryo complementation. *Proc Natl Acad Sci USA* 2001, 98, (11), 6209–6214.

143. Boland, M. J.; Hazen, J. L.; Nazor, K. L.; Rodriguez, A. R.; Gifford, W.; Martin, G.; Kupriyanov, S.; Baldwin, K. K., Adult mice generated from induced pluripotent stem cells. *Nature* 2009, 461, (7260), 91–94.

144. Loh, Y. H.; Wu, Q.; Chew, J. L.; Vega, V. B.; Zhang, W.; Chen, X.; Bourque, G.; George, J.; Leong, B.; Liu, J.; Wong, K. Y.; Sung, K. W.; Lee, C. W.; Zhao, X. D.; Chiu, K. P.; Lipovich, L.; Kuznetsov, V. A.; Robson, P.; Stanton, L. W.; Wei, C. L.; Ruan, Y.; Lim, B.; Ng, H. H., The Oct4 and Nanog transcription network regulates pluripotency in mouse embryonic stem cells. *Nat Genet* 2006, 38, (4), 431–440.

145. Kim, J.; Chu, J.; Shen, X.; Wang, J.; Orkin, S. H., An extended transcriptional network for pluripotency of embryonic stem cells. *Cell* 2008, 132, (6), 1049–1061.

146. Yamanaka, S., Strategies and new developments in the generation of patient-specific pluripotent stem cells. *Cell Stem Cell* 2007, 1, (1), 39–49.

147. Richards, M.; Tan, S. P.; Tan, J. H.; Chan, W. K.; Bongso, A., The transcriptome profile of human embryonic stem cells as defined by SAGE. *Stem Cells* 2004, 22, (1), 51–64.

148. Viswanathan, S. R.; Daley, G. Q.; Gregory, R. I., Selective blockade of microRNA processing by Lin28. *Science* 2008, 320, (5872), 97–100.

149. Scholer, H. R.; Hatzopoulos, A. K.; Balling, R.; Suzuki, N.; Gruss, P., A family of octamer-specific proteins present during mouse embryogenesis: Evidence for germline-specific expression of an Oct factor. *EMBO J* 1989, 8, (9), 2543–2550.

150. Nichols, J.; Zevnik, B.; Anastassiadis, K.; Niwa, H.; Klewe-Nebenius, D.; Chambers, I.; Scholer, H.; Smith, A., Formation of pluripotent stem cells in the mammalian embryo depends on the POU transcription factor Oct4. *Cell* 1998, 95, (3), 379–391.

151. Pan, G.; Qin, B.; Liu, N.; Scholer, H. R.; Pei, D., Identification of a nuclear localization signal in OCT4 and generation of a dominant negative mutant by its ablation. *J Biol Chem* 2004, 279, (35), 37013–37020.

152. Scholer, H. R.; Ruppert, S.; Suzuki, N.; Chowdhury, K.; Gruss, P., New type of POU domain in germ line-specific protein Oct-4. *Nature* 1990, 344, (6265), 435–439.

153. Pan, G. J.; Chang, Z. Y.; Scholer, H. R.; Pei, D., Stem cell pluripotency and transcription factor Oct4. *Cell Res* 2002, 12, (5–6), 321–329.

154. Yuan, H.; Corbi, N.; Basilico, C.; Dailey, L., Developmental-specific activity of the FGF-4 enhancer requires the synergistic action of Sox2 and Oct-3. *Genes Dev* 1995, 9, (21), 2635–2645.

155. Scaffidi, P.; Bianchi, M. E., Spatially precise DNA bending is an essential activity of the sox2 transcription factor. *J Biol Chem* 2001, 276, (50), 47296–47302.

156. Avilion, A. A.; Nicolis, S. K.; Pevny, L. H.; Perez, L.; Vivian, N.; Lovell-Badge, R., Multipotent cell lineages in early mouse development depend on SOX_2 function. *Genes Dev* 2003, 17, (1), 126–140.

157. Li, J.; Pan, G.; Cui, K.; Liu, Y.; Xu, S.; Pei, D., A dominant-negative form of mouse SOX_2 induces trophectoderm differentiation and progressive polyploidy in mouse embryonic stem cells. *J Biol Chem* 2007, 282, (27), 19481–19492.

158. Masui, S.; Nakatake, Y.; Toyooka, Y.; Shimosato, D.; Yagi, R.; Takahashi, K.; Okochi, H.; Okuda, A.; Matoba, R.; Sharov, A. A.; Ko, M. S.; Niwa, H., Pluripotency governed by Sox2 via regulation of Oct3/4 expression in mouse embryonic stem cells. *Nat Cell Biol* 2007, 9, (6), 625–635.

159. Schmidt, E. V., The role of c-myc in cellular growth control. *Oncogene* 1999, 18, (19), 2988–2996.

160. Murphy, M. J.; Wilson, A.; Trumpp, A., More than just proliferation: Myc function in stem cells. *Trends Cell Biol* 2005, 15, (3), 128–137.

161. Zeller, K. I.; Zhao, X.; Lee, C. W.; Chiu, K. P.; Yao, F.; Yustein, J. T.; Ooi, H. S.; Orlov, Y. L.; Shahab, A.; Yong, H. C.; Fu, Y.; Weng, Z.; Kuznetsov, V. A.; Sung, W. K.; Ruan, Y.; Dang, C. V.; Wei, C. L., Global mapping of c-Myc binding sites and target gene networks in human B cells. *Proc Natl Acad Sci USA* 2006, 103, (47), 17834–17839.

162. Cowling, V. H.; Cole, M. D., Mechanism of transcriptional activation by the Myc oncoproteins. *Semin Cancer Biol* 2006, 16, (4), 242–252.

163. Hooker, C. W.; Hurlin, P. J., Of Myc and Mnt. *J Cell Sci* 2006, 119, (Pt 2), 208–216.

164. Chang, T. C.; Yu, D.; Lee, Y. S.; Wentzel, E. A.; Arking, D. E.; West, K. M.; Dang, C. V.; Thomas-Tikhonenko, A.; Mendell, J. T., Widespread microRNA repression by Myc contributes to tumorigenesis. *Nat Genet* 2008, 40, (1), 43–50.

165. Dews, M.; Homayouni, A.; Yu, D.; Murphy, D.; Sevignani, C.; Wentzel, E.; Furth, E. E.; Lee, W. M.; Enders, G. H.; Mendell, J. T.; Thomas-Tikhonenko, A., Augmentation of tumor angiogenesis by a Myc-activated microRNA cluster. *Nat Genet* 2006, 38, (9), 1060–1065.

166. Brenner, C.; Deplus, R.; Didelot, C.; Loriot, A.; Vire, E.; De Smet, C.; Gutierrez, A.; Danovi, D.; Bernard, D.; Boon, T.; Pelicci, P. G.; Amati, B.; Kouzarides, T.; de Launoit, Y.; Di Croce, L.; Fuks, F., Myc represses transcription through recruitment of DNA methyltransferase corepressor. *EMBO J* 2005, 24, (2), 336–346.

167. Secombe, J.; Li, L.; Carlos, L.; Eisenman, R. N., The Trithorax group protein Lid is a trimethyl histone H3K4 demethylase required for dMyc-induced cell growth. *Genes Dev* 2007, 21, (5), 537–551.

168. Knoepfler, P. S.; Zhang, X. Y.; Cheng, P. F.; Gafken, P. R.; McMahon, S. B.; Eisenman, R. N., Myc influences global chromatin structure. *EMBO J* 2006, 25, (12), 2723–2734.

169. Dominguez-Sola, D.; Ying, C. Y.; Grandori, C.; Ruggiero, L.; Chen, B.; Li, M.; Galloway, D. A.; Gu, W.; Gautier, J.; Dalla-Favera, R., Non-transcriptional control of DNA replication by c-Myc. *Nature* 2007, 448, (7152), 445–451.

170. Eminli, S.; Utikal, J.; Arnold, K.; Jaenisch, R.; Hochedlinger, K., Reprogramming of neural progenitor cells into induced pluripotent stem cells in the absence of exogenous Sox2 expression. *Stem Cells* 2008, 26, (10), 2467–2474.

171. Kim, J. B.; Zaehres, H.; Wu, G.; Gentile, L.; Ko, K.; Sebastiano, V.; Arauzo-Bravo, M. J.; Ruau, D.; Han, D. W.; Zenke, M.; Scholer, H. R., Pluripotent stem cells induced from adult neural stem cells by reprogramming with two factors. *Nature* 2008, 454, (7204), 646–650.

172. Segre, J. A.; Bauer, C.; Fuchs, E., Klf4 is a transcription factor required for establishing the barrier function of the skin. *Nat Genet* 1999, 22, (4), 356–360.

173. Rowland, B. D.; Peeper, D. S., KLF4, p21 and context-dependent opposing forces in cancer. *Nat Rev Cancer* 2006, 6, (1), 11–23.

174. Lin, T.; Chao, C.; Saito, S.; Mazur, S. J.; Murphy, M. E.; Appella, E.; Xu, Y., p53 induces differentiation of mouse embryonic stem cells by suppressing Nanog expression. *Nat Cell Biol* 2005, 7, (2), 165–171.

175. Rowland, B. D.; Bernards, R.; Peeper, D. S., The KLF4 tumour suppressor is a transcriptional repressor of p53 that acts as a context-dependent oncogene. *Nat Cell Biol* 2005, 7, (11), 1074–1082.

176. Sumi, T.; Tsuneyoshi, N.; Nakatsuji, N.; Suemori, H., Apoptosis and differentiation of human embryonic stem cells induced by sustained activation of c-Myc. *Oncogene* 2007, 26, (38), 5564–5576.

177. Nakatake, Y.; Fukui, N.; Iwamatsu, Y.; Masui, S.; Takahashi, K.; Yagi, R.; Yagi, K.; Miyazaki, J.; Matoba, R.; Ko, M. S.; Niwa, H., Klf4 cooperates with Oct3/4 and Sox2 to activate the Lefty1 core promoter in embryonic stem cells. *Mol Cell Biol* 2006, 26, (20), 7772–7782.

178. Li, Y.; McClintick, J.; Zhong, L.; Edenberg, H. J.; Yoder, M. C.; Chan, R. J., Murine embryonic stem cell differentiation is promoted by SOCS-3 and inhibited by the zinc finger transcription factor Klf4. *Blood* 2005, 105, (2), 635–637.

179. Pan, G. J.; Pei, D. Q., Identification of two distinct transactivation domains in the pluripotency sustaining factor nanog. *Cell Res* 2003, 13, (6), 499–502.

180. Pan, G.; Li, J.; Zhou, Y.; Zheng, H.; Pei, D., A negative feedback loop of transcription factors that controls stem cell pluripotency and self-renewal. *FASEB J* 2006, 20, (10), 1730–1732.

181. Chambers, I.; Silva, J.; Colby, D.; Nichols, J.; Nijmeijer, B.; Robertson, M.; Vrana, J.; Jones, K.; Grotewold, L.; Smith, A., Nanog safeguards pluripotency and mediates germline development. *Nature* 2007, 450, (7173), 1230–1234.

182. Moss, E. G.; Lee, R. C.; Ambros, V., The cold shock domain protein LIN-28 controls developmental timing in *C. elegans* and is regulated by the lin-4 RNA. *Cell* 1997, 88, (5), 637–646.

183. Moss, E. G.; Tang, L., Conservation of the heterochronic regulator Lin-28, its developmental expression and microRNA complementary sites. *Dev Biol* 2003, 258, (2), 432–442.

184. Balzer, E.; Moss, E. G., Localization of the developmental timing regulator Lin28 to mRNP complexes, P-bodies and stress granules. *RNA Biol* 2007, 4, (1), 16–25.

185. Polesskaya, A.; Cuvellier, S.; Naguibneva, I.; Duquet, A.; Moss, E. G.; Harel-Bellan, A., Lin-28 binds IGF-2 mRNA and participates in skeletal myogenesis by increasing translation efficiency. *Genes Dev* 2007, 21, (9), 1125–1138.

186. Teich, N. M.; Weiss, R. A.; Martin, G. R.; Lowy, D. R., Virus infection of murine teratocarcinoma stem cell lines. *Cell* 1977, 12, (4), 973–982.

187. Wolf, D.; Goff, S. P., TRIM28 mediates primer binding site-targeted silencing of murine leukemia virus in embryonic cells. *Cell* 2007, 131, (1), 46–57.

188. Lois, C.; Hong, E. J.; Pease, S.; Brown, E. J.; Baltimore, D., Germline transmission and tissue-specific expression of transgenes delivered by lentiviral vectors. *Science* 2002, 295, (5556), 868–872.

189. Pfeifer, A.; Ikawa, M.; Dayn, Y.; Verma, I. M., Transgenesis by lentiviral vectors: Lack of gene silencing in mammalian embryonic stem cells and preimplantation embryos. *Proc Natl Acad Sci USA* 2002, 99, (4), 2140–2145.

190. Yao, S.; Sukonnik, T.; Kean, T.; Bharadwaj, R. R.; Pasceri, P.; Ellis, J., Retrovirus silencing, variegation, extinction, and memory are controlled by a dynamic interplay of multiple epigenetic modifications. *Mol Ther* 2004, 10, (1), 27–36.

191. Ellis, J., Silencing and variegation of gammaretrovirus and lentivirus vectors. *Hum Gene Ther* 2005, 16, (11), 1241–1246.

192. Blelloch, R.; Venere, M.; Yen, J.; Ramalho-Santos, M., Generation of induced pluripotent stem cells in the absence of drug selection. *Cell Stem Cell* 2007, 1, (3), 245–247.

193. He, J.; Yang, Q.; Chang, L. J., Dynamic DNA methylation and histone modifications contribute to lentiviral transgene silencing in murine embryonic carcinoma cells. *J Virol* 2005, 79, (21), 13497–13508.

194. Kouzarides, T., Chromatin modifications and their function. *Cell* 2007, 128, (4), 693–705.

195. Srinivasan, L.; Atchison, M. L., YY1 DNA binding and PcG recruitment requires CtBP. *Genes Dev* 2004, 18, (21), 2596–2601.

196. Mizutani, T.; Ito, T.; Nishina, M.; Yamamichi, N.; Watanabe, A.; Iba, H., Maintenance of integrated proviral gene expression requires Brm, a catalytic subunit of SWI/SNF complex. *J Biol Chem* 2002, 277, (18), 15859–15864.

197. Fan, Y.; Nikitina, T.; Zhao, J.; Fleury, T. J.; Bhattacharyya, R.; Bouhassira, E. E.; Stein, A.; Woodcock, C. L.; Skoultchi, A. I., Histone H1 depletion in mammals alters global chromatin structure but causes specific changes in gene regulation. *Cell* 2005, 123, (7), 1199–1212.

198. Watanabe, T.; Takeda, A.; Tsukiyama, T.; Mise, K.; Okuno, T.; Sasaki, H.; Minami, N.; Imai, H., Identification and characterization of two novel classes of small RNAs in the mouse germline: Retrotransposon-derived siRNAs in oocytes and germline small RNAs in testes. *Genes Dev* 2006, 20, (13), 1732–1743.

199. Azuara, V.; Perry, P.; Sauer, S.; Spivakov, M.; Jorgensen, H. F.; John, R. M.; Gouti, M.; Casanova, M.; Warnes, G.; Merkenschlager, M.; Fisher, A. G., Chromatin signatures of pluripotent cell lines. *Nat Cell Biol* 2006, 8, (5), 532–538.

200. Meshorer, E.; Yellajoshula, D.; George, E.; Scambler, P. J.; Brown, D. T.; Misteli, T., Hyperdynamic plasticity of chromatin proteins in pluripotent embryonic stem cells. *Dev Cell* 2006, 10, (1), 105–116.

201. Wang, G. P.; Ciuffi, A.; Leipzig, J.; Berry, C. C.; Bushman, F. D., HIV integration site selection: Analysis by massively parallel pyrosequencing reveals association with epigenetic modifications. *Genome Res* 2007, 17, (8), 1186–1194.

202. Wiblin, A. E.; Cui, W.; Clark, A. J.; Bickmore, W. A., Distinctive nuclear organisation of centromeres and regions involved in pluripotency in human embryonic stem cells. *J Cell Sci* 2005, 118, (Pt 17), 3861–3868.

203. Perkins, K. J.; Lusic, M.; Mitar, I.; Giacca, M.; Proudfoot, N. J., Transcription-dependent gene looping of the HIV-1 provirus is dictated by recognition of pre-mRNA processing signals. *Mol Cell* 2008, 29, (1), 56–68.

204. Alon, U., Network motifs: Theory and experimental approaches. *Nat Rev Genet* 2007, 8, (6), 450–461.

205. Odom, D. T.; Dowell, R. D.; Jacobsen, E. S.; Nekludova, L.; Rolfe, P. A.; Danford, T. W.; Gifford, D. K.; Fraenkel, E.; Bell, G. I.; Young, R. A., Core transcriptional regulatory circuitry in human hepatocytes. *Mol Syst Biol* 2006, 2, 2006. 0017.

206. Kuroda, T.; Tada, M.; Kubota, H.; Kimura, H.; Hatano, S. Y.; Suemori, H.; Nakatsuji, N.; Tada, T., Octamer and Sox elements are required for transcriptional cis regulation of Nanog gene expression. *Mol Cell Biol* 2005, 25, (6), 2475–2485.

207. Okumura-Nakanishi, S.; Saito, M.; Niwa, H.; Ishikawa, F., Oct-3/4 and Sox2 regulate Oct-3/4 gene in embryonic stem cells. *J Biol Chem* 2005, 280, (7), 5307–5317.

208. Wang, J.; Rao, S.; Chu, J.; Shen, X.; Levasseur, D. N.; Theunissen, T. W.; Orkin, S. H., A protein interaction network for pluripotency of embryonic stem cells. *Nature* 2006, 444, (7117), 364–368.

209. Pan, G.; Thomson, J. A., Nanog and transcriptional networks in embryonic stem cell pluripotency. *Cell Res* 2007, 17, (1), 42–49.

210. Zhou, Q.; Chipperfield, H.; Melton, D. A.; Wong, W. H., A gene regulatory network in mouse embryonic stem cells. *Proc Natl Acad Sci USA* 2007, 104, (42), 16438–16443.

211. Davis, R. L.; Weintraub, H.; Lassar, A. B., Expression of a single transfected cDNA converts fibroblasts to myoblasts. *Cell* 1987, 51, (6), 987–1000.

212. Ivanova, N.; Dobrin, R.; Lu, R.; Kotenko, I.; Levorse, J.; DeCoste, C.; Schafer, X.; Lun, Y.; Lemischka, I. R., Dissecting self-renewal in stem cells with RNA interference. *Nature* 2006, 442, (7102), 533–538.

213. Wu, Q.; Chen, X.; Zhang, J.; Loh, Y. H.; Low, T. Y.; Zhang, W.; Sze, S. K.; Lim, B.; Ng, H. H., Sall4 interacts with Nanog and co-occupies Nanog genomic sites in embryonic stem cells. *J Biol Chem* 2006, 281, (34), 24090–24094.

214. Zhang, J.; Tam, W. L.; Tong, G. Q.; Wu, Q.; Chan, H. Y.; Soh, B. S.; Lou, Y.; Yang, J.; Ma, Y.; Chai, L.; Ng, H. H.; Lufkin, T.; Robson, P.; Lim, B., Sall4 modulates embryonic stem cell pluripotency and early embryonic development by the transcriptional regulation of Pou5f1. *Nat Cell Biol* 2006, 8, (10), 1114–1123.

215. Williams, R. L.; Hilton, D. J.; Pease, S.; Willson, T. A.; Stewart, C. L.; Gearing, D. P.; Wagner, E. F.; Metcalf, D.; Nicola, N. A.; Gough, N. M., Myeloid leukaemia inhibitory factor maintains the developmental potential of embryonic stem cells. *Nature* 1988, 336, (6200), 684–687.

216. Nichols, J.; Chambers, I.; Taga, T.; Smith, A., Physiological rationale for responsiveness of mouse embryonic stem cells to gp130 cytokines. *Development* 2001, 128, (12), 2333–2339.

217. Ying, Q. L.; Nichols, J.; Chambers, I.; Smith, A., BMP induction of Id proteins suppresses differentiation and sustains embryonic stem cell self-renewal in collaboration with STAT3. *Cell* 2003, 115, (3), 281–292.

218. Qi, X.; Li, T. G.; Hao, J.; Hu, J.; Wang, J.; Simmons, H.; Miura, S.; Mishina, Y.; Zhao, G. Q., BMP4 supports self-renewal of embryonic stem cells by inhibiting mitogen-activated protein kinase pathways. *Proc Natl Acad Sci USA* 2004, 101, (16), 6027–6032.

219. Sato, N.; Meijer, L.; Skaltsounis, L.; Greengard, P.; Brivanlou, A. H., Maintenance of pluripotency in human and mouse embryonic stem cells through activation of Wnt signaling by a pharmacological GSK-3-specific inhibitor. *Nat Med* 2004, 10, (1), 55–63.

220. Ogawa, K.; Nishinakamura, R.; Iwamatsu, Y.; Shimosato, D.; Niwa, H., Synergistic action of Wnt and LIF in maintaining pluripotency of mouse ES cells. *Biochem Biophys Res Commun* 2006, 343, (1), 159–166.

221. Park, Y. B.; Kim, Y. Y.; Oh, S. K.; Chung, S. G.; Ku, S. Y.; Kim, S. H.; Choi, Y. M.; Moon, S. Y., Alterations of proliferative and differentiation potentials of human embryonic stem cells during long-term culture. *Exp Mol Med* 2008, 40, (1), 98–108.

222. Oh, S. K.; Kim, H. S.; Park, Y. B.; Seol, H. W.; Kim, Y. Y.; Cho, M. S.; Ku, S. Y.; Choi, Y. M.; Kim, D. W.; Moon, S. Y., Methods for expansion of human embryonic stem cells. *Stem Cells* 2005, 23, (5), 605–609.

223. Bendall, S. C.; Stewart, M. H.; Menendez, P.; George, D.; Vijayaragavan, K.; Werbowetski-Ogilvie, T.; Ramos-Mejia, V.; Rouleau, A.; Yang, J.; Bosse, M.; Lajoie, G.; Bhatia, M., IGF and FGF cooperatively establish the regulatory stem cell niche of pluripotent human cells in vitro. *Nature* 2007, 448, (7157), 1015–1021.

224. Xu, C.; Rosler, E.; Jiang, J.; Lebkowski, J. S.; Gold, J. D.; O'Sullivan, C.; Delavan-Boorsma, K.; Mok, M.; Bronstein, A.; Carpenter, M. K., Basic fibroblast growth factor supports undifferentiated human embryonic stem cell growth without conditioned medium. *Stem Cells* 2005, 23, (3), 315–323.

225. Xu, R. H.; Chen, X.; Li, D. S.; Li, R.; Addicks, G. C.; Glennon, C.; Zwaka, T. P.; Thomson, J. A., BMP4 initiates human embryonic stem cell differentiation to trophoblast. *Nat Biotechnol* 2002, 20, (12), 1261–1264.

226. Vallier, L.; Alexander, M.; Pedersen, R. A., Activin/Nodal and FGF pathways cooperate to maintain pluripotency of human embryonic stem cells. *J Cell Sci* 2005, 118, (Pt 19), 4495–4509.

227. Cartwright, P.; McLean, C.; Sheppard, A.; Rivett, D.; Jones, K.; Dalton, S., LIF/STAT3 controls ES cell self-renewal and pluripotency by a Myc-dependent mechanism. *Development* 2005, 132, (5), 885–896.

228. Suzuki, A.; Raya, A.; Kawakami, Y.; Morita, M.; Matsui, T.; Nakashima, K.; Gage, F. H.; Rodriguez-Esteban, C.; Izpisua Belmonte, J. C., Maintenance of embryonic stem cell pluripotency by Nanog-mediated reversal of mesoderm specification. *Nat Clin Pract Cardiovasc Med* 2006, 3, (Suppl 1), S114–S122.

229. Ura, H.; Usuda, M.; Kinoshita, K.; Sun, C.; Mori, K.; Akagi, T.; Matsuda, T.; Koide, H.; Yokota, T., STAT3 and Oct-3/4 control histone modification through induction of Eed in embryonic stem cells. *J Biol Chem* 2008, 283, (15), 9713–9723.

230. Ko, S. Y.; Kang, H. Y.; Lee, H. S.; Han, S. Y.; Hong, S. H., Identification of Jmjd1a as a STAT3 downstream gene in mES cells. *Cell Struct Funct* 2006, 31, (2), 53–62.

231. Kinoshita, K.; Ura, H.; Akagi, T.; Usuda, M.; Koide, H.; Yokota, T., GABPalpha regulates Oct-3/4 expression in mouse embryonic stem cells. *Biochem Biophys Res Commun* 2007, 353, (3), 686–691.

232. Humphrey, R. K.; Beattie, G. M.; Lopez, A. D.; Bucay, N.; King, C. C.; Firpo, M. T.; Rose-John, S.; Hayek, A., Maintenance of pluripotency in human embryonic stem cells is STAT3 independent. *Stem Cells* 2004, 22, (4), 522–530.

233. Armstrong, L.; Hughes, O.; Yung, S.; Hyslop, L.; Stewart, R.; Wappler, I.; Peters, H.; Walter, T.; Stojkovic, P.; Evans, J.; Stojkovic, M.; Lako, M., The role of PI3K/AKT, MAPK/ERK and NFkappabeta signalling in the maintenance of human embryonic stem cell pluripotency and viability highlighted by transcriptional profiling and functional analysis. *Hum Mol Genet* 2006, 15, (11), 1894–1913.

234. Li, J.; Wang, G.; Wang, C.; Zhao, Y.; Zhang, H.; Tan, Z.; Song, Z.; Ding, M.; Deng, H., MEK/ERK signaling contributes to the maintenance of human embryonic stem cell self-renewal. *Differentiation* 2007, 75, (4), 299–307.

235. Ying, Q. L.; Wray, J.; Nichols, J.; Battle-Morera, L.; Doble, B.; Woodgett, J.; Cohen, P.; Smith, A., The ground state of embryonic stem cell self-renewal. *Nature* 2008, 453, (7194), 519–523.

236. Jenuwein, T.; Allis, C. D., Translating the histone code. *Science* 2001, 293, (5532), 1074–1080.

237. Santos, F.; Dean, W., Epigenetic reprogramming during early development in mammals. *Reproduction* 2004, 127, (6), 643–651.

238. Lee, T. I.; Jenner, R. G.; Boyer, L. A.; Guenther, M. G.; Levine, S. S.; Kumar, R. M.; Chevalier, B.; Johnstone, S. E.; Cole, M. F.; Isono, K.; Koseki, H.; Fuchikami, T.; Abe, K.; Murray, H. L.; Zucker, J. P.; Yuan, B.; Bell, G. W.; Herbolsheimer, E.; Hannett, N. M.; Sun, K.; Odom, D. T.; Otte, A. P.; Volkert, T. L.; Bartel, D. P.; Melton, D. A.; Gifford, D. K.; Jaenisch, R.; Young, R. A., Control of developmental regulators by Polycomb in human embryonic stem cells. *Cell* 2006, 125, (2), 301–313.

239. Boyer, L. A.; Plath, K.; Zeitlinger, J.; Brambrink, T.; Medeiros, L. A.; Lee, T. I.; Levine, S. S.; Wernig, M.; Tajonar, A.; Ray, M. K.; Bell, G. W.; Otte, A. P.; Vidal, M.; Gifford, D. K.; Young, R. A.; Jaenisch, R., Polycomb complexes repress developmental regulators in murine embryonic stem cells. *Nature* 2006, 441, (7091), 349–353.

240. Schuettengruber, B.; Chourrout, D.; Vervoort, M.; Leblanc, B.; Cavalli, G., Genome regulation by polycomb and trithorax proteins. *Cell* 2007, 128, (4), 735–745.

241. Stock, J. K.; Giadrossi, S.; Casanova, M.; Brookes, E.; Vidal, M.; Koseki, H.; Brockdorff, N.; Fisher, A. G.; Pombo, A., Ring1-mediated ubiquitination of H2A restrains poised RNA polymerase II at bivalent genes in mouse ES cells. *Nat Cell Biol* 2007, 9, (12), 1428–1435.

242. Guenther, M. G.; Levine, S. S.; Boyer, L. A.; Jaenisch, R.; Young, R. A., A chromatin landmark and transcription initiation at most promoters in human cells. *Cell* 2007, 130, (1), 77–88.

243. Lan, F.; Bayliss, P. E.; Rinn, J. L.; Whetstine, J. R.; Wang, J. K.; Chen, S.; Iwase, S.; Alpatov, R.; Issaeva, I.; Canaani, E.; Roberts, T. M.; Chang, H. Y.; Shi, Y., A histone H3 lysine 27 demethylase regulates animal posterior development. *Nature* 2007, 449, (7163), 689–694.

244. Lee, J. H.; Hart, S. R.; Skalnik, D. G., Histone deacetylase activity is required for embryonic stem cell differentiation. *Genesis* 2004, 38, (1), 32–38.

245. Kloosterman, W. P.; Plasterk, R. H., The diverse functions of microRNAs in animal development and disease. *Dev Cell* 2006, 11, (4), 441–450.

246. Suh, M. R.; Lee, Y.; Kim, J. Y.; Kim, S. K.; Moon, S. H.; Lee, J. Y.; Cha, K. Y.; Chung, H. M.; Yoon, H. S.; Moon, S. Y.; Kim, V. N.; Kim, K. S., Human embryonic stem cells express a unique set of microRNAs. *Dev Biol* 2004, 270, (2), 488–498.

247. Houbaviy, H. B.; Murray, M. F.; Sharp, P. A., Embryonic stem cell-specific MicroRNAs. *Dev Cell* 2003, 5, (2), 351–358.

248. Wilson, K. D.; Venkatasubrahmanyam, S.; Jia, F.; Sun, N.; Butte, A. J.; Wu, J. C., MicroRNA profiling of human-induced pluripotent stem cells. *Stem Cells Dev* 2009, 18, (5), 749–758.

249. Newman, M. A.; Thomson, J. M.; Hammond, S. M., Lin-28 interaction with the Let-7 precursor loop mediates regulated microRNA processing. *RNA* 2008, 14, (8), 1539–1549.

250. Yu, F.; Yao, H.; Zhu, P.; Zhang, X.; Pan, Q.; Gong, C.; Huang, Y.; Hu, X.; Su, F.; Lieberman, J.; Song, E., let-7 regulates self renewal and tumorigenicity of breast cancer cells. *Cell* 2007, 131, (6), 1109–1123.

251. Kumar, M. S.; Erkeland, S. J.; Pester, R. E.; Chen, C. Y.; Ebert, M. S.; Sharp, P. A.; Jacks, T., Suppression of non-small cell lung tumor development by the let-7 microRNA family. *Proc Natl Acad Sci USA* 2008, 105, (10), 3903–3908.

252. Stead, E.; White, J.; Faast, R.; Conn, S.; Goldstone, S.; Rathjen, J.; Dhingra, U.; Rathjen, P.; Walker, D.; Dalton, S., Pluripotent cell division cycles are driven by ectopic Cdk2, cyclin A/E and E2F activities. *Oncogene* 2002, 21, (54), 8320–8333.

253. Savatier, P.; Lapillonne, H.; van Grunsven, L. A.; Rudkin, B. B.; Samarut, J., Withdrawal of differentiation inhibitory activity/leukemia inhibitory factor up-regulates D-type cyclins and cyclin-dependent kinase inhibitors in mouse embryonic stem cells. *Oncogene* 1996, 12, (2), 309–322.

254. Drayton, S.; Peters, G., Immortalisation and transformation revisited. *Curr Opin Genet Dev* 2002, 12, (1), 98–104.

255. Herbig, U.; Sedivy, J. M., Regulation of growth arrest in senescence: Telomere damage is not the end of the story. *Mech Ageing Dev* 2006, 127, (1), 16–24.

256. Qin, H.; Yu, T.; Qing, T.; Liu, Y.; Zhao, Y.; Cai, J.; Li, J.; Song, Z.; Qu, X.; Zhou, P.; Wu, J.; Ding, M.; Deng, H., Regulation of apoptosis and differentiation by p53 in human embryonic stem cells. *J Biol Chem* 2007, 282, (8), 5842–5852.

257. Wright, W. E.; Shay, J. W., The two-stage mechanism controlling cellular senescence and immortalization. *Exp Gerontol* 1992, 27, (4), 383–389.

258. Ducray, C.; Pommier, J. P.; Martins, L.; Boussin, F. D.; Sabatier, L., Telomere dynamics, end-to-end fusions and telomerase activation during the human fibroblast immortalization process. *Oncogene* 1999, 18, (29), 4211–4223.

259. Wu, K. J.; Grandori, C.; Amacker, M.; Simon-Vermot, N.; Polack, A.; Lingner, J.; Dalla-Favera, R., Direct activation of TERT transcription by c-MYC. *Nat Genet* 1999, 21, (2), 220–224.

260. Kunisato, A.; Wakatsuki, M.; Kodama, Y.; Shinba, H.; Ishida, I.; Nagao, K., Generation of induced pluripotent stem (iPS) cells by efficient reprogramming of adult bone marrow cells. *Stem Cells Dev* 2010, 19, (2), 229–238.

261. Nagata, S.; Toyoda, M.; Yamaguchi, S.; Hirano, K.; Makino, H.; Nishino, K.; Miyagawa, Y.; Okita, H.; Kiyokawa, N.; Nakagawa, M.; Yamanaka, S.; Akutsu, H.; Umezawa, A.; Tada, T., Efficient reprogramming of human and mouse primary extra-embryonic cells to pluripotent stem cells. *Genes Cells* 2009, 14, (12), 1395–1404.

262. Li, W.; Ding, S., Small molecules that modulate embryonic stem cell fate and somatic cell reprogramming. *Trends Pharmacol Sci* 2010, 31, (1), 36–45.

263. Kim, J. B.; Greber, B.; Arauzo-Bravo, M. J.; Meyer, J.; Park, K. I.; Zaehres, H.; Scholer, H. R., Direct reprogramming of human neural stem cells by OCT4. *Nature* 2009, 461, (7264), 649–643.

264. Stadtfeld, M.; Brennand, K.; Hochedlinger, K., Reprogramming of pancreatic beta cells into induced pluripotent stem cells. *Curr Biol* 2008, 18, (12), 890–894.

265. Hanna, J.; Markoulaki, S.; Schorderet, P.; Carey, B. W.; Beard, C.; Wernig, M.; Creyghton, M. P.; Steine, E. J.; Cassady, J. P.; Foreman, R.; Lengner, C. J.; Dausman, J. A.; Jaenisch, R., Direct reprogramming of terminally differentiated mature B lymphocytes to pluripotency. *Cell* 2008, 133, (2), 250–264.

266. Maherali, N.; Hochedlinger, K., Tgfbeta signal inhibition cooperates in the induction of iPSCs and replaces Sox2 and cMyc. *Curr Biol* 2009, 19, (20), 1718–1723.

267. Utikal, J.; Maherali, N.; Kulalert, W.; Hochedlinger, K., Sox2 is dispensable for the reprogramming of melanocytes and melanoma cells into induced pluripotent stem cells. *J Cell Sci* 2009, 122, (Pt 19), 3502–3510.

268. Shao, L.; Feng, W.; Sun, Y.; Bai, H.; Liu, J.; Currie, C.; Kim, J.; Gama, R.; Wang, Z.; Qian, Z.; Liaw, L.; Wu, W. S., Generation of iPS cells using defined factors linked via the self-cleaving 2A sequences in a single open reading frame. *Cell Res* 2009, 19, (3), 296–306.

269. Chang, C. W.; Lai, Y. S.; Pawlik, K. M.; Liu, K.; Sun, C. W.; Li, C.; Schoeb, T. R.; Townes, T. M., Polycistronic lentiviral vector for "hit and run" reprogramming of adult skin fibroblasts to induced pluripotent stem cells. *Stem Cells* 2009, 27, (5), 1042–1049.

270. Gonzalez, F.; Barragan Monasterio, M.; Tiscornia, G.; Montserrat Pulido, N.; Vassena, R.; Batlle Morera, L.; Rodriguez Piza, I.; Izpisua Belmonte, J. C., Generation of mouse-induced pluripotent stem cells by transient expression of a single nonviral polycistronic vector. *Proc Natl Acad Sci USA* 2009, 106, (22), 8918–8922.

271. Liao, J.; Cui, C.; Chen, S.; Ren, J.; Chen, J.; Gao, Y.; Li, H.; Jia, N.; Cheng, L.; Xiao, H.; Xiao, L., Generation of induced pluripotent stem cell lines from adult rat cells. *Cell Stem Cell* 2009, 4, (1), 11–15.

272. Shimada, H.; Nakada, A.; Hashimoto, Y.; Shigeno, K.; Shionoya, Y.; Nakamura, T., Generation of canine induced pluripotent stem cells by retroviral transduction and chemical inhibitors. *Mol Reprod Dev* 2010, 77, (1), 2.

273. Esteban, M. A.; Xu, J.; Yang, J.; Peng, M.; Qin, D.; Li, W.; Jiang, Z.; Chen, J.; Deng, K.; Zhong, M.; Cai, J.; Lai, L.; Pei, D., Generation of induced pluripotent stem cell lines from Tibetan miniature pig. *J Biol Chem* 2009, 284, (26), 17634–17640.

274. Wu, Z.; Chen, J.; Ren, J.; Bao, L.; Liao, J.; Cui, C.; Rao, L.; Li, H.; Gu, Y.; Dai, H.; Zhu, H.; Teng, X.; Cheng, L.; Xiao, L., Generation of pig induced pluripotent stem cells with a drug-inducible system. *J Mol Cell Biol* 2009, 1, (1), 46–54.

275. Wang, Y.; Jiang, Y.; Liu, S.; Sun, X.; Gao, S., Generation of induced pluripotent stem cells from human beta-thalassemia fibroblast cells. *Cell Res* 2009, 19, (9), 1120–1123.

276. Seifinejad, A.; Taei, A.; Totonchi, M.; Vazirinasab, H.; Hasani, S. N.; Aghdami, N.; Shahbazi, E.; Salekdeh, G. H.; Baharvand, H., Generation of human induced pluripotent stem cells from a Bombay individual: Moving towards "universal-donor" red blood cells. *Biochem Biophys Res Commun* 2010, 391, (1), 329–334.

277. Loh, Y. H.; Agarwal, S.; Park, I. H.; Urbach, A.; Huo, H.; Heffner, G. C.; Kim, K.; Miller, J. D.; Ng, K.; Daley, G. Q., Generation of induced pluripotent stem cells from human blood. *Blood* 2009, 113, (22), 5476–5479.

278. Ye, Z.; Zhan, H.; Mali, P.; Dowey, S.; Williams, D. M.; Jang, Y. Y.; Dang, C. V.; Spivak, J. L.; Moliterno, A. R.; Cheng, L., Human induced pluripotent stem cells from blood cells of healthy donors and patients with acquired blood disorders. *Blood* 2009, 114, (27), 5473–5480.

279. Li, C.; Zhou, J.; Shi, G.; Ma, Y.; Yang, Y.; Gu, J.; Yu, H.; Jin, S.; Wei, Z.; Chen, F.; Jin, Y., Pluripotency can be rapidly and efficiently induced in human amniotic fluid-derived cells. *Hum Mol Genet* 2009, 18, (22), 4340–4349.

280. Maehr, R.; Chen, S.; Snitow, M.; Ludwig, T.; Yagasaki, L.; Goland, R.; Leibel, R. L.; Melton, D. A., Generation of pluripotent stem cells from patients with type 1 diabetes. *Proc Natl Acad Sci USA* 2009, 106, (37), 15768–15773.

281. Giorgetti, A.; Montserrat, N.; Aasen, T.; Gonzalez, F.; Rodriguez-Piza, I.; Vassena, R.; Raya, A.; Boue, S.; Barrero, M. J.; Corbella, B. A.; Torrabadella, M.; Veiga, A.; Izpisua Belmonte, J. C., Generation of induced pluripotent stem cells from human cord blood using Oct4 and Sox2. *Cell Stem Cell* 2009, 5, (4), 353–357.

282. Hester, M. E.; Song, S.; Miranda, C. J.; Eagle, A.; Schwartz, P. H.; Kaspar, B. K., Two factor reprogramming of human neural stem cells into pluripotency. *PLoS One* 2009, 4, (9), e7044.

283. Kim, J. B.; Zaehres, H.; Arauzo-Bravo, M. J.; Scholer, H. R., Generation of induced pluripotent stem cells from neural stem cells. *Nat Protoc* 2009, 4, (10), 1464–1470.

284. Mali, P.; Ye, Z.; Hommond, H. H.; Yu, X.; Lin, J.; Chen, G.; Zou, J.; Cheng, L., Improved efficiency and pace of generating induced pluripotent stem cells from human adult and fetal fibroblasts. *Stem Cells* 2008, 26, (8), 1998–2005.

285. Ebert, A. D.; Yu, J.; Rose, F. F., Jr.; Mattis, V. B.; Lorson, C. L.; Thomson, J. A.; Svendsen, C. N., Induced pluripotent stem cells from a spinal muscular atrophy patient. *Nature* 2009, 457, (7227), 277–280.

286. Haase, A.; Olmer, R.; Schwanke, K.; Wunderlich, S.; Merkert, S.; Hess, C.; Zweigerdt, R.; Gruh, I.; Meyer, J.; Wagner, S.; Maier, L. S.; Han, D. W.; Glage, S.; Miller, K.; Fischer, P.; Scholer, H. R.; Martin, U., Generation of induced pluripotent stem cells from human cord blood. *Cell Stem Cell* 2009, 5, (4), 434–441.

287. Zhou, W.; Freed, C. R., Adenoviral gene delivery can reprogram human fibroblasts to induced pluripotent stem cells. *Stem Cells* 2009, 27, (11), 2667–2674.

288. Fusaki, N.; Ban, H.; Nishiyama, A.; Saeki, K.; Hasegawa, M., Efficient induction of transgene-free human pluripotent stem cells using a vector based on Sendai virus, an RNA virus that does not integrate into the host genome. *Proc Jpn Acad Ser B Phys Biol Sci* 2009, 85, (8), 348–362.

289. Yu, J.; Hu, K.; Smuga-Otto, K.; Tian, S.; Stewart, R.; Slukvin, II; Thomson, J. A., Human induced pluripotent stem cells free of vector and transgene sequences. *Science* 2009, 324, (5928), 797–801.

Part II

In Vitro *Studies for Angiogenesis, Vasculogenesis, and Arteriogenesis*

4 Guiding Stem Cell Fate through Microfabricated Environments

Lisa R. Trump, Gregory Timp,
and Lawrence B. Schook

CONTENTS

4.1 INTRODUCTION

The multipotent nature of stem cells provides enormous potential for clinical applications for treatment of disease, cancers, and for organ replacement. Despite decades of research, robust culture techniques that consistently permit isolation, expansion, and directed differentiation of stem and progenitor cells in adequate numbers remains a major hurdle to ensure full clinical usage of stem cell therapies. *In vivo*, stem cell fate is governed by specialized microenvironments termed "niches." The stem cell niche consists of supporting cells, extracellular matrix (ECM), and extrinsic cues such as growth factors and cytokines that are spatially and temporally controlled to direct differentiation and maintain stem cell pools (Figure 4.1).[1,2] When removed from niches and cultured *in vitro*, stem cells rapidly lose self-renewal capabilities and undergo spontaneous differentiation due to the loss of intrinsic and extrinsic

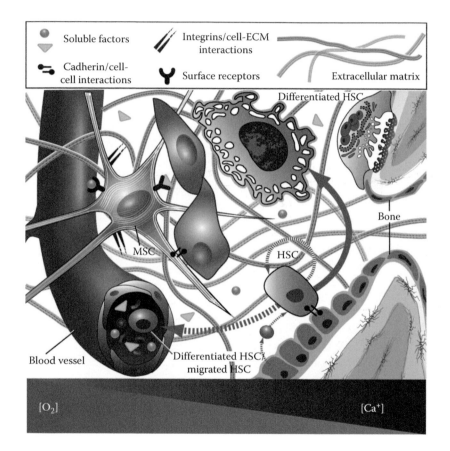

FIGURE 4.1 (See color insert.)   The perivascular and endosteal bone marrow niche: example of signals and cues regulating stem cell function. The stem cell niche is a complex microenvironment that guides stem cell fate through a combination of extracellular matrix (ECM), cell–cell interactions, and extrinsic factors such as growth factors and cytokines. In the bone marrow, hematopoietic stem cells (HSCs) situated in the endosteal niche are physically anchored to the niche by cadherens junctions with osteoblasts (cell–cell interactions). In this location, cells are exposed to high Ca^{2+} concentrations, low oxygen tension, and a variety of autocrine, paracrine, and endocrine signals (extrinsic factors) and are attached to the ECM through integrin receptors. These signals and cues maintain the quiescence and self-renewal of HSCs. In the perivascular niche, however, HSCs are exposed to low Ca^{2+}, high oxygen tension, different cell–cell interactions, and ECM composition that promote migration and differentiation of stem cells. While a majority of these signals are found in a variety of stem cell niches, their utilization and the effects of niche components vary from niche to niche.

cues found in stem cell niches and physiological tissues. This loss of stem cell characteristics in culture *in vitro* severely limits the ability to expand and directly differentiate cells into sufficient numbers for clinical usage.

The ability to recapitulate aspects of physiological tissue environments is key to identifying and understanding the intrinsic and extrinsic cues directing stem cell self-renewal and differentiation. Currently, the understanding of spatial and temporal cues directing stem cell fate is generated from tissue culture systems where the cellular microenvironment is regulated in batch conditions. Typical *in vitro* cell culture techniques rely on the use of two-dimensional (2D), plastic surfaces such as petri dishes and tissue culture flasks to propagate, differentiate, and understand cell behavior in response to various small molecules and chemical stimuli. These conventional cell culture techniques are well established and inexpensive. However, traditional systems poorly recapitulate the complex physiochemical tissue environment and offer little control over cell seeding, cell–cell interactions, and biologically relevant presentation of soluble molecules. Removed from the niche and cultured in *in vitro*, stem cells display altered phenotypes and gene expression and have limited expansion and differentiation capabilities.[3–5] Furthermore, cell isolation techniques are unable to provide homogeneous populations of stem cells. Contaminating cells may secrete soluble molecules that can affect cellular function, select for a subpopulation of cells, or easily proliferate and overtake populations of stem cells. Since cellular responses are mostly measured on a population basis, responses of a small subset or limited population of cells may be masked.

Though recent efforts have increased our knowledge of stem cell biology, little is known about the combinatorial signals that guide stem cell fate. Thorough understanding of the combinatorial microenvironments that direct the behavior and differentiation properties of stem cells require robust culture systems that permit precise control over cell–cell interactions, ECM properties, and extrinsic factor delivery. To circumvent limitations of poorly controlled microenvironments found in traditional batch culture systems, cell biologists are looking toward tissue engineering and microfabrication technologies to design culture systems that more accurately recapitulate *in vivo* cellular microenvironments. These technologies combine biomaterial scaffolds with various engineering strategies that provide the ability to tailor cellular microenvironments and provide signals and cues spatially and temporally (Figure 4.2).[6] This chapter first briefly discusses how the various components of the stem cell niche guide cell behavior and then reviews the various microscale technologies currently used to recreate the stem cell niche *in vitro*.

4.2   THREE-DIMENSIONAL ENVIRONMENTS AND THE STEM CELL NICHE

Physiologic tissue environments are complex, three dimensional (3D) environments that direct cell function through ECM, cell–cell interactions, mechanical stimuli, and soluble factors. The concept of stem cell niches, first proposed by Schofield et al. in 1978, suggests that adult stem cells reside in defined compartments (i.e., "niches"), which balance stem cell self-renewal and differentiation to maintain tissue homeostasis and the stem cell pool.[2] To date, stem cell niches have been identified in a variety of

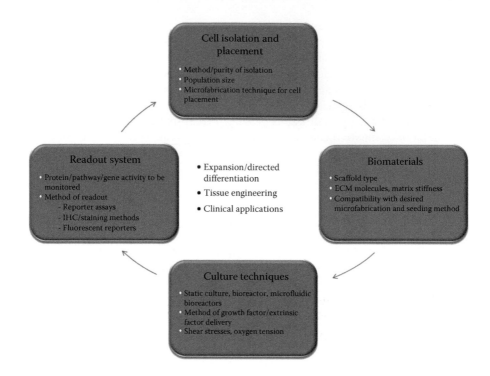

FIGURE 4.2 (See color insert.) Overview of processes to design environments that guide cell fate. First, the cell source or line, isolation, and optimal microfabrication technique for the study or application must be identified. Biomaterials selected for microfabrication techniques must be compatible with the cell and microfabrication technique of interest. Modifications of biomaterials such as the incorporation of ECM molecules or material stiffness can affect cellular processes. Culture techniques provide nutrient delivery and waste removal to micro-fabricated cells. Spatial delivery of growth factors, cytokines, and other extrinsic factors can be controlled by bioreactors and microfluidics systems. The culture technique or bioreactor type can affect nutrient delivery throughout scaffolds, provide shear stresses, affect cell seeding into scaffolds, or modulate ECM deposition by cells. Lastly, systems to monitor cellular activity in the microfabricated system must be identified. Assays can be end point, such as cellular staining, or measured in real time by reporter assays. Each of these factors must be taken into consideration when designing microfabrication experiments.

tissues including the bone marrow,[7,8] skin,[9] hair follicles,[10] intestine,[11] brain,[12] and muscle.[13] The stem cell niche provides signals and cues that balance self-renewal, maintenance, and differentiation as well as protect cells from apoptosis and prevent depletion or overpopulation of stem cells. Cells are physically anchored within the niche by ECM proteins and supporting stromal cells, which in combination with soluble signals regulate the maintenance and self-renewal of stem cells. Figure 4.1 details the bone marrow niche and the various cues that maintain stem cell function. Within the niche, stem cells either undergo symmetric division to give rise to identical progeny (self-renewal), asymmetric diffusion to produce one stem cell and a differentiated progeny, division without differentiation, or remain quiescent.[1] Aberrations within the niche are thought to cause pathologies such as cancer, ageing, and degeneration of tissue function.[14,15]

4.2.1 EXTRACELLULAR MATRIX

The ECM is composed of a combination of proteoglycans, polysaccharides, and proteins that provide structural support to cells. The ECM varies in composition and stiffness from tissue to tissue and plays an integral role in maintaining cellular phenotypes and cell fate decisions. Cells attach to the matrix through integrin receptors on the cell surface that, when bound to their ECM ligands, activate cellular signaling cascades. Loss of cell–ECM interactions results in a specialized form of detachment-induced cell death termed "anoikis," which is derived from the Greek word for homelessness.[16] For example, when cultured in PEG matrices that do not permit cellular attachment, mesenchymal stem cells (MSCs) undergo anoikis. Restoration of cellular attachment by the cell-adhesive peptide Arg-Gly-Asp (RGD) increases viability of encapsulated cells by engaging cell integrin receptors.[17,18] Furthermore, several studies have reported that the stiffness and elasticity of the ECM affect stem cell processes. When cultured on stiff surfaces mimicking bone tissue, MSCs display hallmarks of osteoprogenitor differentiation, while culture on soft surfaces promotes adipose differentiation.[19] ECM interactions also govern cell shape and size, which affect cellular survival, proliferation,[20] and differentiation.[21] McBeath et al. patterned fibronectin ECM in various geometries onto tissue culture substrates and seeded human MSCs onto the ECM. Large islands of fibronectin moieties promoted cell spreading, whereas cells had a rounded phenotype on small ECM islands. Cells allowed to spread on large islands displayed osteoprogenitor commitment, while rounded cells differentiated into adipocytes. A recent study by Chowdhury et al. explored the effects of cyclic strain on embryonic stem cells and embryonic differentiated cells.[22] Cyclic stress induced cell spreading and down regulation of the stemness marker Oct 3/4, whereas embryonic differentiated cells demonstrated no genotypic or phenotypic changes from the cyclic stressors. It is hypothesized that the cell softness, defined as the ratio of strain to stress on the cells, affects a cells response to stress. As embryonic stem cells are significantly softer than embryonic differentiated cells, it is concluded that the ES cells showed responses to stressors due to greater cyclic strain.

4.2.2 EXTRINSIC FACTORS

Within the niche, stem cells are exposed to a mixture of extrinsic factors that influence cell fate decisions. Such factors include growth factors, cytokines, small proteins, and ions. The spatial and temporal presentation of extrinsic factors within the niche affects stem cell self-renewal and differentiation fates. Secreted factors arise from adjacent cells, from diffusion throughout the niche, or immobilized to ECM proteins. Soluble proteins that affect stem cells include Wnts, hedgehog proteins, fibroblast growth factors (FGFs), and the BMP/TGFβ superfamily. In the neural stem cell niche, for example, TGFβ secreted by nearby differentiated neurons suppress the division of neural stem cells (NSCs) within the niche. It is important to note, however, that the spatial and temporal presentation of soluble molecules can also affect stem cell activity. Immobilization of growth factors and small proteins by ECM proteins affect concentrations, stability, and bioavailability to niche cells. FGF-2 tethered to fibrinogen increases endothelial cell (EC) proliferation relative to

FGF-2 in solution.[23] Similarly, bone marrow MSCs exposed to biomaterial surfaces with tethered EGF promotes cell spreading and survival more strongly than soluble EGF.[24] Inorganic ion concentrations and gradients within the niche also affect stem cell behavior. Hematopoietic stem cells (HSCs) situated near the endosteal surface are exposed to high calcium levels from nearby osteoblasts and low oxygen tension. These conditions are thought to help maintain HSCs in the quiescent state. In contrast, HSCs situated closer to microvasculature are exposed to higher oxygen tensions and lower calcium ion levels, which promotes HSC division and differentiation.[25]

4.2.3 CELL–CELL INTERACTIONS

Stem cells represent a very small portion of adult tissues and exist as single cells or small clusters of cells and are in contact and respond to a variety of differentiated cell types within the niche. These interactions, mediated by adherens and gap junctions, influence stem cell fate. Chondrogenic differentiation of MSCs is facilitated *in vitro* by increasing cell–cell interactions via pellet culture.[26] Supporting cells such as stromal cells, vasculature, and basal lamina anchor stem cells within the niche and may direct cellular placement to soluble signals secreted by surrounding cells. Osteoblasts anchor HSCs to the perivascular niche through N-cadherins that are involved in maintaining the quiescent state. The proximity of HSCs to osteoblasts places them in high Ca^{2+} and low oxygen tension microenvironments as discussed above, as well as induces production of cytokines and growth factors. However, exposing HSCs to cocktails of these cytokines is not sufficient to maintain stemness, suggesting that direct HSC–osteoblast contact is required for maintenance of stem cell properties.[27] Changes in cell density or loss of adherens junctions initiate cell division or migration out of the niche. Loss of cadherin junctions between HSCs and osteoblasts induces loss of HSC and migration out of the niche.

4.3 ENGINEERING TECHNOLOGIES TO GUIDE STEM CELL FATE

To understand the requirements for cell microenvironments, cell biologists and tissue engineers developed microfabrication techniques that enable precise control over cell seeding onto substrates and biomaterials, as well as control spatial and temporal cues within the culture microenvironment. Borrowed from semiconductor and microelectronics industries, microfabrication technologies are able to pattern ECM proteins onto 2D substrates such as glass and 3D substrates and scaffolds to control cell adhesion and cell–cell interactions. Other microfabrication techniques offer the unique ability to mold 3D biomaterials into desired shapes and precisely place cells within biomaterial scaffolds. Either way, such techniques are reproducible and able to create objects from tens of microns to millimeters in size with high resolution. These microscale technologies promise advances in elucidating the *in vivo* function of stem cells and niche components, generation of tissue engineering constructs and for development of high throughput platforms for drug discovery and cell-based biosensors. The following sections will first briefly discuss properties of biomaterial scaffolds and then various 2D and 3D microfabrication technologies to recapitulate aspects of the stem cell niche microenvironments within cell culture surfaces or biomaterial scaffolds. Examples of various microfabrication technologies and their applications are further outlined in Table 4.1.

TABLE 4.1
Strategies to Engineer Various Components of the Stem Cell Niche

Niche Component	Engineering Strategies	Examples	References
Extracellular matrix			
Substrate stiffness	Scaffold/ substrate type and design	Human ESCs cultured on PDMS surfaces of varying stiffness affected cellular spreading, growth rate, and osteogenic differentiation. Culture of cells on stiff surfaces increases the degree of cell spreading, attachment, and osteogenic differentiation as compared to softer substrates	Evans et al. (87, 2009, p. 1)
Ligand presentation and gradients	Inkjet printing	Patterns of collagen printed onto agarose films directed smooth muscle cell and primary neuron attachment in predefined patterns	Roth et al. [85, p. 3707]
	Microcontact printing	Microcontact printing techniques have been used to specifically place ECM ligands onto cell repellant surfaces to determine effects of ECM on cellular activity	McBeath et al. [21, p. 483], Offenhäusser et al. (88, 2007, p. 290)
	Microfluidic patterning	Microfluidic chips create gradients of Fc-tagged fusion proteins through laminar flow deposition	Cosson et al. (89, 2009, 3411)
	Photolithography	Two photon laser scanning photolithography micropatterned RGDs onto 3D hydrogel scaffolds to direct cell migration	Lee et al. (90, 2008, p. 2962)
Topography	Laser-guided direct writing	Direct writing techniques fabricate biomaterial scaffolds with precise 3D architecture and composition to guide cell patterning and behavior	Lewis et al. (91, 2004, p. 32)
	Photolithography	Photolithographic masks precisely pattern poly (ethylene glycol) diacrylate (PEGDA) scaffolds into desired architectures. Sequential patterning allows for development of 3D architectures	Hahn et al. (92, 2006, p. 2679)
Cell–cell interactions			
Direct cell placement in 3D	Optical tweezers	Time-shared optical tweezers used in conjunction with microfluidic devices allow precise 2D and 3D placement of *E. coli* bacterium within hydrogel scaffolds	Mirsaidov et al. [53, p. 2174]

(continued)

TABLE 4.1 (Continued)
Strategies to Engineer Various Components of the Stem Cell Niche

Niche Component	Engineering Strategies	Examples	References
	Dielectrophoresis	Dielectrophoresis techniques enable trapping of single cells and cell-laden hydrogels within 3D scaffolds	Albrecht et al. [66, 2007, p. 702]
	Plasmonic trapping	*S. cerevisiae* were arranged into arrays of defined architecture using plasmonic traps and microfluidic devices	Huang et al. [57, p. 6018]
	Bioreactors	Rotating wall vessel (RWV) bioreactors improve cell seeding density and uniformity within 3D scaffolds	Martin et al. [69, p. 80]
Direct cell placement in 2D	Microcontact printing	Microcontact printing of ECM modulates placement and cell–cell interactions by selective adhesion of cells to defined substrates	Ruiz et al. (93, 2008, p. 2921)
	Laser-guided direct writing	Optical forces directly placed chick neuronal cells onto glass surfaces in various 2D patterns with minimal loss in cellular viability	Odde et al. [58, p. 312]
	Inkjet printing	Chinese Hamster Ovary (CHO) Cells were specifically placed onto biomaterial substrates in predefined patterns via inkjet printing technologies	Xu et al. (94, 2005, p. 0210131)
Extrinsic factors			
Growth factor, culture medium, and inorganic ion delivery	Inkjet printing	Muscle-derived stem cells (MDSCs) cultured on patterns of BMP-2 printed onto fibrin substrates. Cells cultured on BMP-2-patterned substrates in myogenic conditions differentiate into osteoblasts, while unpatterned cells differentiate into myoblasts	Phillippi et al. [68, p. 127]
	Microfluidic bioreactors	Oxygen gradients of differing size and shape were created in specially designed microfluidics, where fluid flow was controlled by a computer-controlled gas mixer	Allen et al. (95, 2010) Adler et al. (96, 2010, p. 388)
		Pressure-driven laminar flow quickly switches solution streams presented to cell, enabling rapid microenvironmental changes and growth factor delivery	

TABLE 4.1 (Continued)
Strategies to Engineer Various Components of the Stem Cell Niche

Niche Component	Engineering Strategies	Examples	References
	Bioreactors	Mass transport of growth factors, ions, and oxygen is increased in several types of bioreactors, leading to increased cellular proliferation, matrix deposition, and differentiation	Martin et al. [69, p. 80]
Shear stresses	Bioreactors and microfluidic bioreactors	Shear stresses are modulated through changes in design, fluid flow, and fluid velocity throughout microfluidics and bioreactors	Martin et al. [69, p. 80]

4.3.1 Biomaterial Scaffolds

Biomaterial scaffolds serve as the foundation for many tissue engineering and microfabrication technologies. Made of natural or synthetic materials, such as alginate, collagen, poly(ethylene glycol) diacrylate (PEGDA), and polyesters, these materials form biocompatible networks that provide structural support to the cells, allow rapid diffusion of nutrients, metabolites, and small molecules to and away from encapsulated cells and are resistant to protein absorption. Many biomaterial scaffolds can be modified to include ECM molecules, vary mechanical stiffness, and tune degradation properties. Properties of biomaterial scaffolds vary based on application and have been shown to enhance osteogenic, neural, and adipose differentiation. When selecting biomaterial scaffolds for cell seeding or microfabrication technologies, cell type, fluid dynamics within the scaffold, material stiffness and surface chemistries, method of polymerization, delivery of bioactive molecules, and matrix degradation properties need to be taken into consideration. For extensive review of the properties and types of biomaterial scaffolds, see Chapter 10 of this book or a review by Dawson et al.[28] Several of the microfabrication technologies discussed below utilizes the tunable properties of biomaterial scaffolds to precisely engineer the cellular microenvironment.

4.3.2 Microfabrication Technologies

4.3.2.1 Photolithography

One of the first techniques used to pattern cells and substrates, photolithography utilizes materials that harden or soften in response to light irradiation. A schematic of the photolithography process is presented in Figure 4.3. In most cases, photolithographic micropatterns are created by spin coating glass, silicon, or quartz with a thin layer of liquid prepolymer solution termed photoresist. The spun photoresist is patterned by exposing and hardening the photoresist to light irradiation through

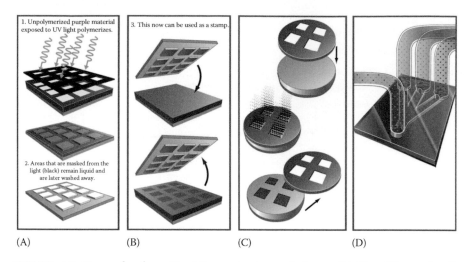

1. Unpolymerized purple material exposed to UV light polymerizes.

2. Areas that are masked from the light (black) remain liquid and are later washed away.

3. This now can be used as a stamp.

(A)   (B)   (C)   (D)

FIGURE 4.3 (See color insert.) Micropatterning techniques. (A) Photolithography. To create patterns using photolithography, a thin film of photopolymerizable material such as photoresist or poly (ethylene glycol) diacrylate hydrogel (PEGDA) is spun onto a glass substrate. The material is photopolymerized through a mask of the desired pattern. The material only hardens where exposed to light, and unexposed material is washed away. The resulting material can be used to directly culture cells, as a master to mold and shape 3D scaffolds, as a stamp for soft lithography (B) as a stencil (C), or as a microfluidic device (D). (B) Soft lithography. A master is formed in the desired pattern using photolithographic techniques and then filled with material such as PDMS, a soft elastomeric material that is commonly used for soft lithographic techniques. This stamp can be dipped directly into ECM molecules such as fibronectin, or functionalized with alkanethiols, and stamped onto substrates. (C) Stencil. The master created by photolithographic techniques can be used as a stencil to limit exposure of the substrate to molecules. (D) Microfluidic devices. Generally, microfluidics devices are formed in the same manner as stamps for soft lithography but have continuous channels to allow for fluid flow. These microfluidic devices can be used as guides to deposit ECM or cells, and then are peeled off the substrate. Additionally, they are used as bioreactors to control fluid flow, soluble molecule presentation, and cell deposition.

an opaque photomask of the desired pattern. Areas unexposed to light irradiation remain liquid and are removed. The resulting photoresist "master" is filled with the material of interest (e.g., ECM proteins), and the photoresist is removed from the substrate.[29] Cells are then patterned by selective adhesion to ECM proteins stamped on the culture substrate. Cell adhesion proteins utilized in photolithographic techniques include fibronectin, collagen, laminins, and Matrigel.[30]

Photolithography is able to accurately pattern substrates with a resolution from 2 to 500 μm.[30] Photomasks of the desired pattern are cheaply and easily fabricated with freely available computer software and printed onto transparencies, microfiche films, or quartz/chromium surfaces, depending on the resolution desired.[31] However, the photolithography process requires expensive clean room facilities, and many of the solvents used to process the photoresist can easily denature biological molecules and are toxic to living cells. Once created, the photoresist can be reused for several experiments, but changing patterns or design requires fabrication of new photoresist masters that can be cost limiting.

4.3.2.2 Soft Lithography

As a cheaper and more biocompatible alternative to photolithography, Whitesides and colleagues developed a patterning process termed soft lithography.[32] This method is termed soft lithography since it uses "soft" elastomeric materials such as poly (dimethylsiloxane) or PDMS. PDMS is a durable, biocompatible elastomer that is permeable to gases, optically transparent, and permissive for culturing cells. Once a PDMS master is formed, it can be used over an extended period of time with little degradation. In soft lithography techniques, a silicon master of the desired pattern is generated, usually by photolithographic techniques. PDMS is cast onto the silicon master and hardened. The resulting PDMS mold can then be used to directly culture cells or as a template to form microchannels filled with material or cells of interest.[33] Soft lithography techniques are generally able to pattern structures that are 500 nm or larger. Odom et al. improved the resolution of soft lithography patterns into the 50–100 nm range by using composite PDMS stamps.[34]

One of the most widely used soft lithography techniques, microcontact printing, utilizes PDMS molds created by soft lithography to stamp patterns onto tissue culture substrates. The simplest studies simply absorb ECM molecules onto the PDMS stamp and transfer them onto substrates. While this type of microcontact printing has been successful in patterning poly-L lysine,[35] laminin,[35] immunoglobulins,[36,37] and even lipid bilayers,[37] it may not be suitable for long-term biological studies due to the loose linkage of the stamped material and the substrate. To form stronger bonds between the protein and substrate, self-assembled monolayers (SAMs) of alkalinethiols are deposited onto gold surfaces. While non-patterned areas are rendered protein resistant, ECM that is added only absorbs to the SAMs. In most cases, the resolution of this technique is $\sim$100 μm. However, using stamps made of polyolefin plastomers, Csucs et al. were able to stamp fibrinogen proteins using microcontact printing with nanometer-scale resolution.[38] Microcontact printing is cost effective and flexible, allowing various substrates and printing material. However, ligand density can vary from experiment to experiment since transfer efficiency of the stamp can vary. Furthermore, physioabsorbed ECM proteins can degrade from the substrate when in contact with the culture medium.

The elastomeric properties of PDMS stamps provide a unique ability to quickly and reversibly seal surfaces to form microfluidic devices. Microchannels formed by the bonding of PDMS to glass substrates can be used to selectively deliver ECM or cells onto the tissue culture substrate through capillary action. In addition, etching of culture surfaces can be achieved by flowing etching solutions through the microfluidic channels to form grooves that guide cell placement. In recent studies, biomaterial scaffolds have been patterned by flowing prepolymer solution into the microchannels, then polymerized to form 3D structures. Once patterned, the microfluidic device can be used as a culture vessel or is easily removed from the substrate. Drawbacks of microfluidic technology include limited spacing between microchannels—too little spacing between channels can compromise the structural integrity of the PDMS stamp. Due to the requirement of fluid flow, patterns with continuous features can only be implemented. To overcome this problem, Khademhosseni et al. used microfluidics in combination with patterned substrates to trap cells in specific location within the microchannel. They successfully used this technique to pattern embryonic stem cells and MEF feeder cells.[39]

4.3.2.3 Optical Fabrication

Developed by Ashkin et al. in 1987, optical trapping techniques utilize tightly focused laser beams to manipulate dielectric particles.[40] Optical tweezers allow precise manipulation and placement of objects in both 2D and 3D. They have found many uses in biological applications, such as measuring molecular forces,[41] manipulating DNA,[42] cellular organelles,[43] viruses,[44] bacterium,[44] and more recently, mammalian cells with minimal damage to cell viability.[45]

A typical set up of optical tweezers consists of Nd:YAG or Ti:Sapphire lasers, beam steering optics, and an inverted microscope with high numerical aperture (NA) objectives. To form optical traps, a tightly focused laser beam is directed through a high NA lens onto a dielectric particle. Photon from the point of focus from the laser beam creates an electrical field that attracts dielectrical particles and traps the object near the focal point of the laser. Trapping forces depend on the size and shape of the particle in question as well as properties of the surrounding medium.[46–48] Average trapping forces are ~1 nN, which are sufficient for manipulating most bacterium and mammalian cells. Often, the manipulation of biological molecules requires the use of one or more optical traps. Acoustic optical deflectors (AODs) allows for time sharing of the laser beam between different positions in a planar field. The laser beam dwells at a position for a predetermined period of time before moving to the next position. As long as the "dark time" is faster than the Brownian motion and diffusion of the cell, the beam is able to fix the position of the object and is as effective as a continuous beam.[49]

In addition to directly manipulating cells, optical tweezers have been used to study the effects of mechanical forces on cells. Microbeads are attached to the cell surface through ligands and act as handles for the optical tweezers. Displacement of the microbeads on the cell membrane by optical tweezers generates stretching or bending forces. These techniques can recreate physiological forces from stretching, compression, and ECM stiffness. Wang et al. used this technique in combination with fluorescent resonance energy transfer (FRET) to study the mechanoactivation of Src, an important signal transduction molecule playing an important role in self-renewal and differentiation of stem cells, as well as in many cancers.[50]

Minimizing damage to cells during manipulation is a major factor in optical trapping design and optimization. Cellular viability is dependent on laser wavelength, laser power, and duration of exposure to the traps. Biological specimens typically absorb light in the near infrared range, and thus most lasers are often tuned in the 800–1000 nm range.[51] Vorobjev et al. reported faulty mitosis and abnormal chromosome bridges in PtK2 cells exposed to continuous wave optical traps in the 760–765 nm range and minimal damage to 700 and 800–820 nm light.[51] Similarly, Liang et al. demonstrated wavelength dependence on the growth of Chinese hamster ovary (CHO) cells. When exposed to 740–760 and 900 nm light, CHO cells had poor growth and cell division characteristics compared to nonirradiated controls. Light in the 950–990 nm wavelengths resulted in the highest clonablilty of all wavelengths tested. In all cases, shorter exposure and lower power of traps result in increased cellular function.[52] Time-shared optical traps, as described above, reduce the exposure of biological samples to laser light as compared to continuous wave (CW) technology. Mirisaidov et al. discovered that, under the same wavelength and duration of trapping, *E. coli* bacterium displayed higher viability with time-shared optical traps than with CW traps.[53]

Optical trapping provides stringent control of the cell and its placement within an environment with a resolution of ~19 nm. To permanently fix cells in position for long-term studies, Jordan et al. and Akselrod et al. entrapped cells into 3D scaffolds with minimal effect on cellular viability.[54,55] Optical tweezers, however, require some knowledge of optical technologies for set up and use. As of yet, most cellular studies utilizing optical trapping are relatively low throughput and monitoring only 10–100 cells in a single experiment.

In recent years, a new field of optical trapping termed plasmonic trapping has emerged. When light is applied to metal nanoparticles, photons excite the electrons in the nanoparticles that form energy waves and strong electromagnetic fields. Plasmonic trapping was first coupled with optical tweezers for nanotechnology applications as a method to enhance optical gradient forces from optical tweezers, and therefore reducing the Brownian motion of nanoparticles in traps. Combining optical tweezers and plasmonic traps, the power required to manipulate and trap biological objects can be greatly reduced.[56] More recently, Huang et al. designed a plasmonic trapping device in a microfluidic system for lab-on-a-chip applications. Whereas cell viability was not explored, the team was able to successfully trap single nanoparticles and *S. cerevisiae* cells in plasmonic traps without the complex optical setup required for optical tweezers.[57] Still in its infancy, plasmonic trapping holds great potential as a new cell-patterning technique or to augment biological optical trapping setups.

Another form of optical trapping technologies to pattern cells is laser-guided direct writing (LGDW). Utilizing the same principles as optical trapping, a weakly focused laser beam is used to trap and direct cells down hollow fibers onto cell culture surfaces. This method provides single-cell manipulation with ~1 μm scale resolution.[58] Nahmias et al. have used LGDW to create vascular and sinusoid-like structures onto collagen scaffolds.[59] It is unknown what effects laser power has on cell viability. Unless the substrate is patterned with adhesion molecules, the cells will randomly spread on the substrate after patterning.

4.3.2.4 Dielectrophoresis

Dielectrophoresis (DEP) has emerged as a promising technique to identify and place cells and microparticles through their electrical properties, size, and shape of the entrapped specimen. When presented to a nonuniform electric field, all objects exert some dielectrical forces that can change the motion of the particle. The strength of the force and movement depends on the size, shape, and electrical properties of the object and the surrounding medium.[60,61] DEP technology has mainly been utilized in cell-sorting applications, as no modification or manipulation is required prior to sorting. Recently, DEP has undergone resurgence for micromanipulation and patterning of DNA, viruses, proteins, and cell applications.[62,63]

To pattern cells using DEP, electrodes are microfabricated into a microfluidic chip or other culture devices. Thousands of individual electrodes can be placed on a centimeter of surface area using common microfabrication techniques. Cells are introduced into the system and pulled toward the electrode surface through DEP forces. Fluid flow across the surface removes unpatterned cells. After trapping, cells can then either be encapsulated in 3D scaffolds or adherent cell lines can be cultured on the surface. As with optical trapping technology, the duration and intensity of electrical stimulation can affect biological activity of living cells. Grey et al. demonstrated DEP patterning of mammalian cells using bovine aortic endothelial cells.[64] The group patterned a $1 \times 1\,cm^2$ array within 5 min with minimal affects on cell viability. Suzuki et al. further modified the procedures by exposing C2C12 cells to DEP forces for 5 min to allow cell adherence, flushed the device, and electropatterned again with a second cell type.[65] Albrect et al. successfully patterned cell-laden alginate beads with DEP technologies.[66]

With DEP technology, the precise location of cells and microparticles can be patterned onto various substrates. The technology is rapid and easily scaled for larger experiments. However, there is little control over the exact cells that are patterned, and coculture experiments so far have only been established by engaging one set of electrodes, removing the cells, then flowing in the next cell type. Exposure to high power traps must be limited as they may result in cell death or local heating of the medium. In DEP applications, the cells must be suspended in low conductivity medium, as physiological medium has high electrical conductivity and will not allow DEP to occur. This medium may be toxic to living cells, so exposure to the medium must be limited. Alternatively, negative DEP occurs when the object is less polarizable than the surrounding medium, allowing patterning in physiological medium.[67]

4.3.2.5 Inkjet Printing

Another microfabrication technique that adapted technologies from an electronics industry is inkjet printing of biomaterials, scaffolds, and cells. Commercially available inkjet printers reproduce electronic images by depositing nanoliter-sized drops of ink onto the paper substrate. Inkjet patterning technologies utilize these same commercial inkjet printers and ink cartridges to deposit small drops of "ink" (i.e., proteins, alkanethiols, scaffold materials, and more recently, cells) onto "paper" (i.e., tissue culture substrate) into desired configurations. The resolution of inkjet printing is approximately 350 μm. While the resolution of inkjet printing is significantly lower than other microfabrication techniques, the configuration of deposited patterns is easily changed without the costs and time constraints of fabricating new masters[68] (Figure 4.4).

4.4 CULTURE HANDLING SYSTEMS

4.4.1 BIOREACTORS

Expansion of progenitor cells in traditional static cultures leads to a loss of proliferation and differentiation potential of stem and progenitor cells, therefore severely limiting the number of cells available for tissue engineering and stem cell therapies.

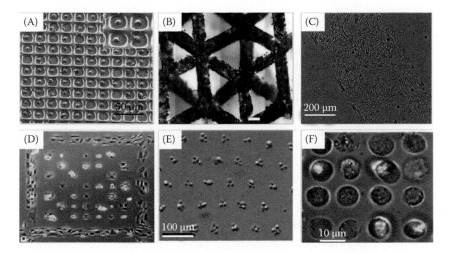

FIGURE 4.4 Microfabrication techniques to engineer the cellular microenvironment. (A) and (B) 2D and 3D photolithography. (A) Mouse 3T3 fibroblast cells seeded into poly (ethylene glycol) (PEG) wells created by photolithography. The PEG wells guided and patterned cellular adhesion to the glass surface. (Reprinted with permission from Revzin, A. et al., Surface engineering with poly (ethylene glycol) photolithography to create high-density cell arrays on glass, *Langmuir*, 19(23), 9855–9862. Copyright 2003 American Chemical Society.) (B) Hepatocytes were patterned into three dimensions by additive photolithography of photopolymerizable poly (ethylene glycol) diacrylate (PEGDA) hydrogels. (From Tsang, L. et al., *FASEB J.*, 21, 798, 2007. With permission.) (C) Inkjet Printing. Defined patterns of collagen were deposited onto cell-repellant agarose surfaces by inkjet printing. Smooth muscle cells (SMCs) were seeded onto the surface and adhered only to patterned collagen surfaces. (Reprinted from *Biomaterials*, 25(17), Roth, E.A. et al., Inkjet printing for high-throughput cell patterning, 3707–3715. Copyright 2004. With permission from Elsevier.) (D) Soft lithography/microcontact printing. Hexadecanethiolate and tri (ethylene glycol) were printed onto gold surfaces by microcontact printing. The ECM molecule fibronectin absorbs to the hexadecanethiolate but not onto tri (ethylene glycol). Bovine capillary endothelial cells (BCE) were patterned by selective adhesion to the fibronectin coated areas. (Reprinted from *Biomaterials*, 20(23–24), Kane, R.S., Takayama, S., Ostuni, E., Ingber, D.E., and Whitesides, G.M., Patterning proteins and cells using soft lithography, 2363–2376. Copyright 1999. With permission from Elsevier.) (E) Dielectrophoresis. Arrays of fibroblasts patterned through dielectrophoresis methodology and encapsulated in PEGDA hydrogels. (From Albrecht, D.R. et al., *Lab Chip— Roy. Soc. Chem.*, 5, 111–118, 2005. With permission.) (F) Optical Tweezers. Human monocytic U937 cells manipulated into a 4×4 array by optical tweezers and encapsulated in PEGDA hydrogels. (Trump, unpublished data.)

Once seeded into culture, cells require a proper balance of nutrients, oxygen, soluble molecules, and waste removal that is typically provided *in vivo* by the vasculature system. The static culture of cells in either 2D or 3D results in gradients of nutrients and small molecules. Further limiting the application of 3D scaffolds is that nutrients and oxygen can only penetrate the scaffold for a few hundred microns, leaving large constructs with a hypoxic and necrotic center surrounded by viable cells.[69] Static culture conditions so do not recapitulate the laminar flow and shear stress features found in physiological tissue environments. These limitations severely inhibit the continued expansion of cells, the size of scaffolds available for use, and the differentiation capabilities of seeded stem cells. Another limitation to the clinical usage of 3D scaffolds and stem cells is the isolation, expansion, and differentiation of stem cells in sufficient quantities to seed into scaffolds or for cell based therapies. Bulk cultures of cells and scaffolds also result in local microenvironmental changes, which form concentration gradients that can affect cell behavior.

Bioreactors are cell culture vessels designed to provide strict control over culture conditions such as temperature, pH, oxygen levels, and for the perfusion of medium in large cultures of cells and 3D constructs. Many different types of bioreactors have been used for the culture and expansion of stem cells and 3D scaffolds, such as stirred flask bioreactors, rotating wall vessels, perfusion chambers, and microfluidic bioreactors. These bioreactors offer a distinct advantage over traditional cell culture as they provide automation, the ability to control and change culture parameters, and offer a more homogeneous environment for cell culture. Parameters can easily be changed from one experiment to another and are highly dependent on the objective of the experiment (i.e., expansion or differentiation). Continuous mixing of oxygen and nutrients in bioreactors reduces concentration gradients and increases nutrient diffusion throughout cellular colonies and constructs.[70]

Initial studies of stem cells in bioreactors reported the increased expansion and long-term maintenance of HSC over static culture systems. Murine HSCs cultured in stirred flask bioreactors showed a fivefold increase of the stemness marker Sca1+ and a fourfold increase in long-term culture initiating cells (LTC-IC) over 21 days of culture. Expansion of murine ESCs increased without the need for feeder layers or the loss of differentiation potential.[71] Bioreactors have also enabled the expansion of embryoid body culture to be scaled up. Cameron et al. reported increased expansion, more uniform morphology, and maintenance of differentiation potential of embryoid bodies cultured in stirred vessel bioreactors.[72]

Bioreactors have also been extensively used to seed progenitor cells onto 3D scaffolds. Important for the development of functional tissue engineering constructs, cell seeding of scaffolds remains a highly variable process. Bioreactor-based seeding methodologies have resulted in increased seeding densities and efficiencies and more uniform distributions of cells within the scaffold.[73] Seeding efficiency depends on cell type and density, scaffold type, and flow rates of culture medium. Both murine and human MSCs have been efficiently seeded onto a variety of scaffolds using stirred flask and perfusion bioreactors. In general, most studies have reported increased seeding efficiency, density, cell penetration, and overall more uniform distributions of cells throughout the scaffold using spinner and perfusion bioreactors. MSCs seeded and cultured onto 3D scaffolds increased expansion while maintaining differentiation capacity as compared to cells cultured in traditional culture vessels.[74]

High flow rates needed for efficient seeding and nutrient diffusion, however, can greatly affect cellular processes. Structural integrity of seeded scaffolds can be compromised by fluid channel formation at high flow rates. Even at low velocities, shear stresses imparted onto cells can be significant.[75] Often, the fluid flow required for efficient cell seeding and nutrient mixing is significantly higher than what is found in physiological tissues (100 and 0.1–10 μL/s, respectively).[70] Shear tolerance of cells and scaffolds depends on the cell type, scaffold used, and experimental parameters. Mechanical loading imparted by shear can affect cellular differentiation. Higher fluid rates are conductive to osteogenic differentiation of MSCs, while lower rates facilitated expansion. Fluid rates also affect ECM distribution. Zhao et al. reported decreased collagen and laminin I deposition in cell-seeded scaffold.[70] Differentiation into osteoblasts was promoted at high flow rates. Similarly, bovine chondrocytes seeded on PGA scaffolds had increased GAG formation and synthesis, while net GAG accumulation throughout the scaffolds was reduced, most likely by release of GAG into culture medium in stirred flask bioreactors.[76] To protect cells from shear stressors in bulk culture systems, cells have been encapsulated in alginate microbeads and cultured in bioreactors.

4.4.2 MICROFLUIDICS

In vivo cellular microenvironments are highly dynamic, and soluble factor concentrations can vary drastically on a scale of microns. In traditional cultures, soluble factors and medium exchange require the physical removal of culture medium and bolus delivery of soluble factors, resulting in a homogeneous mixture that is difficult to control in real time on a microscale level. Microfluidic bioreactors, on the other hand, allow soluble factor delivery and replacement in a matter of seconds, allowing real-time control over the cellular microenvironment. Fluid flow rates, pressures, and soluble factor concentration and delivery are easily manipulated using these devices.[77] Microfluidic bioreactors are formed by creating a master of desired pattern with soft lithography techniques and stamps are most commonly made of PDMS. The PDMS stamps contain one or more channel systems to direct nutrient, oxygen, and soluble factor flow on a microscale level. Fluid flow is typically controlled by syringe pumps that allow for rapid and pulsatile delivery of stimuli and can also maintain cultures for weeks at a time. Microfluidic devices designed with two or more channels permit the controlled mixing of soluble factors. Laminar flow in microchannels allows one or more streams of fluid to combine with limited mixing of streams. Thus, a single cell can be exposed to multiple microenvironments by placement at an intersection of streams carrying different soluble molecules.[78]

4.5 READOUT SYSTEMS

Once cells are exposed to various microenvironments, the activity of the cells in response to its stimuli can be assessed. There are several well-established techniques to monitor cellular activity. On a large scale, dynamic responses of the culture in whole can be measured by degree of expansion and apoptosis, morphological changes, migration, and differentiation. Stem cell differentiation has been readily identified through histochemistries such as alkaline phosphatase/von Kossa staining, oil red O, and safranin staining to elucidate bone, fat, and cartilage differentiation of MSCs.[26] Molecular characteristics can be measured by western blotting, RT-PCR, ELISA, and other well-established techniques. However, such measurements are end-point assays that require fixation or cell lysis. Since these methods measure the population as a whole, the signature of stem cells may be lost due to contaminating cell types. The discovery of green fluorescent protein (GFP) and its variants has enabled single and live cell imaging to visualize cellular processes. GFP can be expressed as a fusion tag to proteins of interest to explore protein–protein interactions and gene activation and expression.[79,80] Specific organelles can be targeted, such as the staining of the actin cytoskeleton. Other assays have been developed using variants of GFP, such as FRET and fluorescent resonance after photobleaching (FRAP), which uses fluorescent protein pairs to monitor cellular interactions in real time.[81] Recently, to improve dynamic sensing of cellular activity, a mutated form of GFP has been developed. This GFP mutant shifts color spectrum as the protein matures. When the protein is first synthesized, the cell is green in color, but shifts to red fluorescence in a time dependent manner. Thus, readouts of gene expression dynamics and protein synthesis can be monitored in real time by the ratio of green to red fluorescence.[82]

4.6 CONCLUSIONS AND FUTURE DIRECTIONS

Tissue engineering and microfabrication technologies have arisen as invaluable tools to probe the spatial and temporal cues that govern stem cell fate. The ability to precisely pattern cells and external signals in 2D and 3D enables investigations into the roles of niche components on stem cell plasticity and differentiation. These approaches require (1) biocompatible scaffolds with defined mechanical properties, (2) microfabrication of scaffolds and signals into specific geometries, (3) controlled seeding and placement of cells and signals, (4) effective culture systems for nutrient delivery, and (5) readout systems to monitor cellular activity during culture. While extensive research has been conducted in establishing these technologies, the generation of functional stem cell niches and culture systems will require a multidisciplinary approach that combines and applies these technologies into functional stem cell environments. Combination of engineering approaches with traditional cell biology approaches will facilitate recapitulation of functional stem cell microenvironments and advance our knowledge of functional stem cell niches.

ACKNOWLEDGMENTS

L. Trump was funded by the Illinois Regenerative Medicine Institute (IRMI) grant no. IDPH 2006-05481 and the Initiative for Future Agriculture and Food Systems grant no. 2001-52100-11527 from the UDSA Cooperative State Research, Education, and Extension Service. We also thank Janet Sinn-Hanlon from the Beckman Imaging Technology Group and Visualization Laboratory for her assistance in creating figure art.

REFERENCES

1. Watt, F. M.; Hogan, B. L. M., Out of eden: Stem cells and their niches. *Science* 2000, *287* (5457), 1427–1430.
2. Schofield, R., The relationship between the spleen colony-forming cell and the haemopoietic stem cell. A hypothesis. *Blood Cells* 1978, *4* (1–2), 7–25.
3. Birgersdotter, A.; Sandberg, R.; Ernberg, I., Gene expression perturbation in vitro—A growing case for three-dimensional (3D) culture systems. *Seminars in Cancer Biology* 2005, *15* (5 SPEC. ISS.), 405–412.

4. Doane, K. J.; Birk, D. E., Fibroblasts retain their tissue phenotype when grown in three-dimensional collagen gels. *Experimental Cell Research* 1991, *195* (2), 432–442.
5. Kale, S.; Biermann, S.; Edwards, C.; Tarnowski, C.; Morris, M.; Long, M. W., Three-dimensional cellular development is essential for ex vivo formation of human bone. *Nature Biotechnology* 2000, *18* (9), 954–958.
6. Khademhosseini, A.; Langer, R.; Borenstein, J.; Vacanti, J. P., Microscale technologies for tissue engineering and biology. *Proceedings of the National Academy of Sciences of the United States of America* 2006, *103* (8), 2480–2487.
7. Shi, S.; Gronthos, S., Perivascular niche of postnatal mesenchymal stem cells in human bone marrow and dental pulp. *Journal of Bone and Mineral Research* 2003, *18* (4), 696–704.
8. Zhang, J.; Niu, C.; Ye, L.; Huang, H.; He, X.; Tong, W. G.; Ross, J.; Haug, J.; Johnson, T.; Feng, J. Q.; Harris, S.; Wiedemann, L. M.; Mishina, Y.; Li, L., Identification of the haematopoietic stem cell niche and control of the niche size. *Nature* 2003, *425* (6960), 836–841.
9. Tumbar, T.; Guasch, G.; Greco, V.; Blanpain, C.; Lowry, W. E.; Rendl, M.; Fuchs, R., Defining the epithelial stem cell niche in skin. *Science* 2004, *303* (5656), 359–363.
10. Ohyama, M.; Terunuma, A.; Tock, C. L.; Radonovich, M. F.; Pise-Masison, C. A.; Hopping, S. B.; Brady, J. N.; Udey, M. C.; Vogel, J. C., Characterization and isolation of stem cell-enriched human hair follicle bulge cells. *Journal of Clinical Investigation* 2006, *116* (1), 249–260.
11. Marshman, E.; Booth, C.; Potten, C. S., The intestinal epithelial stem cell. *BioEssays* 2002, *24* (1), 91–98.
12. Palmer, T. D.; Willhoite, A. R.; Gage, F. H., Vascular niche for adult hippocampal neurogenesis. *Journal of Comparative Neurology* 2000, *425* (4), 479–494.
13. Collins, C. A.; Olsen, I.; Zammit, P. S.; Heslop, L.; Petrie, A.; Partridge, T. A.; Morgan, J. E., Stem cell function, self-renewal, and behavioral heterogeneity of cells from the adult muscle satellite cell niche. *Cell* 2005, *122* (2), 289–301.
14. Corre, J.; Mahtouk, K.; Attal, M.; Gadelorge, M.; Huynh, A.; Fleury-Cappellesso, S.; Danho, C.; Laharrague, P.; Klein, B.; Rème, T.; Bourin, P., Bone marrow mesenchymal stem cells are abnormal in multiple myeloma. *Leukemia* 2007, *21* (5), 1079–1088.
15. Conboy, I. M.; Conboy, M. J.; Wagers, A. J.; Girma, E. R.; Weismann, I. L.; Rando, T. A., Rejuvenation of aged progenitor cells by exposure to a young systemic environment. *Nature* 2005, *433* (7027), 760–764.
16. Frisch, S. M.; Francis, H., Disruption of epithelial cell-matrix interactions induces apoptosis. *Journal of Cell Biology* 1994, *124* (4), 619–626.
17. Nuttelman, C. R.; Tripodi, M. C.; Anseth, K. S., Synthetic hydrogel niches that promote hMSC viability. *Matrix Biology* 2005, *24* (3), 208–218.
18. Benoit, D. S. W.; Tripodi, M. C.; Blanchette, J. O.; Langer, S. J.; Leinwand, L. A.; Anseth, K. S., Integrin-linked kinase production prevents anoikis in human mesenchymal stem cells. *Journal of Biomedical Materials Research Part A* 2007, *81* (2), 259–268.
19. Winer, J. P.; Janmey, P. A.; McCormick, M. E.; Funaki, M., Bone marrow-derived human mesenchymal stem cells become quiescent on soft substrates but remain responsive to chemical or mechanical stimuli. *Tissue Engineering Part A* 2009, *15* (1), 147–154.
20. Chen, C. S.; Mrksich, M.; Huang, S.; Whitesides, G. M.; Ingber, D. E., Geometric control of cell life and death. *Science* 1997, *276* (5317), 1425–1428.
21. McBeath, R.; Pirone, D. M.; Nelson, C. M.; Bhadriraju, K.; Chen, C. S., Cell shape, cytoskeletal tension, and RhoA regulate stem cell lineage commitment. *Developmental Cell* 2004, *6* (4), 483–495.
22. Chowdhury, F.; Na, S.; Li, D.; Poh, Y. C.; Tanaka, T. S.; Wang, F.; Wang, N., Material properties of the cell dictate stress-induced spreading and differentiation in embryonic stem cells. *Nature Materials* 2010, *9* (1), 82–88.

23. Sahni, A.; Sporn, L. A.; Francis, C. W., Potentiation of endothelial cell proliferation by fibrin(ogen)-bound fibroblast growth factor-2. *Journal of Biological Chemistry* 1999, *274* (21), 14936–14941.

24. Fan, V. H.; Tamama, K.; Au, A.; Littrell, R.; Richardson, L. B.; Wright, J. W.; Wells, A.; Griffith, L. G., Tethered epidermal growth factor provides a survival advantage to mesenchymal stem cells. *Stem Cells* 2007, *25* (5), 1241–1251.

25. Adams, G. B.; Chabner, K. T.; Alley, I. R.; Olson, D. P.; Szczepiorkowski, Z. M.; Poznansky, M. C.; Kos, C. H.; Pollak, M. R.; Brown, E. M.; Scadden, D. T., Stem cell engraftment at the endosteal niche is specified by the calcium-sensing receptor. *Nature* 2006, *439* (7076), 599–603.

26. Pittenger, M. F.; Mackay, A. M.; Beck, S. C.; Jaiswal, R. K.; Douglas, R.; Mosca, J. D.; Moorman, M. A.; Simonetti, D. W.; Craig, S.; Marshak, D. R., Multilineage potential of adult human mesenchymal stem cells. *Science* 1999, *284* (5411), 143–147.

27. Shiozawa, Y.; Takenouchi, H.; Taguchi, T.; Saito, M.; Katagiri, Y. U.; Okita, H.; Shimizu, T.; Yamashiro, Y.; Fujimoto, J.; Kiyokawa, N., Human osteoblasts support hematopoietic cell development in vitro. *Acta Haematologica* 2009, *120* (3), 134–145.

28. Dawson, E.; Mapili, G.; Erickson, K.; Taqvi, S.; Roy, K., Biomaterials for stem cell differentiation. *Advanced Drug Delivery Reviews* 2008, *60* (2), 215–228.

29. Park, T. H.; Shuler, M. L., Integration of cell culture and microfabrication technology. *Biotechnology Progress* 2003, *19* (2), 243–253.

30. Lee, J. Y.; Shah, S. S.; Zimmer, C. C.; Liu, G. Y.; Revzin, A., Use of photolithography to encode cell adhesive domains into protein microarrays. *Langmuir* 2008, *24* (5), 2232–2239.

31. Whitesides, G. M.; Ostuni, E.; Takayama, S.; Jiang, X.; Ingber, D. E., Soft lithography in biology and biochemistry. *Annual Review of Biological Engineering* 2001, *3*, 335–373.

32. Xia, Y.; Whitesides, G. M., Soft lithography. *Annual Review of Materials Science* 1998, *28* (1), 153–184.

33. Kane, R. S.; Takayama, S.; Ostuni, E.; Ingber, D. E.; Whitesides, G. M., Patterning proteins and cells using soft lithography. *Biomaterials* 1999, *20* (23–24), 2363–2376.

34. Odom, T. W.; Love, J. C.; Wolfe, D. B.; Paul, K. E.; Whitesides, G. M., Improved pattern transfer in soft lithography using composite stamps. *Langmuir* 2002, *18* (13), 5314–5320.

35. James, C. D.; Davis, R. C.; Kam, L.; Craighead, H. G.; Isaacson, M.; Turner, J. N.; Shain, W., Patterned protein layers on solid substrates by thin stamp microcontact printing. *Langmuir* 1998, *14* (4), 741–744.

36. Bernard, A.; Delamarche, E.; Schmid, H.; Michel, B.; Bosshard, H. R.; Biebuyck, H., Printing patterns of proteins. *Langmuir* 1998, *14* (9), 2225–2229.

37. Groves, J. T.; Boxer, S. G., Micropattern formation in supported lipid membranes. *Accounts of Chemical Research* 2002, *35* (3), 149–157.

38. Csucs, G.; Künzler, T.; Feldman, K.; Robin, F.; Spencer, N. D., Microcontact printing of macromolecules with submicrometer resolution by means of polyolefin stamps. *Langmuir* 2003, *19* (15), 6104–6109.

39. Khademhosseini, A.; Ferreira, L.; Blumling III, J.; Yeh, J.; Karp, J. M.; Fukuda, J.; Langer, R., Co-culture of human embryonic stem cells with murine embryonic fibroblasts on microwell-patterned substrates. *Biomaterials* 2006, *27* (36), 5968–5977.

40. Ashkin, A.; Dziedzic, J. M.; Yamane, T., Optical trapping and manipulation of single cells using infrared laser beams. *Nature* 1987, *330* (6150), 769–771.

41. Visscher, K.; Schnltzer, M. J.; Block, S. M., Single kinesin molecules studied with a molecular force clamp. *Nature* 1999, *400* (6740), 184–189.

42. Wang, M. D.; Yin, H.; Landick, R.; Gelles, J.; Block, S. M., Stretching DNA with optical tweezers. *Biophysical Journal* 1997, *72* (3), 1335–1346.

43. Shelby, J. P.; Edgar, J. S.; Chiu, D. T., Monitoring cell survival after extraction of a single subcellular organelle using optical trapping and pulsed-nitrogen laser ablation. *Photochemistry and Photobiology* 2005, *81* (4), 994–1001.

44. Ashkin, A.; Dziedzic, J. M., Optical trapping and manipulation of viruses and bacteria. *Science* 1987, *235* (4795), 1517–1520.

45. Uchida, M.; Sato-Maeda, M.; Tashiro, H., Micromanipulation: Whole-cell manipulation by optical trapping. *Current Biology* 1995, *5* (4), 380–382.

46. Mazolli, A.; Maia Neto, P. A.; Nussenzveig, H. M., Theory of trapping forces in optical tweezers. *Proceedings of the Royal Society A Mathematical, Physical and Engineering Sciences* 2003, *459* (2040), 3021–3041.

47. Berns, M. W., Optical tweezers: Tethers, wavelengths, and heat. *Methods in Cell Biology* 2007, 82, 457–466.

48. Sun, B.; Roichman, Y.; Grier, D. G., Theory of holographic optical trapping. *Optics Express* 2008, *16* (20), 15765–15776.

49. Brouhard, G. J.; Schek III, H. T.; Hunt, A. J., Advanced optical tweezers for the study of cellular and molecular biomechanics. *IEEE Transactions on Biomedical Engineering* 2003, *50* (1), 121–125.

50. Wang, Y.; Lu, S., The application of FRET biosensors to visualize Src activation. In *Small Animal Whole-Body Optical Imaging Based on Genetically Engineered Probes*, The International Society for Optical Engineering: San Jose, CA, 2008.

51. Vorobjev, I. A.; Liang, H.; Wright, W. H.; Berns, M. W., Optical trapping for chromosome manipulation: A wavelength dependence of induced chromosome bridges. *Biophysical Journal* 1993, *64* (2), 533–538.

52. Liang, H.; Vu, K. T.; Krishnan, P.; Trang, T. C.; Shin, D.; Kimel, S.; Berns, M. W., Wavelength dependence of cell cloning efficiency after optical trapping. *Biophysical Journal* 1996, *70* (3), 1529–1533.

53. Mirsaidov, U.; Scrimgeour, J.; Timp, W.; Beck, K.; Mir, M.; Matsudaira, P.; Timp, G., Live cell lithography: Using optical tweezers to create synthetic tissue. *Lab on a Chip—Miniaturisation for Chemistry and Biology* 2008, *8* (12), 2174–2181.

54. Akselrod, G. M.; Timp, W.; Mirsaidov, U.; Zhao, Q.; Li, C.; Timp, R.; Timp, K.; Matsudaira, P.; Timp, G. L., Laser-guided assembly of heterotypic three-dimensional living cell microarrays. *Biophysical Journal* 2006, *91* (9), 3465–3473.

55. Jordan, P.; Leach, J.; Padgett, M.; Blackburn, P.; Isaacs, N.; Goksör, M.; Hanstorp, D.; Wright, A.; Girkin, J.; Cooper, J., Creating permanent 3D arrangements of isolated cells using holographic optical tweezers. *Lab on a Chip—Miniaturisation for Chemistry and Biology* 2005, *5* (11), 1224–1228.

56. Righini, M.; Zelenina, A. S.; Girard, C.; Quidant, R., Parallel and selective trapping in a patterned plasmonic landscape. *Nature Physics* 2007, *3* (7), 477–480.

57. Huang, L.; Maerkl, S. J.; Martin, O. J. F., Integration of plasmonic trapping in a microfluidic environment. *Optics Express* 2009, *17* (8), 6018–6024.

58. Odde, D. J.; Renn, M. J., Laser-guided direct writing of living cells. *Biotechnology and Bioengineering* 2000, *67* (3), 312–318.

59. Nahmias, Y.; Schwartz, R. E.; Verfaillie, C. M.; Odde, D. J., Laser-guided direct writing for three-dimensional tissue engineering. *Biotechnology and Bioengineering* 2005, *92* (2), 129–136.

60. Chiou, P. Y.; Ohta, A. T.; Wu, M. C., Massively parallel manipulation of single cells and microparticles using optical images. *Nature* 2005, *436* (7049), 370–372.

61. Morgan, H.; Hughes, M. P.; Green, N. G., Separation of submicron bioparticles by dielectrophoresis. *Biophysical Journal* 1999, *77* (1), 516–525.

62. Chou, C.-F.; Tegenfeldt, J. O.; Bakajin, O.; Chan, S. S.; Cox, E. C.; Darnton, N.; Duke, T.; Austin, R. H., Electrodeless dielectrophoresis of single- and double-stranded DNA. *Biophysical Journal* 2002, *83* (4), 2170–2179.

63. Pethig, R., Dielectrophoresis: Using inhomogeneous AC electrical fields to separate and manipulate cells. *Critical Reviews in Biotechnology* 1996, *16* (4), 331–348.
64. Gray, D. S.; Tan, J. L.; Voldman, J.; Chen, C. S., Dielectrophoretic registration of living cells to a microelectrode array. *Biosensors and Bioelectronics* 2004, *19* (7), 771–780.
65. Suzuki, M.; Yasukawa, T.; Shiku, H.; Matsue, T., Negative dielectrophoretic patterning with different cell types. *Biosensors and Bioelectronics* 2008, *24* (4), 1043–1047.
66. Albrecht, D. R.; Underhill, G. H.; Mendelson, A.; Bhatia, S. N., Multiphase electropatterning of cells and biomaterials. *Lab chip* 2007, *7*, 702–709.
67. Thomas, R. S.; Morgan, H.; Green, N. G., Negative DEP traps for single cell immobilisation. *Lab on a Chip — Miniaturisation for Chemistry and Biology* 2009, *9* (11), 1534–1540.
68. Phillippi, J. A.; Miller, E.; Weiss, L.; Huard, J.; Waggoner, A.; Campbell, P., Microenvironments engineered by inkjet bioprinting spatially direct adult stem cells toward muscle- and bone-like subpopulations. *Stem Cells* 2008, *26* (1), 127–134.
69. Martin, I.; Wendt, D.; Heberer, M., The role of bioreactors in tissue engineering. *Trends in Biotechnology* 2004, *22* (2), 80–86.
70. Zhao, F.; Grayson, W. L.; Ma, T.; Irsigler, A., Perfusion affects the tissue developmental patterns of human mesenchymal stem cells in 3D scaffolds. *Journal of Cellular Physiology* 2009, *219* (2), 421–429.
71. Zandstra, P. W.; Eaves, C. J.; Piret, J. M., Expansion of hematopoietic progenitor cell populations in stirred suspension bioreactors of normal human bone marrow cells. *Biotechnology* 1994, *12* (9), 909–914.
72. Cameron, C. M.; Hu, W. S.; Kaufman, D. S., Improved development of human embryonic stem cell-derived embryoid bodies by stirred vessel cultivation. *Biotechnology and Bioengineering* 2006, *94* (5), 938–948.
73. Wendt, D.; Marsano, A.; Jakob, M.; Heberer, M.; Martin, I., Oscillating perfusion of cell suspensions through three-dimensional scaffolds enhances cell seeding efficiency and uniformity. *Biotechnology and Bioengineering* 2003, *84* (2), 205–214.
74. Braccini, A.; Wendt, D.; Jaquiery, C.; Jakob, M.; Heberer, M.; Kenins, L.; Wodnar-Filipowicz, A.; Quarto, R.; Martin, I., Three-dimensional perfusion culture of human bone marrow cells and generation of osteoinductive grafts. *Stem Cells* 2005, *23* (8), 1066–1072.
75. Tada, S.; Tarbell, J. M., Interstitial flow through the internal elastic lamina affects shear stress on arterial smooth muscle cells. *American Journal of Physiology—Heart and Circulatory Physiology* 2000, *278* (5), H1589–H1597.
76. Martin, I.; Obradovic, B.; Freed, L. E.; Vunjak-Novakovic, G., Method for quantitative analysis of glycosaminoglycan distribution in cultured natural and engineered cartilage. *Annals of Biomedical Engineering* 1999, *27* (5), 656–662.
77. Yarmush, M. L.; King, K. R., Living-cell microarrays. *Annual Review of Biological Engineering* 2009, 11, 235–257.
78. Eriksson, E.; Sott, K.; Lundqvist, F.; Sveningsson, M.; Scrimgeour, J.; Hanstorp, D.; Goksör, M.; Granéli, A., A microfluidic device for reversible environmental changes around single cells using optical tweezers for cell selection and positioning. *Lab on a Chip—Miniaturisation for Chemistry and Biology* 2010, *10* (5), 617–625.
79. Chalfie, M.; Tu, Y.; Euskirchen, G.; Ward, W. W.; Prasher, D. C., Green fluorescent protein as a marker for gene expression. *Science* 1994, *263* (5148), 802–805.
80. van Roessel, P.; Brand, A. H., Imaging into the future: Visualizing gene expression and protein interactions with fluorescent proteins. *Nature Cell Biology* 2002, *4* (1), E15–E20.
81. Ha, T.; Enderle, T.; Ogletree, D. F.; Chemla, D. S.; Selvin, P. R.; Weiss, S., Probing the interaction between two single molecules: Fluorescence resonance energy transfer between a single donor and a single acceptor. *Proceedings of the National Academy of Sciences of the United States of America* 1996, *93* (13), 6264–6268.

82. Terskikh, A.; Fradkov, A.; Ermakova, G.; Zaraisky, A.; Tan, P.; Kajava, A. V.; Zhao, X.; Lukyanov, S.; Matz, M.; Kim, S.; Weissman, I.; Siebert, P., "Fluorescent timer": Protein that changes color with time. *Science* 2000, *290* (5496), 1585–1588.

83. Revzin, A. et al., Surface engineering with poly (ethylene glycol) photolithography to create high-density cell arrays on glass. *Langmuir* 2003, *19* (23), 9855–9862.

84. Tsang, L. et al., Fabrication of 3D hepatic tissues by additive photopatterning of cellular hydrogels. *FASEB Journal* 2007, *21* (3), 798.

85. Roth, E. A. et al., Inkjet printing for high-throughput cell patterning. *Biomaterials* 2004, *25* (17), 3707–3715.

86. Albrecht, D. R. et al., Photo- and electropatterning of hydrogel-encapsulated living cell arrays. *Lab on a Chip—The Royal Society of Chemistry* 2005, *5*, 111–118. http://dx.doi.org/10.1039/b406953f

87. Evans, N. D.; Minelli, C.; Gentleman, E.; LaPointe, V.; Patankar, S. N.; Kallivretaki, M.; Chen, X.; Roberts, C. J.; Stevens, M. M., Substrate stiffness affects early differentation events in embryonic stem cells. *European Cells and Materials.* 2009, *18*, 1–14.

88. Offenhausser, A.; Bocker-Meffert, S.; Decker, T.; Helpenstein, R.; Gasteier, P.; Groll, J.; Moller, M.; Reska, A.; Schafer, S.; Schulte, P.; Vogt-Eisele, A., Microcontact printing of proteins for neuronal cell guidance. *Soft Matter* 2007, *3*, 290–298.

89. Cosson, S.; Kobel, S. A.; Lutolf, M. P., Capturing complex protein gradients on biomimetic hydrogels for cell-based assays *Advanced Functional Materials.* 2009, *19*, 3411–3419.

90. Lee, S.; Moon, J. J.; West, J. L., Three-dimensional micropatterning of bioactive hydrogels via two-photon laser scanning photolithography for guided 3D cell migration. *Biomaterials* 2008, *20*, 2962–2968.

91. Lewis, J. A.; Gratson, G. M., Direct writing in three dimensions. *Langumir* 2004, *7*, 32–39.

92. Hahn, M. S.; Taite, L. J.; Moon, J. J.; Rowland, M. C.; Ruffino, K. A.; West, J. L., Photolithographic patterning of polyethylene glycol hydrogels. *Biomaterials* 2006, *27*, 2519–2524.

93. Ruiz, S. A.; Chen, C. S., Emergence of patterned stem cell differentiation within multicellular structures. *Stem Cells* 2008, *26*, 2921–2927.

94. Xu, Tao, Joyce Jin, Cassie Gregory, JJ Hickman, Thomas Boland. Inkjet printing of viable mammalian cells. *Biomaterials* 2005, *26*, 93–99.

95. Allen, P. B.; Milne, G.; Doepker, B. R.; Chiu, D. T., Pressure-driven laminar flow switching for rapid exchange of solution environment around surface adhered biological particles. *Lab chip* 2010, *10*, 727–733.

96. Adler, Micha, Mark Polinkovsky, Edgar Gutierrez, Alex Groisman. Generation of oxygen gradients with arbitrary shapes in a microfluidic device. *Lab chip* 2010, *10*, 388–391.

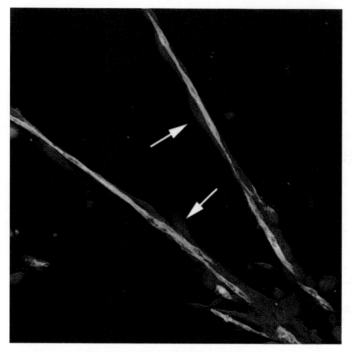

EPCs (green), MPCs (red), nuclei (blue)

FIGURE 2.1 EPCs and MPCs stained in three-dimensional culture *in vitro*. MPCs were labeled with CellTracker Red (Invitrogen), then suspended with EPCs in Matrigel (BD Biosciences), and cultured overnight in EGM-2, 1% serum, and 3 ng/mL bFGF. EPCs were then stained for CD31 (Dako), green. The figure shows linear, multicellular network segments with MPCs aligned lengthwise along EPC cords.

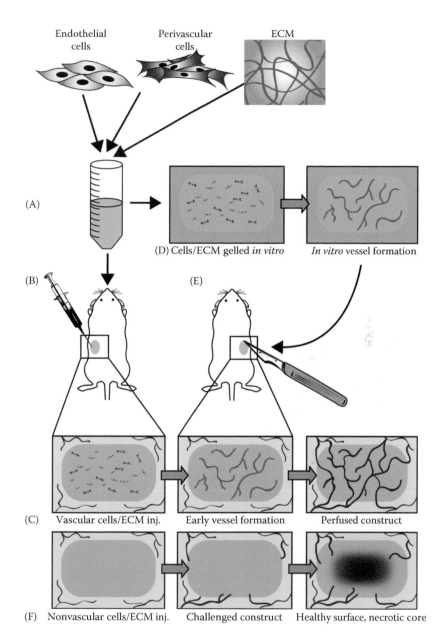

Endothelial cells   Perivascular cells   ECM

(A)

(D) Cells/ECM gelled *in vitro*   *In vitro* vessel formation

(B)

(E)

(C)   Vascular cells/ECM inj.   Early vessel formation   Perfused construct

(F)   Nonvascular cells/ECM inj.   Challenged construct   Healthy surface, necrotic core

FIGURE 2.3 Two emerging paradigms for creation of cell-mediated vascular networks. (A) Endothelial-lineage cells, perivascular cells, and a solubilized extracellular matrix (ECM) are the essential constituents of engineered blood vessel networks. These two cell types are suspended in the ECM and, in one method, (B) injected directly into the animal, where (C) the individual cells form a provisional network and anastomose with the systemic circulation, resulting in a perfused construct. (D) In a second method, the cell/ECM suspension is polymerized and incubated *in vitro*, allowing initial network formation to occur. Upon surgical implantation (E), this network anastomoses with the host circulation to achieve perfusion. (F) It is hypothesized that implantation of a metabolically demanding cell population (cells not depicted) in the absence of vascular cells will result in necrosis within the construct.

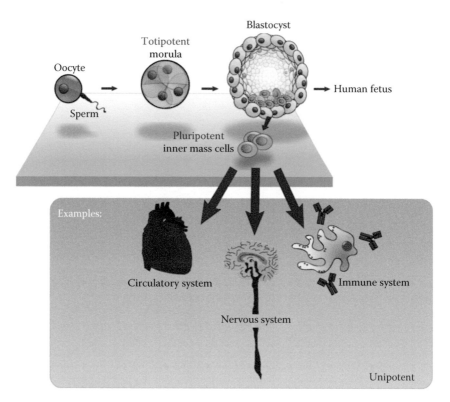

FIGURE 3.1 Pluripotent ES cells originate as inner mass cells within a blastocyst. The stem cells can become any tissue in the body, excluding a placenta. Only the morula's cells are totipotent, able to become all tissues and a placenta. (Image retrieved from Wikipedia. Jones, M., http://commons.wikimedia.org/wiki/File:Stem_cells_diagram.png. With permission.)

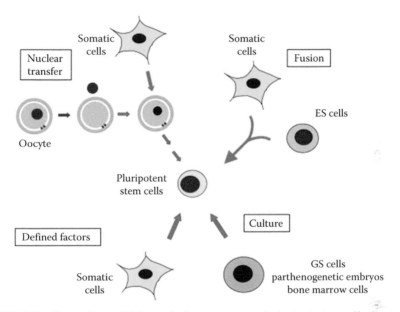

FIGURE 3.2 Currently available methods to generate pluripotent stem cells from adult somatic or germ cells. In mouse models, three methods have been reported to generate pluripotent stem cells from somatic cells: NT, fusion, and forced expression of defined factors. Also reported is the generation of chimera-competent pluripotent stem cells after the long-term culture of bone marrow cells. In addition, pluripotent stem cells can be established from mouse adult germ cells: multipotent GS cells and parthenogenetic ES cells. (From *Cell Stem Cell*, 1(1), Yamanaka, S., Strategies and new developments in the generation of patient-specific pluripotent stem cells, 39–49. Copyright 2007. With permission from Elsevier.)

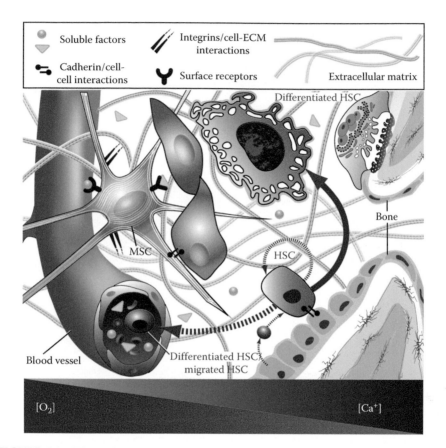

FIGURE 4.1 The perivascular and endosteal bone marrow niche: example of signals and cues regulating stem cell function. The stem cell niche is a complex microenvironment that guides stem cell fate through a combination of extracellular matrix (ECM), cell–cell interactions, and extrinsic factors such as growth factors and cytokines. In the bone marrow, hematopoietic stem cells (HSCs) situated in the endosteal niche are physically anchored to the niche by cadherens junctions with osteoblasts (cell–cell interactions). In this location, cells are exposed to high Ca^{2+} concentrations, low oxygen tension, and a variety of autocrine, paracrine, and endocrine signals (extrinsic factors) and are attached to the ECM through integrin receptors. These signals and cues maintain the quiescence and self-renewal of HSCs. In the perivascular niche, however, HSCs are exposed to low Ca^{2+}, high oxygen tension, different cell–cell interactions, and ECM composition that promote migration and differentiation of stem cells. While a majority of these signals are found in a variety of stem cell niches, their utilization and the effects of niche components vary from niche to niche.

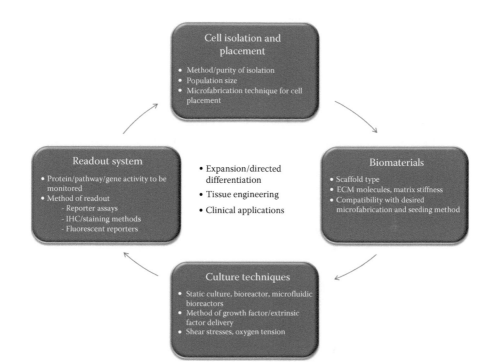

FIGURE 4.2 Overview of processes to design environments that guide cell fate. First, the cell source or line, isolation, and optimal microfabrication technique for the study or application must be identified. Biomaterials selected for microfabrication techniques must be compatible with the cell and microfabrication technique of interest. Modifications of biomaterials such as the incorporation of ECM molecules or material stiffness can affect cellular processes. Culture techniques provide nutrient delivery and waste removal to microfabricated cells. Spatial delivery of growth factors, cytokines, and other extrinsic factors can be controlled by bioreactors and microfluidics systems. The culture technique or bioreactor type can affect nutrient delivery throughout scaffolds, provide shear stresses, affect cell seeding into scaffolds, or modulate ECM deposition by cells. Lastly, systems to monitor cellular activity in the microfabricated system must be identified. Assays can be end point, such as cellular staining, or measured in real time by reporter assays. Each of these factors must be taken into consideration when designing microfabrication experiments.

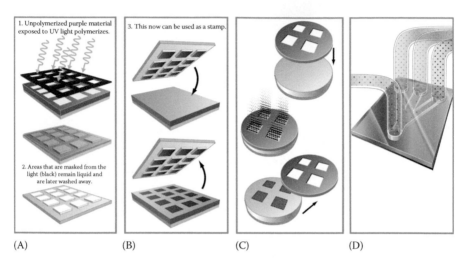

1. Unpolymerized purple material exposed to UV light polymerizes.

2. Areas that are masked from the light (black) remain liquid and are later washed away.

3. This now can be used as a stamp.

(A)   (B)   (C)   (D)

FIGURE 4.3 Micropatterning techniques. (A) Photolithography. To create patterns using photolithography, a thin film of photopolymerizable material such as photoresist or poly (ethylene glycol) diacrylate hydrogel (PEGDA) is spun onto a glass substrate. The material is photopolymerized through a mask of the desired pattern. The material only hardens where exposed to light, and unexposed material is washed away. The resulting material can be used to directly culture cells, as a master to mold and shape 3D scaffolds, as a stamp for soft lithography (B) as a stencil (C), or as a microfluidic device (D). (B) Soft lithography. A master is formed in the desired pattern using photolithographic techniques and then filled with material such as PDMS, a soft elastomeric material that is commonly used for soft lithographic techniques. This stamp can be dipped directly into ECM molecules such as fibronectin, or functionalized with alkanethiols, and stamped onto substrates. (C) Stencil. The master created by photolithographic techniques can be used as a stencil to limit exposure of the substrate to molecules. (D) Microfluidic devices. Generally, microfluidics devices are formed in the same manner as stamps for soft lithography but have continuous channels to allow for fluid flow. These microfluidic devices can be used as guides to deposit ECM or cells, and then are peeled off the substrate. Additionally, they are used as bioreactors to control fluid flow, soluble molecule presentation, and cell deposition.

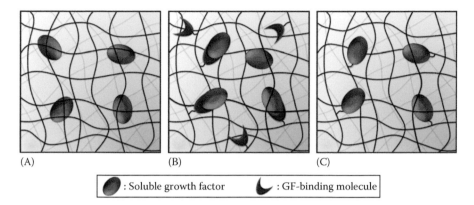

: Soluble growth factor   : GF-binding molecule

FIGURE 5.1 Localization of growth factors within polymeric materials using (A) encapsulation, (B) sequestration via growth factor–binding molecules such as heparin or peptide ligands, or (C) covalent immobilization.

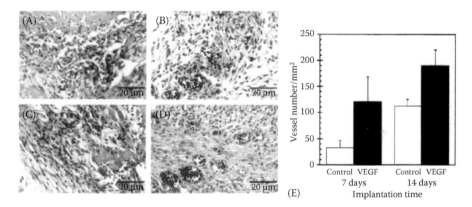

FIGURE 5.2 Growth factors can be encapsulated in porous scaffolds and released to promote localized neovascularization. PLG-alginate scaffold containing encapsulated VEGF were subcutaneously implanted in SCID mice. Scaffolds were retrieved 1 (A and B) or 2 (C and D) weeks after implantation, and histological sections were stained with hematoxylin and eosin to visualize blood vessels within scaffolds containing no VEGF (A and C) and VEGF-releasing (B and D) scaffolds. (E) Quantification of blood vessel density indicated that neovascularization in VEGF conditions was significantly higher ($p < 0.05$) than controls at both time points. (From Peters, M.C., Polverini, P.J., and Mooney, D.J.: Engineering vascular networks in porous polymer matrices. *J Biomed Mater Res*. 2002. 60(4). 668–678. Copyright Wiley-VCH Verlag GmbH & Co. KGaA. Adapted with permission.)

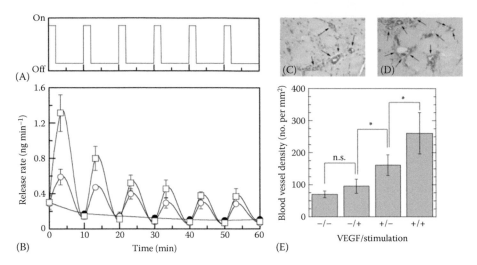

FIGURE 5.3 Encapsulated growth factor release from hydrogels can be controlled via mechanical stimulation and used to direct localized neovascularization within tissues. Mechanical stimulation of VEGF-loaded alginate hydrogels with (A) cyclic compressive loading led to (B) discrete bursts of VEGF release under 10% (open circles) and 25% (open squares) strain amplitude, while control conditions with no compression (filled circles) did not exhibit changes in VEGF release. Implantation of scaffolds into femoral artery ligation sites of non-obese diabetic (NOD) mice lead to increased neovascularization in response to released VEGF with and without daily mechanical stimulation. Tissue sections depict localized neovascularization in implanted hydrogels containing VEGF (C) under static conditions (+/−) and (D) when exposed to daily mechanical stimulation (+/+). (E) Quantification of local blood vessel density for control hydrogels without VEGF under static conditions (−/−) and under daily mechanical stimulation (−/+) as well as VEGF containing hydrogels. (Original magnification 400X. Arrows indicate CD31-stained blood vessels. n.s. indicates no statistical difference. Asterisk indicates statistical significance, $p < 0.05$). (Adapted by permission from Macmillan Publishers Ltd. *Nature* Lee, K.Y., Peters, M.C., Anderson, K.W., and Mooney, D.J., Controlled growth factor release from synthetic extracellular matrices. 408(6815), 998–1000. Copyright 2000.)

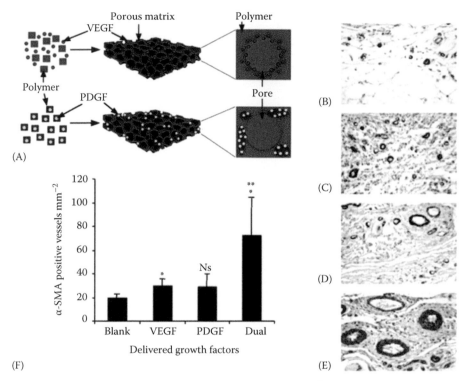

(A)

(B)

(C)

(D)

(E)

α-SMA positive vessels mm^{-2}

120

100

80

60

40

20

0

Blank   VEGF   PDGF   Dual

Delivered growth factors

(F)

FIGURE 5.5 Scaffolds with multiple material components can be used to spatially localize multiple growth factors. Dual release of VEGF and PDGF-BB from a multicomponent material enhanced localized neovascularization and blood vessel maturity within scaffolds. In this approach, VEGF was directly incorporated into the porous polymer scaffold material, while PDGF-BB was pre-encapsulated within polymer microspheres before incorporation into the porous scaffold (schematic, A). Following subcutaneous implantation in Lewis rats, α-smooth muscle actin (α-SMA) staining of tissue sections from (B) blank scaffolds, (C) scaffolds containing VEGF (D), scaffolds containing PDGF-BB-loaded microspheres, and (E) scaffolds containing both VEGF and PDGF-BB-loaded microspheres indicated significant increases in blood vessel maturity from spatial localization of both growth factors. (F) Implantation of scaffolds releasing both VEGF and PDGF-BB into non-obese (NOD) mice following femoral ligation also resulted in significant increases in vessel maturity as indicated by the density of vessels staining positive for α-SMA. (Asterisk indicates statistical significance relative to blank scaffold, $p < 0.05$; two asterisks indicate statistical significance relative to VEGF or PDGF-BB alone $p < 0.05$; ns indicates no statistical significance, $p > 0.05$. Magnification for photomicrographs 400X. (Adapted by permission from Macmillan Publishers Ltd. *Nat. Biotechnol.*, Richardson, T.P., Peters, M.C., Ennett, A.B., and Mooney, D.J., Polymeric system for dual growth factor delivery. 19(11), 1029–1034. Copyright 2001.)

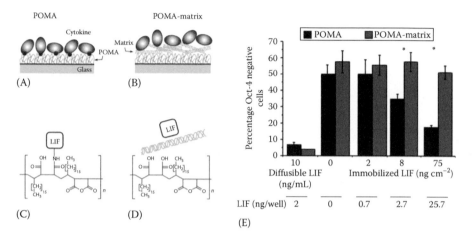

FIGURE 5.6 Surface-immobilized growth factors can serve as a sustained, surface localized biological signal to regulate stem cell activity. Covalent immobilization of LIF to a polymer film substrate was used to maintain mouse embryonic stem cell (mESC) pluripotency *in vitro*. LIF was (A and C) covalently immobilized to poly(octadecene-alt-maleic anhydride) (POMA) or (B and D) noncovalently adsorbed to an ECM coating deposited on top of hydrolyzed POMA (POMA-matrix). Covalently immobilized LIF maintained Oct-4 expression in a surface density–dependent manner while adsorbed LIF did not as indicated by (E) expression of Oct-4 in mESCs cultured for 72 h. Diffusible LIF was replaced during medium change every 24 h (Error bars are standard error of the mean. Asterisk indicates significant difference between POMA and POMA-matrix conditions, $p < 0.05$). (Adapted by permission from Macmillan Publishers Ltd. *Nat. Methods* Alberti, K., Davey, R.E., Onishi, K., George, S., Salchert, K., Seib, F.P., Bornhauser, M., Pompe, T., Nagy, A., Werner, C., and Zandstra, P.W., Functional immobilization of signaling proteins enables control of stem cell fate. 5(7), 645–650. Copyright 2008.)

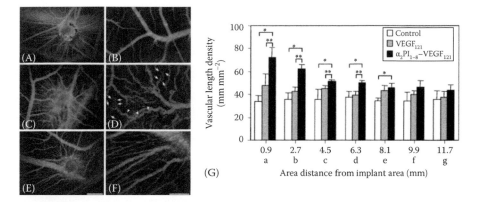

FIGURE 5.7 "Cell-demanded" release of growth factors can be achieved using enzymatically degradable linking chemistries. Plasmin-sensitive linking chemistries were used to create fibrin hydrogels that released VEGF in response to infiltrating cells. Implantation of these scaffolds into chick chorioallantoic membranes led to enhanced localized neovascularization as indicated by *in vivo* fluorescence microscopy of vascular networks: (A and B) Control grafts made of fibrin alone, (C and D) fibrin formulated with soluble VEGF, and (E and F) fibrin formulated with enzymatically (plasmin) released VEGF (α_2PI_{1-8}-VEGF$_{121}$). Both VEGF forms increased neovascularization as indicated by blood vessel density; however, vascular hierarchy as well as neovessel morphology induced by enzymatically released VEGF (E and F) appeared more normal than neovascularization induced by freely diffusible VEGF. Bars = 1 mm (A, C, and E); 0.5 mm (B, D, and F). (G) Quantification of mean vascular length density also indicated enhanced neovascularization via enzymatically released VEGF. (Data are mean ± standard deviation. One asterisk indicates significance versus control fibrin alone, $p < 0.05$; two asterisks indicate significance versus soluble VEGF, $p < 0.05$). (Adapted from Ehrbar, M. et al., *Circ. Res.,* 94(8), 1124, 2004. With permission.)

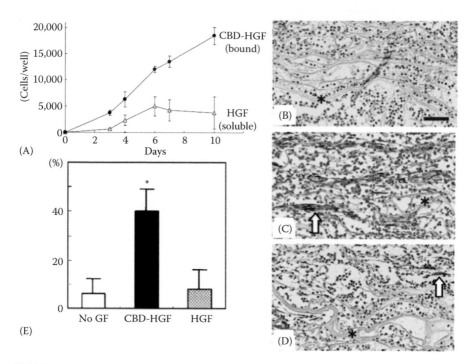

FIGURE 5.8 Specific material-binding domains can be used to localize growth factors on surfaces or within scaffolds via sequestration. A fusion protein of hepatocyte growth factor containing a collagen-binding domain (CBD-HGF) sequestered to collagen materials and locally regulated neovascularization. Sequestration of CBD-HGF to collagen-coated substrates lead to significant increases in (A) human umbilical vein endothelial cell (HUVEC) proliferation compared to HGF conditions. CBD-HGF sequestered within collagen sponges promoted increased localized neovascularization as indicated by (B–D) α-smooth muscle actin stained histological sections and (E) mean blood vessel density measurements of implanted collagen sponges soaked in buffer containing (B) no growth factor, (C) CBD-HGF, or (D) HGF before implantation. (Error bars represent standard deviation, n = 15. Asterisk indicates significance compared to no growth factor, $p < 0.01$). (Adapted from *Biomaterials*, 28(11), Kitajima, T., Terai, H., and Ito, Y., A fusion protein of hepatocyte growth factor for immobilization to collagen, 1989–1997. Copyright (2007). With permission from Elsevier.)

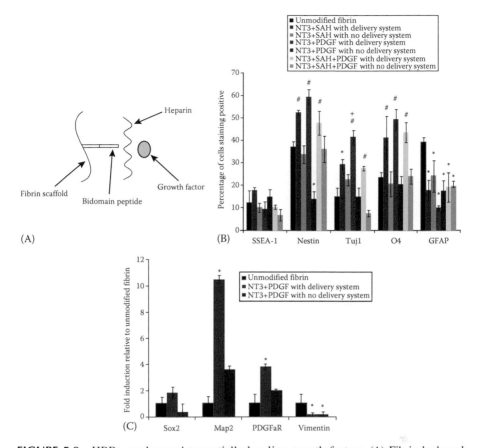

FIGURE 5.9 HBPs can be used to spatially localize growth factors. (A) Fibrin hydrogel scaffolds functionalized with HBPs can be used to sequester heparin and concomitantly modulate the release of heparin-binding growth factors such as neurotrophin 3 (NT-3), sonic headgehog (SHH), and platelet-derived growth factor-AA (PDGF-AA). Spatial localization of growth factors regulated the activity of mouse embryonic stem cell–derived neural progenitor cells (ESNPCs) seeded within fibrin scaffolds. ESNPCs were cultured within scaffolds containing different growth factors for 14 days and then analyzed via (B) fluorescence-activated cell sorting using the markers SSEA-1 (undifferentiated ES cells), nestin (neural progenitors), Tuj1 (early neurons), O4 (oligodendrocytes), and GFAP (astrocytes). Results were verified via (C) real-time RT-PCR analysis using the markers Sox2 (neural progenitors), microtubule-associated protein-2 (Map2; early neurons), platelet-derived growth factor α receptor (PDGFαR; early oligodendrocytes), and vimentin (astrocytes) and confirmed that combined localization of NT-3 and PDGF-AA increased the level of progenitor differentiation into neurons and oligodendrocytes while suppressing astrocyte differentiation. (Adapted from *Stem Cell Res.*, 1(3), Willerth, S.M., Rader, A., and Sakiyama-Elbert, S.E., The effect of controlled growth factor delivery on embryonic stem cell differentiation inside fibrin scaffolds, 205–218. Copyright 2008. With permission from Elsevier.)

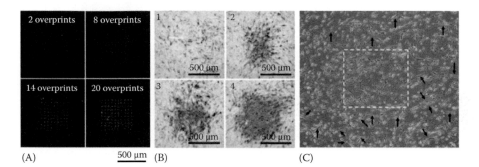

FIGURE 5.10 Spatially patterned growth factors can be used to locally regulate stem cell differentiation. Spatially patterned BMP-2 was used to localize muscle-derived stem cell (MDSC) osteogenesis. Patterns with different amounts of deposited BMP-2 were created by printing and "overprinting" concentrated drops of BMP-2 on fibrin-coated slides and (A) were visualized using Cy3-labeled BMP-2. MDSCs were cultured on printed surfaces for 72 h and then (B) stained for alkaline phosphatase activity (red) indicating localized osteogenesis. Overprints of BMP-2 were as follows: (1) 2 overprints, (2) 8 overprints, (3) 14 overprints, and (4) 20 overprints. (C) Additionally, MDSCs cultured on regions containing 14 overprint were analyzed for myotube formation, indicating that spatially patterned BMP-2 prevented myogenic differentiation (Dotted line indicates edge of pattern, arrows indicate myotubes). (From Phillippi, J.A., Miller, E., Weiss, L., Huard, J., Waggoner, A., and Campbell, P.: Microenvironments engineered by inkjet bioprinting spatially direct adult stem cells toward muscle- and bone-like subpopulations. *Stem Cells*. 2008. 26(1). 127–134. Copyright Wiley-VCH Verlag GmbH & Co. KGaA. Adapted with permission.)

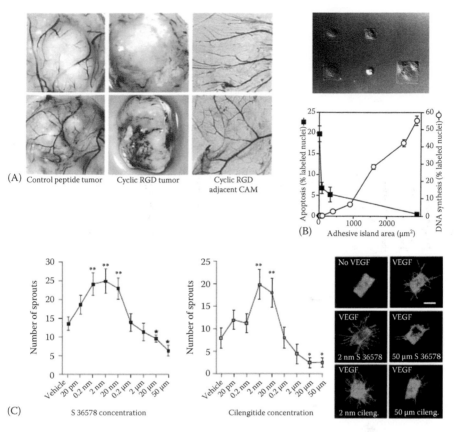

(A) Control peptide tumor Cyclic RGD tumor Cyclic RGD adjacent CAM

(B) Adhesive island area (μm²)

(C) S 36578 concentration   Cilengitide concentration

FIGURE 6.1 Cell adhesion to the ECM regulates angiogenesis. (A) Tumor angiogenesis on the CAM is inhibited by cyclic RGD peptide, which antagonizes integrin αvβ3 binding to the ECM. (From *Cell*, 79(7), Brooks, P.C., Montgomery, A.M., Rosenfeld, M., Reisfeld, R.A., Hu, T., Klier, G., and Cheresh, D.A., Integrin alpha v beta 3 antagonists promote tumor regression by inducing apoptosis of angiogenic blood vessels, 1157–1164. Copyright 1994. With permission from Elsevier.) (B) The degree of cell spreading, controlled by varying sizes of micropatterned islands of fibronectin, regulates endothelial cell proliferation versus apoptosis, with greater proliferation in more highly spread cells. (From Chen, C.S., Mrksich, M., Huang, S., Whitesides, G.M., and Ingber, D.E., Geometric control of cell life and death, *Science*, 1997, 276(5317), 1425–1428. With permission of AAAS.) (C) Low concentrations of the αvβ3/αvβ5 inhibitors S 36578 and cilengitide promote VEGF-mediated angiogenesis from mouse aortic rings, compromising antiangiogenic effects. (By permission from Macmillan Publishers Ltd. *Nat Med*, Reynolds, A.R., Hart, I.R., Watson, A.R., Welti, J.C., Silva, R.G., Robinson, S.D., Da Violante, G., Gourlaouen, M., Salih, M., Jones, M.C., Jones, D.T., Saunders, G., Kostourou, V., Perron-Sierra, F., Norman, J.C., Tucker, G.C., and Hodivala-Dilke, K.M., Stimulation of tumor growth and angiogenesis by low concentrations of RGD-mimetic integrin inhibitors, 15(4) 392–400. Copyright 2009.)

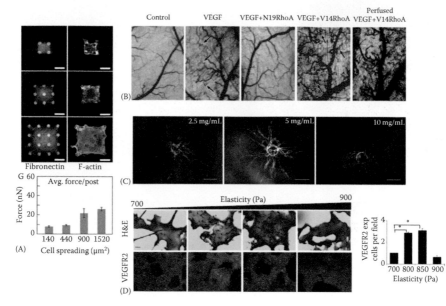

FIGURE 6.2 Cell contractility and ECM mechanical properties regulate angiogenesis. (A) Increased cell spreading increases cell contractility, as measured by culture on arrays of flexible microposts. (From *Proceedings of the National Academy of the United States of America,* 100(4), Tan, J. L.; Tien, J.; Pirone, D. M.; Gray, D. S.; Bhradriraju, K.; Chen, C. S., Cells Lying on a bed of microneedles: An approach to isolate mechanical force, 1484–1489. Copyright 2003. With permission from Elsevier.) (B) RhoA manipulations regulate VEGF-driven angiogenesis in mouse skin. (From Hoang, M.V., Whelan, M.C., and Senger, D.R., Rho activity critically and selectively regulates endothelial cell organization during angiogenesis. *Proc Natl Acad Sci USA* 2004, 101(7), 1874–1879. With permission.) (C) Fibrin density regulates endothelial cell sprouting from microcarrier beads, with higher densities inhibiting sprouting. (From *Biophys J*, 94(5), Ghajar, C.M., Chen, X., Harris, J.W., Suresh, V., Hughes, C.C., Jeon, N.L., Putnam, A.J., and George, S.C., The effect of matrix density on the regulation of 3-D capillary morphogenesis, 1930–1941. Copyright 2008. With permission from Elsevier.) (D) Intermediate stiffness of Matrigel promotes angiogenesis *in vivo* as measured by cellular infiltration and VEGFR2 expression. (By permission from Macmillan Publishers Ltd. *Nature,* Mammoto, A., Connor, K.M., Mammoto, T., Yung, C.W., Huh, D., Aderman, C.M., Mostoslavsky, G., Smith, L.E., and Ingber, D.E., A mechanosensitive transcriptional mechanism that controls angiogenesis, 457(7233), 1103–1138. Copyright 2009.)

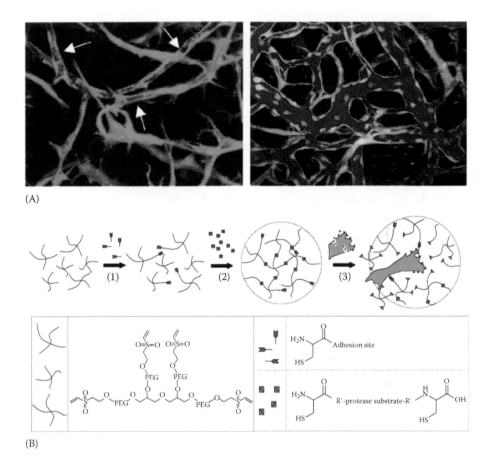

(A)

(1)   (2)   (3)

O=S=O O=S=O

PEG PEG

PEG   PEG

H_2N Adhesion site

HS

H_2N R'-protease substrate-R'

HS   HS

(B)

FIGURE 6.3 Engineering materials to promote tissue vascularization. (A) Endothelial cells and pericytes co-seeded in collagen gel implants begin to form vascular networks at 4 days (left) and are stable and fully perfused at 4 months (right). (By permission from Macmillan Publishers Ltd. *Nature*, Koike, N., Fukumura, D., Gralla, O., Au, P., Schechner, J.S., and Jain, R.K., Tissue engineering: Creation of long-lasting blood vessels, 428(6979), 138–139. Copyright 2004.) (B) One synthetic scheme used to create biomimetic PEG materials. Monocysteine adhesive peptides and bis-cysteine MMP-sensitive peptides are coupled to multiarm PEGs via Michael-type addition reaction to allow cell migration through the matrix. (From Lutolf, M.P., Raeber, G.P., Zisch, A.H., Tirelli, N., and Hubbell, J.A.: Cell-responsive synthetic hydrogels. *Adv Mater.* 2003. 15(11). 888. Copyright Wiley-VCH Verlag GmbH & Co. KGaA. With permission.)

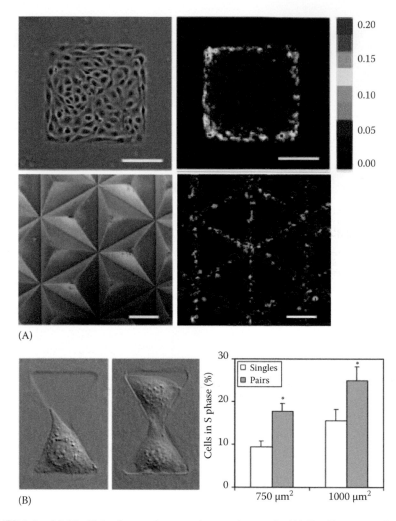

FIGURE 6.4 Multicellular interactions regulate angiogenesis. (A) Confluent monolayers of endothelial cells proliferate at regions of higher stress: at monolayer edges (top) or in valleys of undulating monolayers without edges (bottom). (From Nelson, C.M. et al., *Proc Natl Acad Sci USA*, 102(33), 11594, 2005.) (B) Cell–cell contact increases proliferation when cell spreading is controlled. (From Nelson, C.M. and Chen, C.S., *J Cell Sci*, 116(Pt 17), 3571, 2003. With permission.)

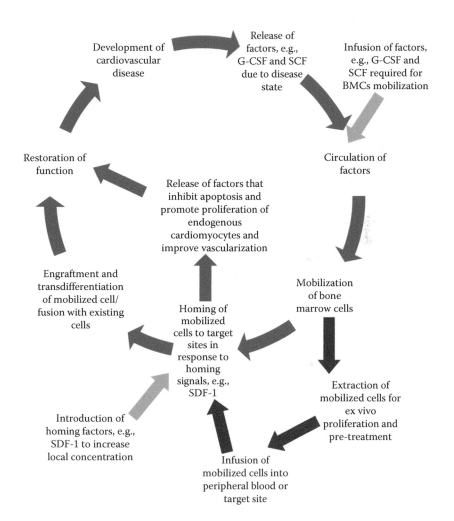

FIGURE 7.1 Overview of bone marrow cell (BMC) mobilization for treatment of cardio-vascular diseases. Blue arrows represent naturally occurring processes, red arrows represent conventional treatment involving extraction and reimplantation of cells, and green arrows represent strategies to directly recruit mobilized cells endogenously.

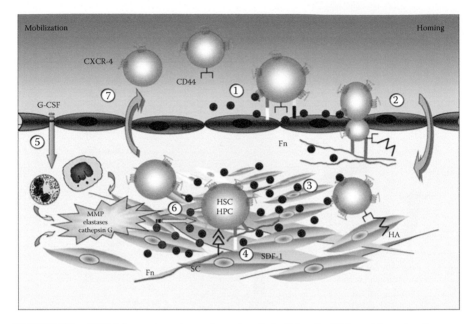

FIGURE 7.2 SDF-1 located on the surface of endothelial cells (EC) and bound to heparan sulfates can optimally interact with its counter-receptor CXCR-4 expressed on the rolling HSC hematopoietic progenitor cells (HPC). CXCR-4 engagement induces an activation of integrins VLA-4 and LFA-1 by inside-out signaling, which converts their interactions with VCAM-1 and ICAM-1 constitutively expressed on EC to firm adhesion. This activation induces the arrest of HSC/HPC and stimulates actin polymerization, which results in transendothelial migration mediated by VLA-4 and VLA-5 in the presence of fibronectin (Fn), these interactions being enhanced by SDF-1. HSC/HPC then polarizes, migrates toward a local gradient of SDF-1 continuously produced by stromal cells (SC), and reaches the hematopoietic niche. SDF-1-induced migration is related to the presence of CD44, the receptor for hyaluronic acid (HA). The final anchoring of HSC/HPC within the niche depends mainly on interactions with SC and extracellular matrix (Fn, HA). Anchoring is maintained by the continuous production of SDF-1 by SC. *In vivo* administration of G-CSF induces local production of proteases (MMP, elastases, and cathepsin G) from leucocytes or SC. These molecules are able to degrade the ECM and to disrupt VLA-4/VCAM-1, c-Kit/SCF, and also CXCR-4/SDF-1 interactions by degrading both CXCR-4 and SDF-1. The loss of attachment to stromal cells and to the ECM, together with the loss of SDF-1 activity, favors the release of HSC/HPC into the peripheral blood. (From Lataillade, J.J. et al., *Eur. Cytokine Netw.*, 15(3), 177, 2004. With permission.)

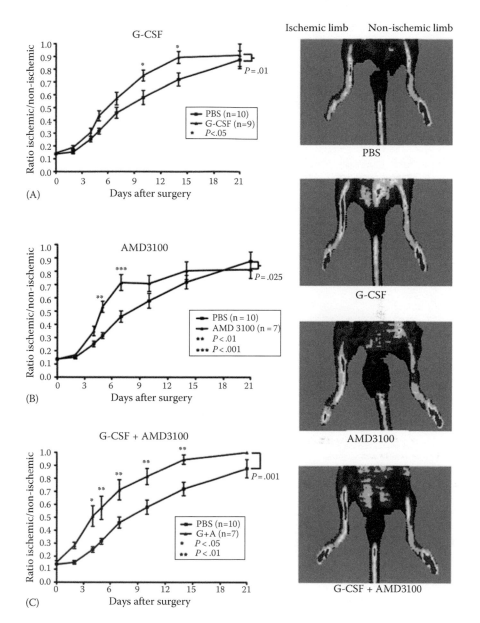

FIGURE 7.3 Cytokine treatment and neovascularization. C57BL/6 mice were treated with saline alone (PBS), G-CSF, AMD3100, or G-CSF in combination with AMD3100 (G + A), and perfusion of the ischemic hindlimb relative to the nonischemic hindlimb was measured. Representative laser Doppler images taken 14 days after surgery are shown. Low perfusion is displayed as blue, while the highest level of perfusion is displayed as red. (A) G-CSF versus PBS. (B) AMD-3100 versus PBS. (C) G-CSF + AMD-3100 versus PBS. Data represent the mean ± standard deviation. (From Capoccia, B.J. et al., *Blood*, 108(7), 2438, 2006. With permission.)

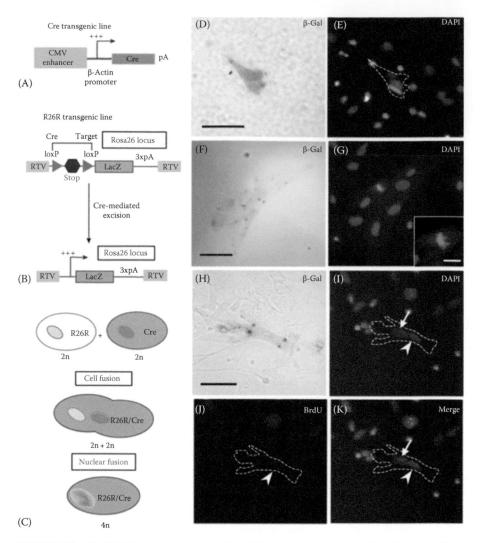

FIGURE 7.4 (A, B) Schematic representation of the transgenes expressed by the mouse lines used. (C) When a cell expressing Cre recombinase (A) fuses with a cell bearing the LacZ reporter transgene (B), the floxed stop cassette is excised and the LacZ reporter is expressed in the fused cell. LacZ expression can be detected by the generation of a blue precipitate after X-gal staining. RTV, Integration retroviral sequence (LTR). (D–G) Cocultures of R26R BMSCs with Cre⁺ neurospheres stained for β-gal and the nuclear dye DAPI (4,6-diamidino-2-phenylindole). (D, E) Multinucleated β-gal⁺ fused cell (4 DIV). (F, G) Colony of β-gal⁺ (15 DIV). These cells were mononucleated and mitotically active, as demonstrated by a cell in metaphase (inset in (G)). (H–K) Coculture of R26R BMCs with BrdU-labeled Cre⁺ neurospheres. Binucleated β-gal⁺ fused cells (H) with one nucleus immunopositive for BrdU (J, K, arrowhead) and the second nucleus negative for BrdU (I, K, arrow). Scale bars: (D), (F), (H) 20 μm; (g-inset) 5 μm. (By permission from Macmillan Publishers Ltd., *Nature*, Alvarez-Dolado, M., Pardal, R., Garcia-Verdugo, J.M., Fike, J.R., Lee, H.O., Pfeffer, K., Lois, C., Morrison, S.J., Alvarez-Buylla, A., Fusion of bone-marrow-derived cells with Purkinje neurons, cardiomyocytes and hepatocytes, 425(6961), 968–973. Copyright 2003.)

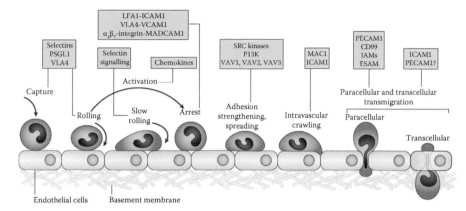

FIGURE 8.1 General scheme of the leukocyte homing cascade, which includes rolling (mediated by selectins), activation (mediated by chemokines), arrest (mediated by integrins), intravascular crawling, and paracellular and transcellular transmigration. Key molecules involved in each step are indicated in boxes. (Adapted by permission from Macmillan Publishers Ltd. *Nat. Rev. Immunol.*, Ley, K., Laudanna, C., Cybulsky, M.I., Nourshargh, S., Getting to the site of inflammation: The leukocyte adhesion cascade updated, 7(9), 678–689. Copyright 2007.)

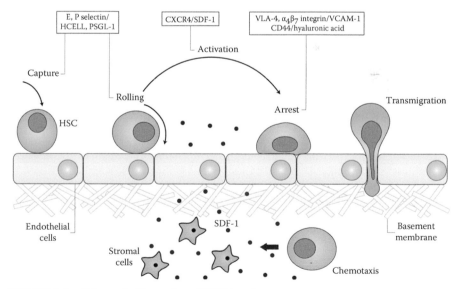

FIGURE 8.2 Schematic illustration of HSC homing.

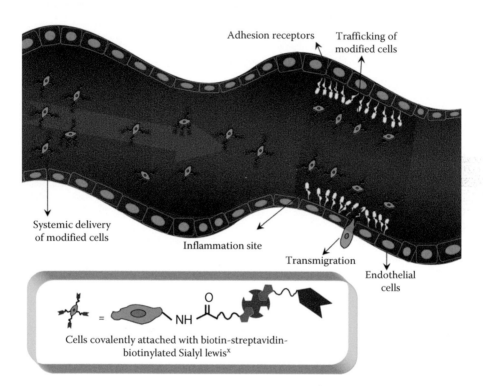

FIGURE 8.3 Schematic of the homing process of systemically infused SLeX-engineered MSCs to P-selectin-expressing endothelium under inflammatory conditions. (Reproduced with permission from Sarkar, D., Vemula, P.K., Teo, G.S.L., Spelke, D., Karnik, R., Wee, L.Y., Karp, J.M., Chemical engineering of mesenchymal stem cells to induce a cell rolling response, *Bioconj. Chem.*, 19(11), 2105–2109. Copyright 2008 American Chemical Society.)

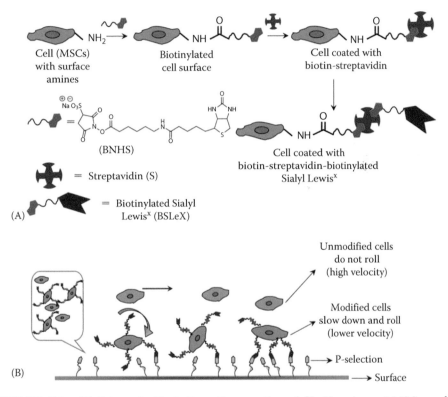

FIGURE 8.4 (A) Schematic illustration of preparation of SLeX-engineered MSCs and (B) SLeX modification induces a MSC rolling response on P-selectin coated surface. (Reproduced with permission from Sarkar, D., Vemula, P.K., Teo, G.S.L., Spelke, D., Karnik, R., Wee, L.Y., Karp, J.M., Chemical engineering of mesenchymal stem cells to induce a cell rolling response, *Bioconj Chem.*, 19(11), 2105–2109. Copyright 2008 American Chemical Society.)

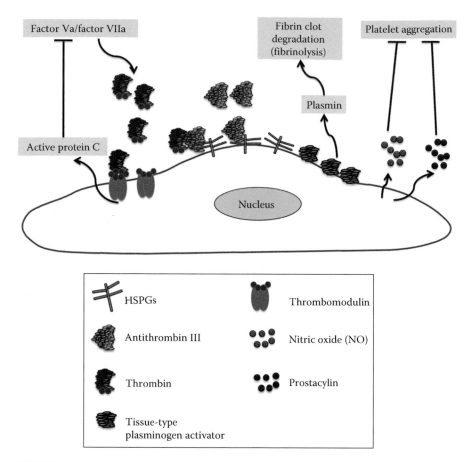

FIGURE 9.1 Antithrombogenic mechanisms of ECs. Membrane-associated thrombomodulin is activated upon thrombin binding. Activation of thrombomodulin leads to activation of protein C. Active protein C inactivates factors Va and VIIa that are responsible for upstream thrombin production. Heparan sulfate proteoglycans (HSPG) on the EC surface bind to antithrombin III (ATIII), which inactivates thrombin. ECs also secrete nitric oxide (NO) and prostacyclin that inhibit platelet aggregation. EC expression of tissue-type-plasminogen activator (t-PA) causes plasminogen conversion to produce plasmin to degrade fibrin in the clot.

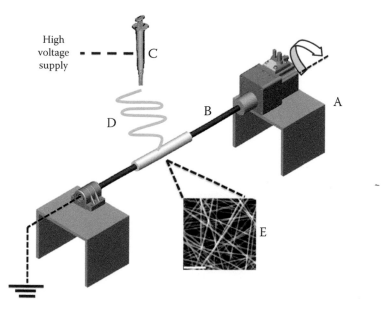

FIGURE 9.2 Making nanofibrous scaffolds by using an electrospinning technique. An electric field is created between charged capillary tip carrying a polymer solution and a grounded mandrel. (A) Motor; (B) rotating mandrel (grounded); (C) polymer solution delivered from a syringe pump to a charged capillary; the capillary tip is connected to a high voltage source (~15kV); (D) nanofibers exiting charged capillary to be collected on rotating mandrel; (E) nanofibers collected on the rotating mandrel.

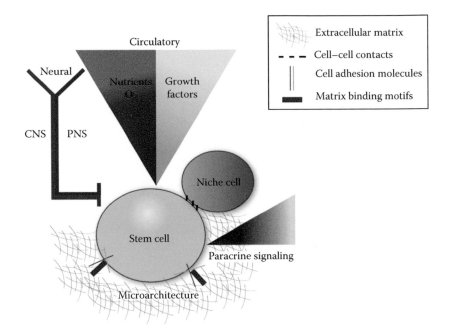

FIGURE 10.1 Microenvironmental inputs regulating stem cell niche. Local environmental cues regulating the stem cell life cycle are depicted. These include mechanical cues, physical interactions with microarchitectural features of ECM, feedback signaling loops from supporting cells, metabolic regulation and control by the circulatory system, neutral input from the surrounding tissues, and the central nervous system.

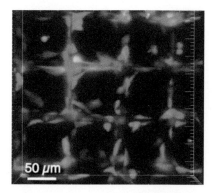

FIGURE 10.2 Fluorescence image showing isometric 3D-rendered view of HT1080 cells inside a 110 mm pore-sized scaffold 24 h after seeding cells. (Modified from Tayalia, P., et al., *Adv. Mater.*, 20 (23), 4494, 2008.)

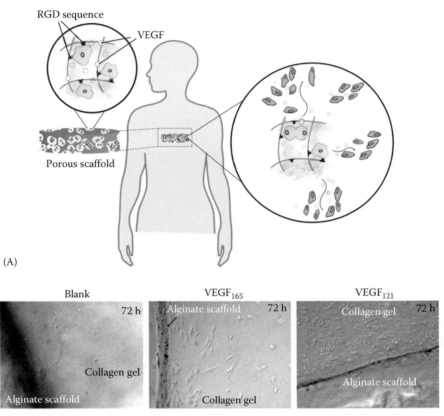

(A)

Blank 72 h
VEGF$_{165}$ 72 h
VEGF$_{121}$ 72 h

(B)

FIGURE 10.3 Proposed cell delivery approach, characterization of cell migration from macroporous alginate scaffolds. (A) Diagram of approach to present cell adhesion ligands (RGD-containing peptides) and local morphogens (VEGF) in the material to maintain cell viability and to activate and induce cell migration out of scaffold. (B) Phase-contrast micrographs of OECs that have migrated out from scaffolds that contain no VEGF (blank), VEGF121, or VEGF165 and populated the surrounding tissue mimic (collagen gel) after 72 h.

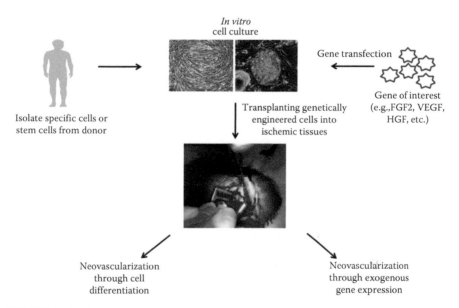

In vitro
cell culture

Gene transfection

Gene of interest
(e.g.,FGF2, VEGF,
HGF, etc.)

Isolate specific cells or
stem cells from donor

Transplanting genetically
engineered cells into
ischemic tissues

Neovascularization
through cell
differentiation

Neovascularization
through exogenous
gene expression

FIGURE 11.1 Schematic representation of genetically engineered cell delivery in ischemic tissue.

5 Spatial Localization of Growth Factors to Regulate Stem Cell Fate

Justin T. Koepsel and William L. Murphy

CONTENTS

5.1   INTRODUCTION

5.1.1   REGULATION OF STEM CELL FATE IN TISSUE ENGINEERING

Tissue engineering approaches that use stem cells to modify, regenerate, or replace diseased tissue will likely require precise control over stem cell fate to create functional tissues. Included in this precise control is a need for mechanisms that locally regulate stem cell activity to generate complex tissue architectures. While many of the mature cell types needed to repair a tissue can in principle be generated by implanting a population of stem cells, the potential for uncontrolled differentiation and tumorigenesis may ultimately hinder this type of approach. For example, in a recent attempt to treat a patient with the neurodegenerative disorder ataxia telangiectasia, repeated local injections of human fetal neural stem cells (NSCs) resulted in tumor formation 4 years later.[1] Although approaches using adult-derived stem cells may decrease the likelihood of such negative outcomes, the ultimate performance of a tissue engineering approach that utilizes stem cells will likely depend on the efficiency with which a stem cell population can be locally differentiated into a specific type of mature cell. Therefore, a major focus of many tissue engineering approaches has been to incorporate biological signals that locally regulate stem cell activity.

5.1.2   GROWTH FACTOR–MEDIATED REGULATION OF STEM CELL FATE

Growth factors are integral to the regulation of stem cell activity and can be used to promote specific stem cell fates. Stem cell behavior is a function of underlying gene expression, which dictates activities such as stem cell survival, self-renewal, and differentiation. Growth factors are a class of signaling proteins that bind to receptors on the cell surface and thereby stimulate changes in gene expression that regulate stem cell activities. In some cases, such as leukemia inhibitory factor (LIF) stimulation of mouse embryonic stem cell self-renewal, sustained growth factor signaling can be used to maintain stem cell pluripotency.[2] In other cases, sequential signals provided by a specific sequence of stem cell–growth factor interactions can be used to drive stem cell differentiation toward a specific mature cell type. For example, motor neurons can be derived from human embryonic stem cells via sequential exposure to different growth factors including basic fibroblast growth factor (bFGF), brain-derived neurotrophic factor (BDNF), glial cell line–derived neurotrophic factor (GDNF), and insulin-like growth factor-1 (IGF-1).[3] In either case, whether promoting self-renewal or directing differentiation, growth factors provide a basis for regulating stem cell activity. Therefore, it follows that mechanisms to localize growth factor stimulation may be used in tissue engineering approaches to locally regulate specific populations of stem cells.

5.1.3 *In Vivo* Regulation of Growth Factor Signals via the Extracellular Matrix

The extracellular matrix (ECM) provides several levels of growth factor regulation, including modulation of growth factor stability, activity, availability, and spatial localization. Growth factors synthesized and secreted into the extracellular space are often physically entrapped or sequestered into the nearby ECM, causing growth factor localization within a tissue.[4] This growth factor sequestration can be facilitated through heparin-binding domains (HBD) found in many growth factors, including vascular endothelial growth factor (VEGF) and bFGF, that specifically bind heparin or heparan sulfate glycosaminoglycan (GAG) side chains of proteoglycans present throughout the ECM.[4,5] Growth factors bound to these GAGs are localized within the ECM and, in some cases, also protected from proteolytic degradation.[5,6] This effectively creates a store of growth factors, which are locally released into the environment of nearby stem cells over time or in response to specific ECM-degrading enzymes. Growth factors can also be sequestered within the ECM through specific ECM protein–growth factor interactions. Multiple forms of collagen, for example, have been shown to specifically bind bFGF and serve as a reservoir for local release of bFGF.[4] Whether physically entrapped or sequestered, growth factors are localized within the ECM of tissues and can locally regulate stem cell fate as well as other processes such as neovascularization. Therefore, many tissue engineering approaches aim to mimic aspects of ECM–growth factor interactions to locally regulate cell activity.

5.1.4 Designing Control over Growth Factors to Regulate Stem Cell Fate

Many of the mechanisms and concepts derived from *in vivo* localization of growth factors can be mimicked when designing tissue engineering approaches that use stem cells. In particular, phenomena observed in ECM-mediated localization of growth factors within tissues, such as growth factor encapsulation and sequestration, can be designed into engineered tissue constructs (Figure 5.1A and B). Additionally, mechanisms such as covalent immobilization of growth factors onto scaffold material can be used to explicitly control growth factor location and subsequent interactions with cells (Figure 5.1C). These different methodologies can potentially be used to localize multiple growth factors within a scaffold as well as create two- (2D) and three-dimensional (3D) patterns of growth factors within a single scaffold to precisely regulate stem cell activity. Throughout this chapter, many different approaches will be surveyed to provide the reader with an understanding of how spatial localization of growth factors can be engineered to locally regulate stem cell fate.

5.1.5 Chapter Scope and Key Definitions

Many strategies have been developed to localize growth factors for *in vitro* and *in vivo* applications. Only a limited number of strategies have been applied to locally regulate stem cell fate. To increase the number and quality of examples, and to more

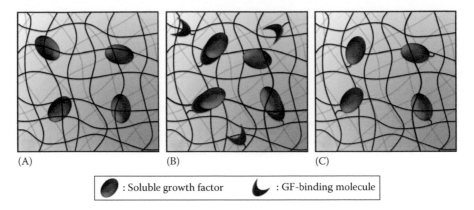

(A)   (B)   (C)

● : Soluble growth factor   ◖ : GF-binding molecule

FIGURE 5.1 (See color insert.) Localization of growth factors within polymeric materials using (A) encapsulation, (B) sequestration via growth factor–binding molecules such as heparin or peptide ligands, or (C) covalent immobilization.

closely align with this book's focus on both stem cells and neovascularization in tissue engineering, this chapter will highlight approaches in which localization of growth factors is used to locally regulate either stem cell fate or neovascularization. Furthermore, in the context of this chapter, growth factor localization within a tissue or throughout a 3D scaffold is classified as *spatial localization*, growth factor localization throughout a 2D surface is classified as *surface localization*, and methods that create distinct regions of concentrated growth factor within a scaffold or on a surface will be classified as *spatial patterning*. For instance, growth factor release from a polymeric delivery vehicle into a surrounding tissue will be considered spatial localization, since the growth factor is locally concentrated around the implanted material; growth factor immobilized on a cell culture substrate will be considered surface localization since the growth factor is locally concentrated on the surface; and growth factor embedded in discrete layers of a scaffold or printed at specific sites on a surface will be considered spatial patterning, since the growth factors are localized at discrete positions within the scaffold or upon the surface. Additionally, for clarity, this chapter will focus on examples of approaches in which growth factor localization is achieved through polymeric materials–based approaches. Our goal is not to provide a comprehensive review of previous work in this area, but instead to introduce and explain key illustrative examples. For detailed discussions of other approaches to locally regulate growth factor–mediated stem cell activity and neovascularization, please refer to reviews on microfluidics,[7] microelectromechanical systems,[8] viral gene delivery,[9] and nonviral gene delivery.[10]

5.2 ENCAPSULATION OF GROWTH FACTORS

Many tissue engineering approaches have used polymeric materials–based platforms to spatially localize growth factors within tissues and tissue engineering scaffolds. Some of the simplest approaches have used natural or synthetic polymers to encapsulate or entrap growth factors within a polymer network (Figure 5.1A).

When these materials are implanted into a tissue, encapsulated growth factors can be released to locally enrich the environment with growth factors that regulate cell activity. This local enrichment can increase growth factor concentrations in the surrounding tissue as well as increase the local growth factor concentration within the interior of a scaffold, depending on the design of the material. In either case, the performance and spatial localization afforded by an encapsulation approach rely on methods to tune the rate at which growth factor releases from the polymer network. For example, materials that release growth factor too quickly may not sustain high enough local levels of growth factor over time to significantly influence cell activity. On the other hand, materials in which growth factors are tightly bound within the solid phase of the scaffold may not release enough growth factor to generate a significant effect on local cell populations. Additionally, the success of spatial localization approaches also depends on the responsiveness of a cell or tissue type to a particular growth factor, as well as the properties of the surrounding environment that dictate the how rapidly growth factors degrade or diffuse away. Fortunately, growth factor release from an encapsulating network can be controlled through a wide range of processing methods and polymer material properties including polymer concentration, crystallinity, copolymer ratio, charge, porosity, cross-linking density, and degradation characteristics.[11] Tuning each of these parameters can lead to changes in growth factor diffusivity within the polymer network and thus modulate the amount of growth factor that is spatially localized around a material. This section will survey several examples of polymer encapsulation approaches that have been developed to spatially localize growth factors and will highlight strategies using *porous scaffolds*, *hydrogel matrices*, and *microspheres* to locally regulate stem cell fate or promote neovascularization. Each subsection will begin with a general description of the approach, followed by specific examples related to neovascularization and stem cells.

5.2.1 POROUS SCAFFOLDS

In general, porous polymer scaffolds are constructed from degradable polymer materials with interconnecting pores on the size scale of 10–1000 µm.[12] Porous architecture allows for cell infiltration, while degradation allows for eventual replacement of the scaffold by regenerated tissue. Growth factors can be encapsulated within the polymer phase of the porous scaffold, and release is governed by growth factor transport through the degrading polymer matrix.[11] In this manner, a porous scaffold can serve as a synthetic ECM that supports tissue ingrowth and spatially localizes growth factors.

Tissue engineering approaches that use porous scaffolds to regulate stem cell activities or neovascularization rely on a variety of polymers and fabrication strategies to encapsulate and spatially localize growth factors. Growth factors have been encapsulated in porous scaffolds composed of natural polymers such as gelatin[13] and collagen[14] by incubation in growth factor–containing solutions. Growth factors can also be encapsulated in scaffolds composed of synthetic polymers, such as poly (lactide-co-glycolide) (PLG)[15,16] via gas foaming.[17,18] Using these encapsulation methods, porous scaffolds have been used to spatially localize growth factors

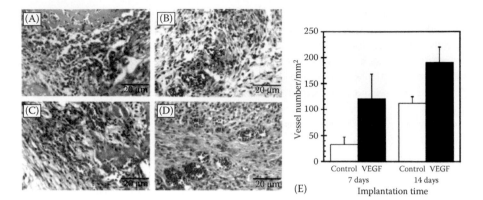

FIGURE 5.2 (See color insert.)   Growth factors can be encapsulated in porous scaffolds and released to promote localized neovascularization. PLG-alginate scaffolds containing encapsulated VEGF were subcutaneously implanted in SCID mice. Scaffolds were retrieved 1 (A and B) or 2 (C and D) weeks after implantation, and histological sections were stained with hematoxylin and eosin to visualize blood vessels within scaffolds containing no VEGF (A and C) and VEGF-releasing (B and D) scaffolds. (E) Quantification of blood vessel density indicated that neovascularization in VEGF conditions was significantly higher ($p < 0.05$) than controls at both time points. (From Peters, M.C., Polverini, P.J., and Mooney, D.J.: Engineering vascular networks in porous polymer matrices. *J Biomed Mater Res*. 2002. 60(4). 668–678. Copyright Wiley-VCH Verlag GmbH & Co. KGaA. Adapted with permission.)

including VEGF,[15,16] transforming growth factor β1 (TGF-β1),[19] acidic fibroblast growth factor (aFGF),[13] bFGF,[14] hepatocyte growth factor (HGF),[14] platelet-derived growth factor-BB (PDGF-BB),[14] IGF-1,[14] and heparin-binding epidermal growth factor (HB-EGF).[14] Therefore, encapsulation can be used to provide several different types of porous scaffolds that spatially localize growth factors, and only a limited subset of previous approaches in this area will be discussed here.

Porous, growth factor–containing polymeric scaffolds can be used to promote local neovascularization *in vivo*. One of the first examples of this concept used porous gelatin sponges to spatially localize aFGF. Implantation of sponges in the neck and peritoneal cavity of rats resulted in local release of aFGF that promoted extensive neovascularization both within implanted sponges and in millimeters of surrounding tissue.[13] In a more recent approach, PLG-alginate porous scaffolds were used to spatially localize VEGF within subcutaneous pockets of SCID mice and locally promote neovascularization within scaffolds (Figure 5.2).[15] In addition to promoting localized neovascularization within the mouse tissue, spatially localized VEGF led to significant increases in human-derived blood vessel formation when similar porous scaffolds were seeded with human microvascular endothelial cells before implantation. Therefore, porous scaffold can be used locally regulate angiogenesis within a tissue as well as neovascularization by a transplanted endothelial cell population.

Porous polymeric scaffolds can also be used to continually release growth factors and regulate stem cell fate *in vitro*. Similar to spatially localizing growth factors in tissues, porous scaffolds can also be used to regulate the amount of growth factor present in cell culture systems. For example, porous poly(ethylene glycol)-terephthalate/poly(butylene terephthalate) multiblock copolymer scaffolds coated with a layer of polymer containing encapsulated TGF-β1 were used to locally release TGF-β1 into cultures of bone marrow–derived mesenchymal stem cells (MSCs).[19] Adjustments in copolymer coating composition were used to tune TGF-β1 release from scaffolds to last up to 12 days or for more than 50 days. Continuous release over 21 days resulted in enhanced MSC chondrogenesis, thus demonstrating that continual release of a growth factor from a porous scaffold can regulate stem cell fate *in vitro*. It is noteworthy that porous scaffolds designed to spatially localize growth factors could also potentially be used to locally regulate stem cells *in vivo*. For example, implantation of a porous scaffold containing an encapsulated growth factor such as TGF-β1 into the site of a chondral defect could be used to locally direct MSC chondrogenesis to aid in the repair of the damaged cartilage.

5.2.2 HYDROGEL MATRICES

Hydrogel matrices have been used extensively in tissue engineering, as they can be designed with structural properties similar to the ECM found in a range of tissues.[20] These materials consist of hydrated polymer networks that can govern the release of encapsulated growth factors through changes in network properties that affect growth factor diffusivity. As a result, growth factor release kinetics can be modulated through polymer formulations and processing methods that control hydrogel network structure, swelling characteristics, and network degradation.[21] Taken together, the structural similarities to native ECM as well as tunable physicochemical properties make hydrogel materials important scaffolds for tissue engineering approaches aimed at spatially localizing growth factors.

Growth factors can be encapsulated in hydrogel materials using straightforward strategies, facilitated by mild cross-linking chemistries. For example, alginate hydrogels can be formed by exposing sodium alginate macromolecules to divalent cations such as Ca^{2+} to create ionic cross-links that result in gelation.[22] Encapsulation can be achieved by simply mixing growth factors within the sodium alginate macromolecule solution before exposure to divalent cations. These mild conditions minimize growth factor denaturation and degradation during cross-linking. Growth factors can also be encapsulated in other hydrogels using relatively mild cross-linking chemistries, such as enzyme-mediated polymerization and photoinitiated free-radical polymerization. Therefore, many different cross-linking chemistries can be used to encapsulate growth factors such as bFGF,[23–26] VEGF,[26–31] aFGF,[25,32] epidermal growth factor (EGF),[25] and transforming growth factor α (TGF-α)[25] in synthetic materials such as dextran[23] and poly(ethylene glycol) (PEG),[33] as well as natural materials such as alginate,[25,28–30] fibrin,[24,26,31,32] and collagen.[27] Furthermore, mild cross-linking conditions can be used to simultaneously encapsulate growth factors and cells, and to facilitate in situ scaffold formation.

Several previous studies have demonstrated that hydrogels can spatially localize pro-angiogenic growth factors to induce localized neovascularization.[24–26,29–32] One particularly interesting hydrogel approach, due to its use of stimulus-based release of growth factor, demonstrated that cyclic mechanical loading of alginate hydrogels

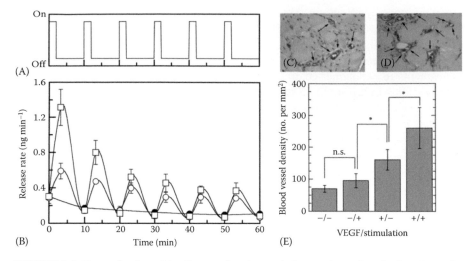

FIGURE 5.3 (See color insert.) Encapsulated growth factor release from hydrogels can be controlled via mechanical stimulation and used to direct localized neovascularization within tissues. Mechanical stimulation of VEGF-loaded alginate hydrogels with (A) cyclic compressive loading led to (B) discrete bursts of VEGF release under 10% (open circles) and 25% (open squares) strain amplitude, while control conditions with no compression (filled circles) did not exhibit changes in VEGF release. Implantation of scaffolds into femoral artery ligation sites of non-obese diabetic (NOD) mice lead to increased neovascularization in response to released VEGF with and without daily mechanical stimulation. Tissue sections depict localized neovascularization in implanted hydrogels containing VEGF (C) under static conditions (+/−) and (D) when exposed to daily mechanical stimulation (+/+). (E) Quantification of local blood vessel density for control hydrogels without VEGF under static conditions (−/−) and under daily mechanical stimulation (−/+) as well as VEGF containing hydrogels. (Original magnification 400X. Arrows indicate CD31-stained blood vessels. n.s. indicates no statistical difference. Asterisk indicates statistical significance, $p < 0.05$). (Adapted by permission from Macmillan Publishers Ltd. *Nature* Lee, K.Y., Peters, M.C., Anderson, K.W., and Mooney, D.J., Controlled growth factor release from synthetic extracellular matrices. 408(6815), 998–1000. Copyright 2000.)

containing encapsulated VEGF resulted in localized bursts of growth factor release following each compression. Daily mechanical stimulation of hydrogels implanted in to the hindlimbs of ischemic mice resulted in enhanced local neovascularization compared to nonmechanically stimulated conditions, and these changes were attributed to daily increases in local VEGF concentration (Figure 5.3).[29] Local VEGF delivery without mechanical stimulation also resulted in significant increases in local neovascularization compared to scaffolds implanted without VEGF. Therefore, while not all hydrogel scaffold approaches to spatially localize growth factors may benefit from repeated mechanical stimulation, this type of approach provides an example of hydrogel-mediated spatial localization over growth factors to regulate neovascularization.

There are many examples in which hydrogels have been implanted to spatially localize growth factors within a tissue. However, very few of these examples have explicitly demonstrated regulation of a specific stem cell population. In one example that clearly regulated stem cell activity, in situ transdermal photopolymerization was used to form poly(ethylene oxide) hydrogel scaffolds containing goat MSCs in the dorsal subcutis of nude mice.[34] In conditions that included encapsulated transforming growth factor β3 (TGF-β3) and high-molecular-weight hyaluronic acid, MSCs expressed increases in the chrondrocyte-associated genes, collagen type II and aggrecan, compared to scaffolds without growth factor. Additionally, safranin-O-staining of implanted scaffolds indicated that TGF-β3 was required for maximum proteoglycans production. Therefore, this example highlights several key capabilities of hydrogel approaches including minimally invasive delivery of stem cells within a scaffold as well as delivery and spatial localization of growth factors via encapsulation.

5.2.3 Microspheres/Microparticles

Encapsulation in dense polymer microparticles or microspheres with diameters ranging from 1 to 1000 μm is a common method to spatially localize growth factors.[35] These microspheres can be injected into a tissue or can be incorporated within scaffolds to locally release growth factor over time. In these systems, release is predominantly dictated by growth factor diffusivity through the polymer network as well as the polymer degradation rate. Additionally, due to their small size and high surface-area-to-volume ratio, distinct growth factor release kinetics can be achieved by controlling microsphere size and surface morphology. Therefore, as growth factors are released from a single microsphere, hundreds of micrometers of tissue or scaffold surrounding a microsphere are locally enriched with growth factor.[36] In this manner, microspheres embedded within a tissue or tissue engineering scaffold can serve as discrete depots that spatially localize growth factors.

Flexible fabrication strategies and a range of polymeric base materials can be used to encapsulate many different growth factors within microspheres. PLG stands out as one of the most commonly used synthetic materials for microsphere encapsulation approaches, mainly due to the fact that small changes in microsphere processing conditions, including the polymer molecular weight and the lactide-to-glycolide ratio, can be used to create polymer microsphere compositions with distinct growth factor release properties.[37] PLG microspheres can also be easily prepared using standard emulsion-solvent extraction or evaporation techniques.[35] As an example of the flexibility afforded by microsphere fabrication strategies, PLG-based microspheres have been used to spatially localize growth factors including bone morphogenic protein 2 (BMP-2),[38,39] GDNF,[40] BDNF,[41] nerve growth factor (NGF),[42,43] VEGF,[44,45] TGF-β1,[46] TGF-β3,[47,48] IGF-1,[49] and PDGF-BB.[45] Several notable natural materials have also been used to create microspheres, including chitosan–albumin microspheres and microfibers to encapsulate endothelial cell growth factor (ECGF),[50] alginate microspheres to encapsulate VEGF,[36] gelatin microspheres to encapsulate TGF-β1[51] and TGF-β2,[52] and type I collagen microspheres to encapsulate bFGF and HGF.[53] Therefore, microspheres can play an integral role in approaches to spatially localize growth factors.

Microspheres have been used to spatially localize growth factors and promote micrometer-scale regulation of neovascularization within tissue. Analogous to growth factors entrapped within the ECM of a tissue, growth factors encapsulated within implanted microspheres can serve as reservoirs that locally regulate cell activity within a tissue. Specifically, release of pro-angiogenic factors, such as ECGF[50] or VEGF,[36] from microspheres can promote localized neovascularization over hundreds of micrometers within a tissue. For example, VEGF release from alginate microspheres in the groin tissue of rats locally increased the VEGF concentration immediately surrounding individual microspheres as indicated by immunoperoxidase staining of the subcutaneous implantation site.[36] Furthermore, the regions staining positive for VEGF exhibited significant increases in neovascularization compared to blank microsphere controls, especially in the region of tissue 0–400 µm away from a microsphere (Figure 5.4).[36]

Polymer microspheres have also been used as a vehicle for *in vitro* release of growth factors that regulate stem cell fate. For example, PLG microspheres have been used to encapsulate and continually release TGF-β3 into human mesenchymal stem cell (hMSC) cultures. Adjustments in the PLA:PGA copolymer ratio were used to tune the rate at which TGF-β3 was released, with 50:50 ratios resulting in more rapid release rates than 75:25 PLA:PGA ratios.[47] Interestingly, release of TGF-β3 from 50:50 PLA:PGA formulations prevented hMSC osteogenesis, even in osteogenic cell culture conditions. Therefore, this example demonstrates a general approach, in which microsphere-based release of an inhibitory growth factor can prevent stem cell differentiation down a specific lineage. This type of approach could be useful to avoid spontaneous differentiation of stem cells into undesired cell types. Conversely, a similar encapsulation approach demonstrated that the stimulatory growth factor TGF-β1 could promote proliferation and osteogenic differentiation of bone marrow stromal cells.[46] In each case, continual release of a growth factor from microspheres provided a sustained signal in the cell culture environment that regulated MSC fate. Microspheres containing inhibitory or stimulatory growth factors could also be implanted to locally regulate native stem cell fate within tissues. Additionally, strategic placement of microspheres containing either stimulatory or inhibitory growth factors could potentially be used to generate complex tissue architectures such as the bone–cartilage interface.

5.2.4 EMBEDDED MICROSPHERES

Microspheres embedded in polymer scaffolds can serve as local depots that release and spatially localize growth factors. Small changes in microsphere formulations that alter the release of growth factors can be made without affecting the bulk properties of a scaffold, since embedded microspheres only constitute a small fraction of a scaffold volume. In this manner, spatial localization of growth factors within a scaffold can be controlled independently of scaffold properties.

Microspheres embedded in scaffolds can locally regulate stem cell differentiation *in vitro*. For example, embedded microspheres have been used to spatially localize growth factors such as TGF-β1[51] and TGF-β3[48] within hydrogel scaffolds. TGF-β1-releasing gelatin microspheres promoted hMSC chondrogenesis when included

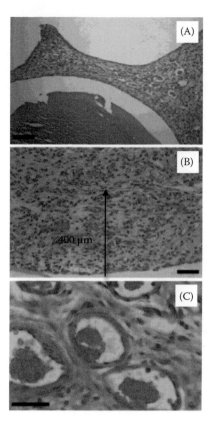

FIGURE 5.4 Growth factors can be encapsulated in microspheres to promote localized neovascularization. Implantation of alginate microspheres containing VEGF lead to localized neovascularization within the groin fascia of Wistar rats. Histological sections stained with hematoxylin and eosin indicated (A) localized neovascularization adjacent to implanted microspheres releasing VEGF and (B) regions of tissue 0–400 μm away from microspheres contained the greatest increases in blood vessel density. (C) High magnification of image of vessels. (Bar is 100 μm). (From Elcin, Y.M., Dixit, V., and Gitnick, G.: Extensive *in vivo* angiogenesis following controlled release of human vascular endothelial cell growth factor: Implications for tissue engineering and wound healing. *Artif Organs.* 2001. 25(7). 558–565. Copyright Wiley-VCH Verlag GmbH & Co. KGaA. Adapted with permission.)

in oligo(poly(ethylene glycol) fumarate) (OPF) hydrogel scaffolds. In particular, encapsulated hMSCs exhibited TGF-β1 dose-dependent upregulation of the chondrocyte-associated genes collagen type II and aggrecan, while gene expression for collagen type I, a marker typically associated with fibroblasts and undifferentiated hMSCs,[51,54] was suppressed. Therefore, this study provides an example of how tunable spatial localization can be used to regulate a population of encapsulated stem cells and demonstrates that embedded microspheres may play a role in approaches requiring precise regulation of stem cell fate.

Embedded microsphere systems have also been used to spatially localize growth factors within tissue-engineered constructs *in vivo*. For example, spatial localization of TGF-β2 from gelatin microspheres was used to regulate bone marrow stromal cells within a PGA scaffold and create an engineered tissue scaffold for tracheal replacement.[52] Localized release of TGF-β2 from gelatin microspheres within the PGA scaffold resulted in localized MSC chondrogenesis that produced a cartilaginous tissue scaffold similar to a natural trachea.[52] Microspheres can be used to expand the capabilities of scaffold approaches by releasing growth factors that are not easily encapsulated in a scaffold alone. Therefore, the adaptability and spatial localization afforded by embedded microsphere approaches is of relevance to tissue engineering approaches aimed at locally regulating stem cell activity and neovascularization.

5.2.5   Dual Growth Factor Release via Co-Encapsulation

Approaches that aim to create increasingly complex tissue architectures will likely require scaffolds that spatially localize multiple growth factors simultaneously. Several pairs of growth factors have been delivered via co-encapsulation, including bFGF and granulocyte colony-stimulating factor (G-CSF),[55] bFGF and HGF,[53,56] VEGF and PDGF-BB,[57] PDGF-BB and TGF-β2,[58] and BMP-2 and TGF-β3.[59] In particular, several studies have shown that dual release of co-encapsulated growth factors *in vivo* leads to increased vascular maturity within scaffolds.[53,55,57] For example, alginate containing both VEGF and PDGF-BB was injected into a myocardial infarct model to promote local neovascularization. Release of both growth factors not only promoted locally increased vascular density within the myocardium compared to delivery of PDGF-BB alone, but also enhanced blood vessel maturity as indicated by smooth muscle cell recruitment.[57] In other approaches, scaffolds containing multiple co-encapsulated growth factors as well as stem cells have been implanted to promote tissue regeneration. For example, alginate hydrogels containing BMP-2, TGF-β3, and bone marrow stromal cells (BMSCs) were subcutaneously implanted into SCID mice.[59] Delivery of both growth factors resulted in localized bone formation by BMSCs, while delivery of either growth factor alone resulted in negligible bone formation as indicated by hematoxylin and eosin staining of harvested implants. Therefore, in both cases, whether promoting neovascularization or regulating stem cell behavior, delivery of multiple growth factors can be used to enhance the performance of a tissue engineering approach.

5.2.6   Dual Growth Factor Release via Dual Component Materials

Scaffolds with multiple material components can also be used to deliver multiple growth factors. For example, microspheres loaded with one type of growth factor can be embedded in a scaffold loaded with another type of growth factor. In this manner, the release of one growth factor is dictated primarily by the properties of the microsphere component, while the release of a second growth factor is dictated primarily by the scaffold properties. For example, PLG microspheres containing encapsulated PDGF-BB were embedded in a PLG porous scaffold containing encapsulated VEGF to create a scaffold that exhibited distinct release kinetics of each growth factor (Figure 5.5A). VEGF was quickly released from the porous PLG

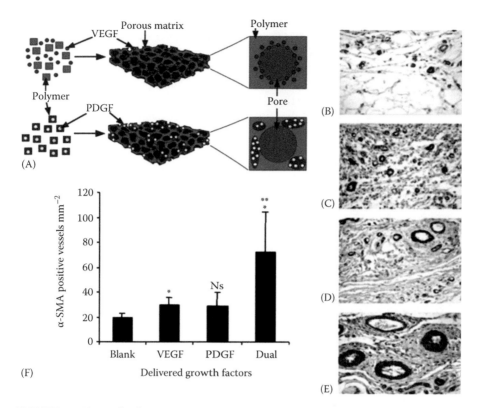

FIGURE 5.5 (See color insert.)   Scaffolds with multiple material components can be used to spatially localize multiple growth factors. Dual release of VEGF and PDGF-BB from a multicomponent material enhanced localized neovascularization and blood vessel maturity within scaffolds. In this approach, VEGF was directly incorporated into the porous polymer scaffold material, while PDGF-BB was pre-encapsulated within polymer microspheres before incorporation into the porous scaffold (schematic, A). Following subcutaneous implantation in Lewis rats, α-smooth muscle actin (α-SMA) staining of tissue sections from (B) blank scaffolds, (C) scaffolds containing VEGF (D), scaffolds containing PDGF-BB-loaded microspheres, and (E) scaffolds containing both VEGF and PDGF-BB-loaded microspheres indicated significant increases in blood vessel maturity from spatial localization of both growth factors. (F) Implantation of scaffolds releasing both VEGF and PDGF-BB into non-obese (NOD) mice following femoral ligation also resulted in significant increases in vessel maturity as indicated by the density of vessels staining positive for α-SMA. (Asterisk indicates statistical significance relative to blank scaffold, $p < 0.05$; two asterisks indicate statistical significance relative to VEGF or PDGF-BB alone $p < 0.05$; ns indicates no statistical significance, $p > 0.05$. Magnification for photomicrographs 400X). (Adapted by permission from Macmillan Publishers Ltd. *Nat. Biotechnol.*, Richardson, T.P., Peters, M.C., Ennett, A.B., and Mooney, D.J., Polymeric system for dual growth factor delivery. 19(11), 1029–1034. Copyright 2001.)

matrix into the pores of the scaffold, while PDGF-BB released more slowly from embedded microspheres. Subcutaneously implanted scaffolds promoted extensive neovascularization within the scaffold, characterized by increased vascular density and more mature vessel formation compared to control conditions with no growth factor or delivery of a single growth factor (Figure 5.5B through F).[45] Therefore, dual component approaches that utilize an embedded microparticle phase offer a straightforward method to achieve release of multiple growth factors and have also been used to stimulate cartilage repair in an osteochondral defect via delivery of IGF-1 and TGF-β1.[60,61] Taken together, multiphase approaches afford the sort of independent spatial localization of growth factors likely necessary to efficiently regulate stem cells and neovascularization, as well as to develop tissue engineering approaches capable of generating complex tissue architectures.

5.2.7 SUMMARY

Encapsulation approaches can be adapted to spatially localize many different growth factors to promote a range of stem cell responses. In these systems, spatial localization of growth factors can be tuned using material processing to change polymer scaffold and microparticle properties as well as using multicomponent materials. As tissue engineering approaches that utilize encapsulation target more complex applications, mechanisms to release growth factors in response to specific stimuli may be of critical importance to regulating stem cells that require multiple sequential signals. Therefore, approaches that use the aforementioned mechanical stimulation or other approaches that employ other stimulus-responsive materials[62] (i.e., pH, heat, ultrasound, magnetic field, electric field, or ligand-induced protein conformational change[63,64]) may aid in the development of complex tissue engineering scaffolds that can temporally trigger release of multiple growth factors to guide stem cell activity.

5.3 COVALENT IMMOBILIZATION OF GROWTH FACTORS

Covalent growth factor linkage to a substrate or scaffold material can limit growth factor diffusion and explicitly define the initial local density or concentration of growth factor, respectively. In this manner, covalently linked growth factors can serve as sustained biological signals that are presented by a scaffold material to locally regulate cell activity. This section begins with a brief discussion of several general methods and considerations related to covalent growth factor immobilization. We then examine several approaches in which covalent immobilization has been used to achieve surface localization and spatial localization of growth factors to regulate neovascularization and stem cell activity (Figure 5.1C). Finally, we discuss emerging methods to immobilize growth factors using linking chemistries that degrade in response specific biological signals.

5.3.1 IMMOBILIZATION STRATEGIES

There are multiple key considerations when a covalent conjugation strategy is used to tether growth factor to a material. First, growth factors interact with cell surface receptors through specific sites or domains present on a growth factor. Therefore, immobilization strategies that prevent these domains from binding a growth factor receptor can decrease the efficacy of a tethered growth factor and interfere with any predicted regulation of cell activity. Crystallographic data for a growth factor bound to its receptor is often useful when designing covalent immobilization strategies, although this information is not always available. Another consideration is the type of biochemical functionality used for immobilization. Many strategies utilize amino acid residues that are ubiquitous across all proteins, such as primary amine groups,[65–70] to covalently bond growth factors to a material. While these immobilization approaches do not necessarily dictate growth factor orientation, some studies have nonetheless indicated that covalently tethered growth factors exhibit better activity when compared to growth factors simply adsorbed to a substrate.[66] Some growth factor immobilization strategies use less common amino acid residues such as cystines[71] or fusion proteins with specific motifs for enzyme-facilitated immobilization[72,73] to potentially control growth factor orientation. However, the issue of orientation and its impact on growth factor signaling remains an unsolved issue in growth factor immobilization strategies. In view of these considerations, many approaches have successfully used tethered growth factors to regulate stem cell fate and neovascularization.

5.3.2 GROWTH FACTOR IMMOBILIZATION TO 2D SURFACES

Surface localization of growth factors can be achieved by covalently immobilizing soluble growth factors to the surface of a cell culture substrate. During adherent cell culture, a major part of the local cell environment is dictated by the cell culture substrate. Therefore, signals conveyed through these cell–substrate interactions can play major roles in regulating stem cell behavior such as proliferation, differentiation, and apoptosis. Covalent growth factor immobilization to a cell culture substrate exploits this relationship between adherent cells and a culture surface to promote growth factor signaling, which can be used to influence stem cell fate. Covalent immobilization on 2D surfaces has been applied to growth factors such as VEGF,[71–73] EGF,[65,66] LIF,[67] stem cell factor (SCF),[67] NGF,[68] and bFGF[69,70] using covalent immobilization strategies previously discussed.

Surface-immobilized growth factors can serve as sustained biological signals to regulate stem cell activity. One particularly important growth factor for maintenance of murine embryonic stem cell (mESC) pluripotency is LIF. Traditionally, mESC cultures require a continual supply of soluble LIF to prevent mESC differentiation. However, immobilization of LIF on poly(octadecene-alt-maleic anhydride) copolymer films has been used to maintain mESC pluripotency for up to 2 weeks in the absence of soluble LIF (Figure 5.6). Furthermore, changes in immobilized LIF density were used to activate multiple signal transduction pathways, including the Jak-STAT and MAP kinase pathways, in a density-dependent manner.[67] Surface localization can also be used to generate intracellular signal levels that exceed those achieved by saturating concentrations of a soluble growth factor, as demonstrated in a study in which EGF was immobilized on a polymer film for hMSC culture.[65] Therefore, surface localization of growth factors through covalent

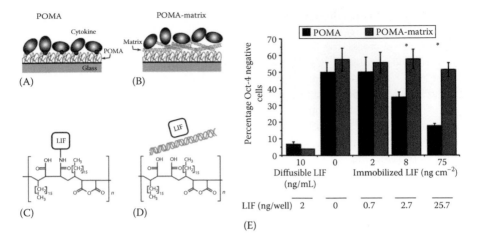

FIGURE 5.6 (See color insert.) Surface-immobilized growth factors can serve as a sustained, surface localized biological signal to regulate stem cell activity. Covalent immobilization of LIF to a polymer film substrate was used to maintain mouse embryonic stem cell (mESC) pluripotency *in vitro*. LIF was (A and C) covalently immobilized to poly(octadecene-alt-maleic anhydride) (POMA) or (B and D) noncovalently adsorbed to an ECM coating deposited on top of hydrolyzed POMA (POMA-matrix). Covalently immobilized LIF maintained Oct-4 expression in a surface density–dependent manner while adsorbed LIF did not as indicated by (E) expression of Oct-4 in mESCs cultured for 72 h. Diffusible LIF was replaced during medium change every 24 h (Error bars are standard error of the mean. Asterisk indicates significant difference between POMA and POMA-matrix conditions, $p < 0.05$). (Adapted by permission from Macmillan Publishers Ltd. *Nat. Methods* Alberti, K., Davey, R.E., Onishi, K., George, S., Salchert, K., Seib, F.P., Bornhauser, M., Pompe, T., Nagy, A., Werner, C., and Zandstra, P.W., Functional immobilization of signaling proteins enables control of stem cell fate. 5(7), 645–650. Copyright 2008.)

immobilization can locally regulate stem cell fate at the cell culture surface using less total growth factor, and thereby convey growth factor signals more efficiently than soluble systems.

5.3.3 GROWTH FACTOR IMMOBILIZATION WITHIN 3D POLYMER SCAFFOLDS

Covalent immobilization strategies can achieve precise spatial localization of growth factors within a scaffold and deliver sustained signals to local cell populations. Many of the covalent immobilization approaches utilized on 2D cell culture substrates can also be used to immobilize growth factors to 3D scaffold materials. Therefore, approaches aimed at regulating stem cell fate have used the aforementioned strategies to immobilize growth factors such as TGF-β2,[74] VEGF,[72,75–77] LIF,[78] or β-nerve growth factor (β-NGF)[73] to polymer scaffolds. For example, LIF immobilized within a synthetic polyester scaffold was used to provide a sustained, spatially localized growth factor signal to mESCs. Interestingly, mESCs seeded in scaffolds with immobilized LIF maintained pluripotency more efficiently than cells on scaffolds with or without soluble LIF.[78] This type of approach is particularly useful when, as described in the case of LIF and mESCs, a continuous growth factor signal is needed to maintain a specific population of stem cells.

5.3.4 IMMOBILIZED GROWTH FACTORS WITH DEGRADABLE LINKERS

Covalent immobilization typically achieves more precise spatial localization of growth factors within scaffolds when compared to growth factor encapsulation approaches. However, this explicit localization is not achieved without limitations— steric hindrance of tethered growth factors or a lack of chemotactic gradients from local growth factor release may ultimately limit the utility of covalent approaches in some tissue engineering applications. One alternative strategy that has been used to achieve spatial localization over growth factors such as β-NGF[73] or VEGF[76,77] employs chemical tethers that degrade, leading to growth factor release into solution. In this manner, the mechanism by which the molecular tether is degraded, such as hydrolysis or enzymatic degradation, modulates growth factor release from a scaffold.

Enzymatically degradable linking chemistries can be used to spatially local- ize growth factor release in response to cellular activity. To achieve this, growth factors can be engineered to contain sites that facilitate release in response to spe- cific cell-secreted enzymes. For example, a mutant fusion protein of β-NGF was designed to contain a plasmin-sensitive cleavage site within the domain used for immobilization. β-NGF immobilized in fibrin hydrogels was released from the net- work in response to plasmin secreted by infiltrating neurons. This method achieved enhanced *in vivo* neurite extension compared to hydrogels with a nondegradable linker or nonimmobilized β-NGF.[73] This type of approach has also been used to locally regulate neovascularization. For example, immobilization of a mutant fusion protein within a fibrin hydrogel scaffold was used to release VEGF in response to plasmin secreted by invading cells. Implantation of these scaffolds into an embry- onic chick chorioallantoic membrane (CAM) resulted in localized neovasculariza- tion at the implantation site, while only diffuse neovessel formation was observed in hydrogels prepared with soluble VEGF (Figure 5.7).[77] Furthermore, subcutane- ous implantation of these scaffolds into rats resulted in formation of fully remod- eled, vascularized tissue similar to native tissue architecture.[77] In both β-NGF and VEGF examples, growth factors were released in response to plasmin secretion from invading cells. These types of "cell-demanded" approaches for regulation of growth factor release may be pivotal when attempting to regulate stem cell fate and neovascularization.

5.3.5 SUMMARY

Covalent immobilization can be used to create tissue engineering scaffolds that pres- ent stem cells with sustained, spatially localized growth factor signals. Additionally, immobilization approaches that employ degradable linking chemistries can be used to design scaffolds that release growth factors in response to specific signals. While examples described in this section highlighted the use of enzymatically degrad- able linking chemistries to create "cell-demanded" release of growth factor, linking

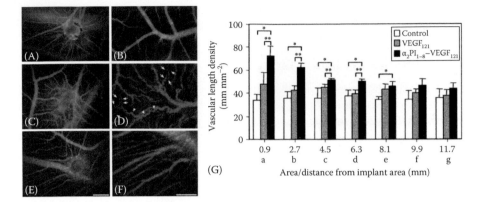

FIGURE 5.7 (See color insert.) "Cell-demanded" release of growth factors can be achieved using enzymatically degradable linking chemistries. Plasmin-sensitive linking chemistries were used to create fibrin hydrogels that released VEGF in response to infiltrating cells. Implantation of these scaffolds into chick chorioallantoic membranes led to enhanced localized neovascularization as indicated by *in vivo* fluorescence microscopy of vascular networks: (A and B) Control grafts made of fibrin alone, (C and D) fibrin formulated with soluble VEGF, and (E and F) fibrin formulated with enzymatically (plasmin) released VEGF ($\alpha_2 PI_{1-8}$-VEGF$_{121}$). Both VEGF forms increased neovascularization as indicated by blood vessel density; however, vascular hierarchy as well as neovessel morphology induced by enzymatically released VEGF (E and F) appeared more normal than neovascularization induced by freely diffusible VEGF. Bars = 1 mm (A, C, and E); 0.5 mm (B, D, and F). (G) Quantification of mean vascular length density also indicated enhanced neovascularization via enzymatically released VEGF. (Data are mean ± standard deviation. One asterisk indicates significance versus control fibrin alone, $p < 0.05$; two asterisks indicate significance versus soluble VEGF, $p < 0.05$). (Adapted from Ehrbar, M. et al., *Circ. Res.*, 94(8), 1124, 2004. With permission.)

chemistries that degraded in response to specific external stimulation such as light[79] or ultrasound[80] could potentially be used for stimulus-responsive growth factor release. Stimulus-based growth factor release may be important in approaches aimed at spatially and temporally regulating stem cell behavior.

5.4   GROWTH FACTOR–MATERIAL AFFINITY INTERACTIONS AND GROWTH FACTOR SEQUESTRATION

As described in the introduction of this chapter, many growth factors exhibit affinity for specific components of the ECM, and these interactions can result in growth factor localization within tissues. These noncovalent, "affinity-based" interactions, as termed by Sakiyama-Elbert et al.,[81] can also be used to localize growth factors in synthetic systems via a number of approaches (Figure 5.1B). One approach involves sequestering a growth factor out of solution onto a surface or within a scaffold via growth factor–material affinity. Another approach uses growth factor–material affinity to tune the spatial localization of an encapsulated growth factor and modulate its transport through a scaffold. In either approach, changes in the amount of growth factor–binding moiety upon or within a material, or changes to the strength of the affinity interaction can potentially be used to fine-tune growth factor localization. Therefore, approaches have used specific material-binding domains,[82–84] metal-ion chelation,[85–88] charge–charge attraction,[89–93] heparin binding,[94–104] and peptide ligands[105–112] to create scaffolds that spatially localize growth factors by either sequestering them out of solution or modulating their transport within a scaffold.

5.4.1 Material Intrinsic Interactions and Engineered Binding Domains

Specific growth factor–material interactions can be used to locally sequester growth factors within a scaffold. In some cases, growth factors have specific binding domains for natural materials such as collagen[113] or fibrin[114,115] that can be exploited to localize growth factors. However, not all growth factor possess these domains. Therefore, one way to achieve a specific growth factor–scaffold interaction is to engineer fusion proteins containing both growth factor and a known material-binding domain.[82–84] For example, HGF has been engineered to specifically bind to collagen-based materials by creating a fusion protein containing both HGF and collagen-binding domains (CBD-HGF). Upon exposure to collagen-based materials, such as collagen-coated surfaces or collagen sponges, the CBD promoted specific growth factor–collagen binding that resulted in surface or spatial localization of HGF, respectively. Importantly, sequestered CBD-HGF lead to enhanced HUVEC proliferation on collagen-coated cell culture substrates compared to HGF without a collagen-binding domain (Figure 5.8A). In the same study, collagen sponges were preincubated in a CBD-HGF solution and then implanted subcutaneously in mice. Results indicated that CBD-HGF was maintained within the sponges for 7 days, leading to significantly enhanced neovascularization within scaffolds when compared to HGF (Figure 5.8D through E).[84] This approach broadly demonstrates that mimicry of natural growth factor–material interactions can be applied to tissue engineering approaches aimed at locally regulating cell activity.

5.4.2 Metal-Ion Chelation

Metal ion–chelating moieties can be engineered into growth factor fusion proteins to promote growth factor–material interactions. In general, these approaches are similar to the those described in the previous section, but instead take advantage of Ni(II) coordination between histidine tags and anionic chemical moieties such as glycidyl methacrylate-iminodiacetic acid[85] or nitrilotriacetic acid (NTA).[86–88] For example, a c-terminal histidine tag has been used to localize EGF on the surface of NTA-Ni(II)-presenting self-assembled monolayer (SAM) substrates. In this approach, variations in NTA surface content controlled the surface density of EGF and were used to optimize substrates for NSCs culture. This optimization resulted in superior NSC expansion and maintenance of multipotency when compared to physically adsorbed or covalently immobilized EGF.[86] A similar method was used to immobilize dimeric EGF, and even greater selective expansion of NSCs was achieved when compared to monomeric EGF.[87] Methods that explicitly define the density and orientation of growth factors on a surface, such as metal-ion chelation, can be tuned to precisely

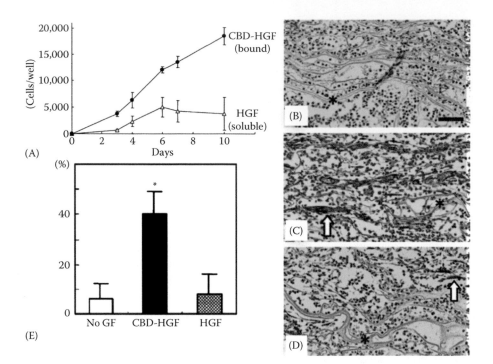

FIGURE 5.8 (See color insert.) Specific material-binding domains can be used to localize growth factors on surfaces or within scaffolds via sequestration. A fusion protein of hepatocyte growth factor containing a collagen-binding domain (CBD-HGF) sequestered to collagen materials and locally regulated neovascularization. Sequestration of CBD-HGF to collagen-coated substrates lead to significant increases in (A) human umbilical vein endothelial cell (HUVEC) proliferation compared to HGF conditions. CBD-HGF sequestered within collagen sponges promoted increased localized neovascularization as indicated by (B–D) α-smooth muscle actin stained histological sections and (E) mean blood vessel density measurements of implanted collagen sponges soaked in buffer containing (B) no growth factor, (C) CBD-HGF, or (D) HGF before implantation. (Error bars represent standard deviation, n = 15. Asterisk indicates significance compared to no growth factor, $p < 0.01$). (Adapted from *Biomaterials*, 28(11), Kitajima, T., Terai, H., and Ito, Y., A fusion protein of hepatocyte growth factor for immobilization to collagen, 1989–1997. Copyright (2007). With permission from Elsevier.)

regulate stem cell fate. Furthermore, approaches capable of regulating growth factor dimerization as well as stem cell–growth factor interactions may potentially offer regulation of stem cell fate

5.4.3 CHARGE INTERACTIONS

Interactions between charged growth factors and ionic scaffold materials can be used to locally sequester and spatially localize growth factors. Scaffolds with acidic functionalities attract growth factors with positive net charges, while scaffolds with basic functionalities attract growth factors with negative net charges. Through these types of charge–charge interactions, materials like acidic gelatin hydrogels have been used to sequester and spatially localize positively charged growth factors like bFGF.[90] For example, when acidic gelatin hydrogels were used to sequester bFGF and then subcutaneously implanted in mice, sequestered bFGF remained spatially localized within the hydrogel and promoted more extensive localized neovascularization within the tissue when compared to basic gelatin hydrogels.[91] Similar approaches using acidic gelatin-PLA copolymer scaffolds have been used to spatially localize bFGF and G-SCF simultaneously and promote localized neovascularization within ischemic hindlimbs of mice.[92] Therefore, while charge–charge interactions are not as specific as specific protein-binding or material-binding domains, they offer a straightforward and broadly adaptable method to sequester growth factors for tissue engineering applications. As such, charge–charge interactions have also been used to design microsphere-based approaches to sequester and fine-tune spatial localization of growth factors such as VEGF or bFGF to promote local neovascularization within tissues.[89,93] Taken together, these examples demonstrate how materials processing can be used to capitalize on the intrinsic charge of specific growth factors.

5.4.4 Heparin Functionalization and Heparin-Mimetic Peptides

As described in the introduction, glycosaminoglycans (GAGs) such as heparin and heparan sulfate are biomacromolecules present throughout the ECM of different tissues. Many growth factors capable of regulating stem cell activity have heparin-binding domains that bind GAG polysaccharide chains of proteoglycans. This binding interaction is characterized by a pattern of negatively charged sulfates found on polysaccharide chains that drive GAG–growth factor HBD interactions.[116] These heparin-binding growth factors include most of the growth factors discussed in this chapter, and a recent survey of known heparin-binding proteins indicated that there are greater than 100 growth factors, cytokines, chemokines, and morphogens known to interact with heparin or heparan sulfates[5]. As a result, GAGs such as heparin have been covalently immobilized to scaffold materials including collagen,[94,95] PEG,[96–99,102] or alginate[100,101,103,104] to sequester as well as tune spatial localization of growth factors, including bFGF[94–101,103,104] and VEGF.[99,102]

In some cases, heparin functionalization has been used to sequester growth factors out of solution into the interior of a scaffold.[96] However, in most cases, heparin functionalization has been instead used to modulate the rate at which an encapsulated growth factor releases from a scaffold. For example, heparin functionalization of hydrogels formed from cross-linked hyaluronic acid and poly(ethylene glycol) was used to tune bFGF release from scaffolds. Hydrogels functionalized with high concentrations of heparin slowly released bFGF while hydrogels functionalized with low concentrations released bFGF at an intermediate rate compared to unfunctionalized scaffolds. When these scaffolds were implanted subcutaneously in mice, the intermediate release rate and subsequent spatial localization achieved by scaffolds with low concentrations of heparin promoted greater localized neovascularization compared to both nonfunctionalized and highly functionalized scaffolds.[98] Therefore, optimum conditions for local neovascularization were achieved by using varied degrees of heparin functionalization to tune bFGF localization. This approach demonstrates that affinity-based interactions can be used to enhance the performance of encapsulation approaches by tuning spatial localization of growth factors. Furthermore, while this specific example demonstrates locally regulated neovascularization, similar approaches could potentially be used to locally regulate stem cell activity.

Short sulfated peptides that mimic heparin and bind growth factors via the HBD[117,118] could potentially be immobilized within scaffolds to directly sequester and spatially localize heparin-binding growth factors without the use of heparin (Figure 5.1B). These sulfated peptides offer several advantages over full-length GAG polysaccharides as they can be made via synthetic strategies and can be designed to contain functionalities for immobilization within scaffolds. Additionally, simple changes to the sulfated peptide amino acid sequence can be used to modulate peptide–growth factor affinity. Although these peptides have not yet been used in tissue engineering approaches, the sequence-based control over affinity offered by these peptides could be used to modulate growth factor spatial localization within scaffolds with greater precision than offered by heparin immobilization. Therefore, these heparin mimics offer a promising mechanism for spatially localizing growth factors that could be broadly applied to tissue engineering approaches aimed at locally regulating cell activity.

5.4.5 Heparin-Binding Peptides

Immobilization of heparin-binding peptides (HBPs) can be used to convey spatial localization over heparin-binding growth factors by specifically binding heparin. Similar to scaffolds functionalized with heparin, heparin sequestered within a scaffold via an immobilized HBP can also be used to spatially localize growth factors (Figure 5.9A).[107] In this type of approach, two different affinity interactions are employed to indirectly regulate growth factors: (1) HBP:heparin affinity, and (2) heparin:growth factor affinity. Practically, these peptides are an attractive alternative to heparin, as they can be synthesized and immobilized using standard techniques without modifying heparin for covalent immobilization. Therefore, this approach has been applied to growth factors such as β-NGF,[108,109] bFGF,[110] VEGF,[111] neurotrophin 3 (NT-3),[107,112] sonic hedgehog (SHH),[107] and platelet-derived growth factor-AA (PDGF-AA).[107]

Spatial localization achieved by immobilized HBP can be used to locally regulate cell activity. For example, HBP functionalization was used to create fibrin hydrogel scaffolds capable of retaining significant concentrations of encapsulated heparin, NT-3, and PDGF-AA over several days. Mouse embryonic stem cell–derived neural progenitor cells seeded within these scaffolds were significantly influenced by growth factors spatially localized within hydrogel scaffolds. In particular, simultaneous spatial localization of NT-3 and PDGF-AA via HBP not only increased the fraction of neural progenitors present in embryoid bodies but also increased the level of progenitor differentiation into neurons and oligodendrocytes while suppressing astrocyte differentiation (Figure 5.9B and C).[107] This approach highlights the utility of immobilized HBP as a means of spatially localizing multiple different growth

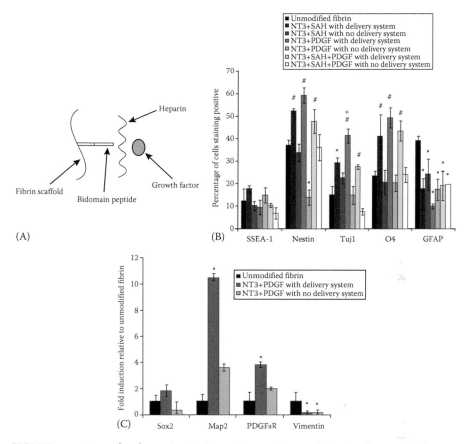

FIGURE 5.9 (See color insert.)   HBPs can be used to spatially localize growth factors. (A) Fibrin hydrogel scaffolds functionalized with HBPs can be used to sequester heparin and concomitantly modulate the release of heparin-binding growth factors such as neurotrophin 3 (NT-3), sonic headgehog (SHH), and platelet-derived growth factor-AA (PDGF-AA). Spatial localization of growth factors regulated the activity of mouse embryonic stem cell–derived neural progenitor cells (ESNPCs) seeded within fibrin scaffolds. ESNPCs were cultured within scaffolds containing different growth factors for 14 days and then analyzed via (B) fluorescence-activated cell sorting using the markers SSEA-1 (undifferentiated ES cells), nestin (neural progenitors), Tuj1 (early neurons), O4 (oligodendrocytes), and GFAP (astrocytes). Results were verified via (C) real-time RT-PCR analysis using the markers Sox2 (neural progenitors), microtubule-associated protein-2 (Map2; early neurons), platelet-derived growth factor α receptor (PDGFαR; early oligodendrocytes), and vimentin (astrocytes) and confirmed that combined localization of NT-3 and PDGF-AA increased the level of progenitor differentiation into neurons and oligodendrocytes while suppressing astrocyte differentiation. (Adapted from *Stem Cell Res.*, 1(3), Willerth, S.M., Rader, A., and Sakiyama-Elbert, S.E., The effect of controlled growth factor delivery on embryonic stem cell differentiation inside fibrin scaffolds, 205–218. Copyright 2008. With permission from Elsevier.)

factors within scaffolds and demonstrates that HBPs can be used in place of covalently immobilized heparin.

5.4.6 Growth Factor–Binding Peptide Ligands

Specific growth factor–material interactions can be designed into scaffolds using peptide ligands that bind specific growth factors (Figure 5.1B). Various methods, such as phage display and intelligent design from growth factor receptor domains, have been used to develop peptide ligands that specifically bind growth factors such as keratinocyte growth factor (KGF),[119] VEGF,[120] NGF,[105] and tumor necrosis factor α (TNFα).[106,121–123] In many cases, these short peptides may offer significant advantages over other affinity approaches as they can be designed to bind a single growth factor with high specificity and can potentially be incorporated into materials using standard techniques. Similar to several examples already described in this chapter, materials functionalized with peptide ligands can be used to maintain high concentrations of growth factor within a scaffold to create localized signals for encapsulated cells.[105]

Although growth factor–specific ligands have yet to be used extensively in tissue engineering scaffolds, their specificity could more closely mimic the spatial localization of growth factors observed in the ECM and precisely regulate stem cell activity or neovascularization. The use of growth factor–binding peptides also offers several other capabilities when immobilized within a scaffold including: (1) The ability to tune local concentrations of growth factor within a scaffold though peptide modifications that affect peptide–growth factor affinity and (2) the ability to specifically localize multiple growth factors within a scaffold by using multiple growth factor–specific binding peptides. In this manner, the ability to control growth factor–peptide binding affinity as well as growth factor specificity could potentially be used to spatially localize a wide range of growth factors within a scaffold and design synthetic materials that precisely regulate stem cell fate or neovascularization.

5.4.7 Summary

Noncovalent, affinity-based approaches can be used to create scaffolds that mimic the spatial localization of growth factor observed in the ECM of tissues. As knowledge and understanding of growth factor–material interactions grow, our ability to regulate growth factors will also expand. Therefore, methods to identify growth factor–binding ligands may emerge as powerful tools for regulating growth factor signaling in tissue engineering and understanding the mechanisms that govern stem cell behavior.

5.5 FUTURE PERSPECTIVE: SPATIALLY PATTERING GROWTH FACTORS

Methods to create spatially patterned microenvironments of growth factors within scaffolds will likely be necessary to engineer fully functional tissues from stem cells. As described in the introduction, the ECM of tissues exerts precise regulation over growth factors and other signals that locally regulate cell activity. However,

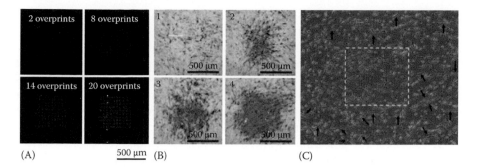

FIGURE 5.10 (See color insert.)   Spatially patterned growth factors can be used to locally regulate stem cell differentiation. Spatially patterned BMP-2 was used to localize muscle-derived stem cell (MDSC) osteogenesis. Patterns with different amounts of deposited BMP-2 were created by printing and "overprinting" concentrated drops of BMP-2 on fibrin-coated slides and (A) were visualized using Cy3-labeled BMP-2. MDSCs were cultured on printed surfaces for 72 h and then (B) stained for alkaline phosphatase activity (red) indicating local-ized osteogenesis. Overprints of BMP-2 were as follows: (1) 2 overprints, (2) 8 overprints, (3) 14 overprints, and (4) 20 overprints. (C) Additionally, MDSCs cultured on regions con-taining 14 overprint were analyzed for myotube formation, indicating that spatially patterned BMP-2 prevented myogenic differentiation (Dotted line indicates edge of pattern, arrows indicate myotubes). (From Phillippi, J.A., Miller, E., Weiss, L., Huard, J., Waggoner, A., and Campbell, P.: Microenvironments engineered by inkjet bioprinting spatially direct adult stem cells toward muscle- and bone-like subpopulations. *Stem Cells.* 2008. 26(1). 127–134. Copyright Wiley-VCH Verlag GmbH & Co. KGaA. Adapted with permission.)

within a tissue, the ECM contains a multitude of different microenvironments, such as stem cell niches, that generate and maintain the hierarchical architecture of func-tioning tissues. Therefore, tissue-engineered scaffolds that contain micrometer scale patterns of localized growth factors may more closely recapitulate the function of the natural ECM within tissues to regulate stem cell fate. Currently, very few approaches to spatially pattern growth factor microenvironments exist, and only a handful have been applied to regulate stem cells or neovascularization. The approaches that have been used to regulate stem cell fate or neovascularization have built upon approaches already described in this chapter.

Substrates can be patterned with growth factors to create discrete microenvi-ronments capable of locally regulating stem cell activity. One approach to achieve this patterning is to deposit or print growth factors onto surfaces.[124,125] For example, microenvironments containing different amounts of growth factor were created by inkjet printing square features of BMP-2 on fibrin-coated glass slides (Figure 5.10A). When these substrates were used to culture muscle-derived stem cells (MDSCs), cells residing on square features expressed increased levels of osteogenic differen-tiation depending on the amount of BMP-2 deposited (Figure 5.10B). Additionally, MDSCs residing on regions of fibrin without BMP-2 formed myotubes indicative of myogenic differentiation (Figure 5.10C).[125] In this case, micrometer-scale patterns of deposited growth factor served as local microenvironments that regulated stem cell activity on a substrate.

Processing methods can also be used to create 3D patterns of growth factors within a scaffold. One method to achieve this type of spatial patterning is to fabricate scaffolds using multiple layers of material to encapsulate different growth factors.[126,127] For example, a bilayered, porous PLG scaffold containing encapsulated VEGF in one layer and co-encapsulated VEGF and PDGF in a second layer was used to spatially pattern growth factor within a scaffold. When this scaffold was implanted in ischemic hindlimbs of SCID mice, layers containing only VEGF resulted in densely formed networks of immature blood vessels, while layers containing both VEGF and PDGF resulted in less dense but significantly more mature blood vessels networks.[126] In this type of approach, each layer independently controlled the local concentration of growth factors at distinct, spatially patterned positions throughout the scaffold. Development of such an approach to spatially pattern multiple growth factors that stimulate different stem cell fates could potentially be used to build complex tissue interfaces and significantly expand tissue engineering capabilities.

Spatial patterning approaches can also be used to screen for the effects of different biological signals, such as growth factors, on stem cell behavior. One particular type of approach has focused on exposing stem cells to defined biochemical signaling environments, with an emphasis on controllable chemistries and simple, automated processing methods. For example, formation of multiple 1 μL volume PEG hydrogels within a macroscale PEG hydrogel was used to create arrays of 3D microenvironments, in which human mesenchymal stem cells (hMSCs) were exposed to various combinations of signals, including bFGF, cell adhesion peptides, and degradable network chemistries. Each component influenced hMSC viability, and microenvironments containing degradable PEG networks, bFGF, and the cell adhesion peptide Arg-Gly-Asp-Ser-Pro (RGDSP) led to a significant increase in hMSC spreading compared to all other conditions.[128] Arrays of chemically defined microenvironments have also been developed by spatially patterning surfaces. In particular, investigators have used the well-defined chemistries of alkanethiolate SAMs combined with simple micropatterning techniques to screen for the effects of immobilized peptides on hMSCs[129] and hESCs.[130] Similar arrays may also be used to screen for the effects of different immobilized growth factors on stem cell activity in emerging studies. Spatial patterning approaches can be used to create 2D or 3D microenvironments to rapidly screen for conditions that promote specific stem cell behaviors, a critical capability in emerging stem cell biology applications.

5.6 CONCLUSION

This chapter surveyed a range of approaches used to spatially localize growth factors within scaffolds, within tissues, and on surfaces to locally regulate stem cell activity and neovascularization. Materials have been designed to encapsulate, covalently immobilize, or sequester growth factors to mimic the spatial localization over growth factors observed in the ECM of tissues. While all of the examples in this chapter explore methods to locally regulate stem cells or neovascularization, very few have been applied to clinical tissue engineering approaches. As methods continually evolve to localize multiple growth factors and spatially pattern growth factor within scaffolds, spatial localization over growth factors will play an increasingly important role in tissue engineering approaches aimed at regulating stem cell fate.

REFERENCES

1. Amariglio, N.; Hirshberg, A.; Scheithauer, B. W.; Cohen, Y.; Loewenthal, R.; Trakhtenbrot, L.; Paz, N.; Koren-Michowitz, M.; Waldman, D.; Leider-Trejo, L.; Toren, A.; Constantini, S.; Rechavi, G., Donor-derived brain tumor following neural stem cell transplantation in an ataxia telangiectasia patient. *PLoS Med* 2009, 6, (2), e1000029.
2. Williams, R. L.; Hilton, D. J.; Pease, S.; Willson, T. A.; Stewart, C. L.; Gearing, D. P.; Wagner, E. F.; Metcalf, D.; Nicola, N. A.; Gough, N. M., Myeloid leukaemia inhibitory factor maintains the developmental potential of embryonic stem cells. *Nature* 1988, 336, (6200), 684–687.
3. Li, X. J.; Du, Z. W.; Zarnowska, E. D.; Pankratz, M.; Hansen, L. O.; Pearce, R. A.; Zhang, S. C., Specification of motoneurons from human embryonic stem cells. *Nat Biotechnol* 2005, 23, (2), 215–221.
4. Oehrl, W.; Panayotou, G., Modulation of growth factor action by the extracellular matrix. *Connect Tissue Res* 2008, 49, (3), 145–148.
5. Ori, A.; Wilkinson, M. C.; Fernig, D. G., The heparanome and regulation of cell function: Structures, functions and challenges. *Front Biosci* 2008, 13, 4309–4338.
6. Bishop, J. R.; Schuksz, M.; Esko, J. D., Heparan sulphate proteoglycans fine-tune mammalian physiology. *Nature* 2007, 446, (7139), 1030–1037.
7. van Noort, D.; Ong, S. M.; Zhang, C.; Zhang, S. F.; Arooz, T.; Yu, H., Stem cells in microfluidics. *Biotechnol Progr* 2009, 25, (1), 52–60.
8. Grayson, A. C. R.; Shawgo, R. S.; Li, Y. W.; Cima, M. J., Electronic MEMS for triggered delivery. *Adv Drug Deliv Rev* 2004, 56, (2), 173–184.
9. Zhang, X. J.; Godbey, W. T., Viral vectors for gene delivery in tissue engineering. *Adv Drug Deliv Rev* 2006, 58, (4), 515–534.
10. De Laporte, L.; Shea, L. D., Matrices and scaffolds for DNA delivery in tissue engineering. *Adv Drug Deliv Rev* 2007, 59, (4–5), 292–307.
11. Sokolsky-Papkov, M.; Agashi, K.; Olaye, A.; Shakesheff, K.; Domb, A. J., Polymer carriers for drug delivery in tissue engineering. *Adv Drug Deliv Rev* 2007, 59, (4–5), 187–206.
12. Hollister, S. J., Porous scaffold design for tissue engineering. *Nat Mater* 2005, 4, (7), 518–524.
13. Thompson, J. A.; Anderson, K. D.; DiPietro, J. M.; Zwiebel, J. A.; Zametta, M.; Anderson, W. F.; Maciag, T., Site-directed neovessel formation in vivo. *Science* 1988, 241, (4871), 1349–1352.
14. Kanematsu, A.; Yamamoto, S.; Ozeki, M.; Noguchi, T.; Kanatani, I.; Ogawa, O.; Tabata, Y., Collagenous matrices as release carriers of exogenous growth factors. *Biomaterials* 2004, 25, (18), 4513–4520.
15. Peters, M. C.; Polverini, P. J.; Mooney, D. J., Engineering vascular networks in porous polymer matrices. *J Biomed Mater Res* 2002, 60, (4), 668–678.
16. Ennett, A. B.; Kaigler, D.; Mooney, D. J., Temporally regulated delivery of VEGF in vitro and in vivo. *J Biomed Mater Res A* 2006, 79, (1), 176–184.
17. Sheridan, M. H.; Shea, L. D.; Peters, M. C.; Mooney, D. J., Bioabsorbable polymer scaffolds for tissue engineering capable of sustained growth factor delivery. *J Control Release* 2000, 64, (1–3), 91–102.
18. Hile, D. D.; Amirpour, M. L.; Akgerman, A.; Pishko, M. V., Active growth factor delivery from poly(D,L-lactide-co-glycolide) foams prepared in supercritical $CO(2)$. *J Control Release* 2000, 66, (2–3), 177–185.

19. Sohier, J.; Hamann, D.; Koenders, M.; Cucchiarini, M.; Madry, H.; van Blitterswijk, C.; de Groot, K.; Bezemer, J. M., Tailored release of TGF-[beta]1 from porous scaffolds for cartilage tissue engineering. *Int J Pharm* 2007, 332, (1–2), 80–89.

20. Drury, J. L.; Mooney, D. J., Hydrogels for tissue engineering: Scaffold design variables and applications. *Biomaterials* 2003, 24, (24), 4337–4351.

21. Lin, C. C.; Metters, A. T., Hydrogels in controlled release formulations: Network design and mathematical modeling. *Adv Drug Deliv Rev* 2006, 58, (12–13), 1379–1408.

22. Augst, A. D.; Kong, H. J.; Mooney, D. J., Alginate hydrogels as biomaterials. *Macromol Biosci* 2006, 6, (8), 623–633.

23. Hiemstra, C.; Zhong, Z.; van Steenbergen, M. J.; Hennink, W. E.; Feijen, J., Release of model proteins and basic fibroblast growth factor from in situ forming degradable dextran hydrogels. *J Control Release* 2007, 122, (1), 71–78.

24. Albes, J. M.; Klenzner, T.; Kotzerke, J.; Thiedemann, K. U.; Schafers, H. J.; Borst, H. G., Improvement of tracheal autograft revascularization by means of fibroblast growth factor. *Ann Thorac Surg* 1994, 57, (2), 444–449.

25. Downs, E. C.; Robertson, N. E.; Riss, T. L.; Plunkett, M. L., Calcium alginate beads as a slow-release system for delivering angiogenic molecules in vivo and in vitro. *J Cell Physiol* 1992, 152, (2), 422–429.

26. Wilcke, I.; Lohmeyer, J. A.; Liu, S.; Condurache, A.; Kruger, S.; Mailander, P.; Machens, H. G., VEGF(165) and bFGF protein-based therapy in a slow release system to improve angiogenesis in a bioartificial dermal substitute in vitro and in vivo. *Langenbecks Arch Surg* 2007, 392, (3), 305–314.

27. Tabata, Y.; Miyao, M.; Ozeki, M.; Ikada, Y., Controlled release of vascular endothelial growth factor by use of collagen hydrogels. *J Biomater Sci Polym Ed* 2000, 11, (9), 915–930.

28. Peters, M. C.; Isenberg, B. C.; Rowley, J. A.; Mooney, D. J., Release from alginate enhances the biological activity of vascular endothelial growth factor. *J Biomater Sci Polym Ed* 1998, 9, (12), 1267–1278.

29. Lee, K. Y.; Peters, M. C.; Anderson, K. W.; Mooney, D. J., Controlled growth factor release from synthetic extracellular matrices. *Nature* 2000, 408, (6815), 998–1000.

30. Silva, E. A.; Mooney, D. J., Spatiotemporal control of vascular endothelial growth factor delivery from injectable hydrogels enhances angiogenesis. *J Thromb Haemost* 2007, 5, (3), 590–598.

31. Kipshidze, N.; Chekanov, V.; Chawla, P.; Shankar, L. R.; Gosset, J. B.; Kumar, K.; Hammen, D.; Gordon, J.; Keelan, M. H., Angiogenesis in a patient with ischemic limb induced by intramuscular injection of vascular endothelial growth factor and fibrin platform. *Tex Heart Inst J* 2000, 27, (2), 196–200.

32. Fasol, R.; Schumacher, B.; Schlaudraff, K.; Hauenstein, K. H.; Seitelberger, R., Experimental use of a modified fibrin glue to induce site-directed angiogenesis from the aorta to the heart. *J Thorac Cardiovasc Surg* 1994, 107, (6), 1432–1439.

33. Lin, C. C.; Anseth, K. S., PEG hydrogels for the controlled release of biomolecules in regenerative medicine. *Pharm Res* 2009, 26, (3), 631–643.

34. Sharma, B.; Williams, C. G.; Khan, M.; Manson, P.; Elisseeff, J. H., In vivo chondrogenesis of mesenchymal stem cells in a photopolymerized hydrogel. *Plast Reconstr Surg* 2007, 119, (1), 112–120.

35. Varde, N. K.; Pack, D. W., Microspheres for controlled release drug delivery. *Expert Opin Biol Ther* 2004, 4, (1), 35–51.

36. Elcin, Y. M.; Dixit, V.; Gitnick, G., Extensive in vivo angiogenesis following controlled release of human vascular endothelial cell growth factor: Implications for tissue engineering and wound healing. *Artif Organs* 2001, 25, (7), 558–565.

37. Cohen, S.; Yoshioka, T.; Lucarelli, M.; Hwang, L. H.; Langer, R., Controlled delivery systems for proteins based on poly(lactic/glycolic acid) microspheres. *Pharm Res* 1991, 8, (6), 713–720.

38. Woo, B. H.; Fink, B. F.; Page, R.; Schrier, J. A.; Jo, Y. W.; Jiang, G.; DeLuca, M.; Vasconez, H. C.; DeLuca, P. P., Enhancement of bone growth by sustained delivery of recombinant human bone morphogenetic protein-2 in a polymeric matrix. *Pharm Res* 2001, 18, (12), 1747–1753.

39. Oldham, J. B.; Lu, L.; Zhu, X.; Porter, B. D.; Hefferan, T. E.; Larson, D. R.; Currier, B. L.; Mikos, A. G.; Yaszemski, M. J., Biological activity of rhBMP-2 released from PLGA microspheres. *J Biomech Eng* 2000, 122, (3), 289–292.

40. Aubert-Pouessel, A.; Venier-Julienne, M. C.; Clavreul, A.; Sergent, M.; Jollivet, C.; Montero-Menei, C. N.; Garcion, E.; Bibby, D. C.; Menei, P.; Benoit, J. P., In vitro study of GDNF release from biodegradable PLGA microspheres. *J Control Release* 2004, 95, (3), 463–475.

41. Mittal, S.; Cohen, A.; Maysinger, D., In vitro effects of brain derived neurotrophic factor released from microspheres. *Neuroreport* 1994, 5, (18), 2577–2582.

42. Pean, J. M.; Menei, P.; Morel, O.; Montero-Menei, C. N.; Benoit, J. P., Intraseptal implantation of NGF-releasing microspheres promote the survival of axotomized cholinergic neurons. *Biomaterials* 2000, 21, (20), 2097–2101.

43. Menei, P.; Pean, J. M.; Nerriere-Daguin, V.; Jollivet, C.; Brachet, P.; Benoit, J. P., Intracerebral implantation of NGF-releasing biodegradable microspheres protects striatum against excitotoxic damage. *Exp Neurol* 2000, 161, (1), 259–272.

44. Kim, T. K.; Burgess, D. J., Pharmacokinetic characterization of 14C-vascular endothelial growth factor controlled release microspheres using a rat model. *J Pharm Pharmacol* 2002, 54, (7), 897–905.

45. Richardson, T. P.; Peters, M. C.; Ennett, A. B.; Mooney, D. J., Polymeric system for dual growth factor delivery. *Nat Biotechnol* 2001, 19, (11), 1029–1034.

46. Peter, S. J.; Lu, L.; Kim, D. J.; Stamatas, G. N.; Miller, M. J.; Yaszemski, M. J.; Mikos, A. G., Effects of transforming growth factor beta1 released from biodegradable polymer microparticles on marrow stromal osteoblasts cultured on poly(propylene fumarate) substrates. *J Biomed Mater Res* 2000, 50, (3), 452–462.

47. Moioli, E. K.; Hong, L.; Guardado, J.; Clark, P. A.; Mao, J. J., Sustained release of TGFbeta3 from PLGA microspheres and its effect on early osteogenic differentiation of human mesenchymal stem cells. *Tissue Eng* 2006, 12, (3), 537–546.

48. Moioli, E. K.; Mao, J. J., Chondrogenesis of mesenchymal stem cells by controlled delivery of transforming growth factor-beta3. *Conf Proc IEEE Eng Med Biol Soc* 2006, 1, 2647–2650.

49. Carrascosa, C.; Torres-Aleman, I.; Lopez-Lopez, C.; Carro, E.; Espejo, L.; Torrado, S.; Torrado, J. J., Microspheres containing insulin-like growth factor I for treatment of chronic neurodegeneration. *Biomaterials* 2004, 25, (4), 707–714.

50. Elcin, Y. M.; Dixit, V.; Gitnick, G., Controlled release of endothelial cell growth factor from chitosan-albumin microspheres for localized angiogenesis: In vitro and in vivo studies. *Artif Cells Blood Substit Immobil Biotechnol* 1996, 24, (3), 257–271.

51. Park, H.; Temenoff, J. S.; Tabata, Y.; Caplan, A. I.; Mikos, A. G., Injectable biodegradable hydrogel composites for rabbit marrow mesenchymal stem cell and growth factor delivery for cartilage tissue engineering. *Biomaterials* 2007, 28, (21), 3217–3227.

52. Kojima, K.; Ignotz, R. A.; Kushibiki, T.; Tinsley, K. W.; Tabata, Y.; Vacanti, C. A., Tissue-engineered trachea from sheep marrow stromal cells with transforming growth factor beta2 released from biodegradable microspheres in a nude rat recipient. *J Thorac Cardiovasc Surg* 2004, 128, (1), 147–153.

53. Marui, A.; Kanematsu, A.; Yamahara, K.; Doi, K.; Kushibiki, T.; Yamamoto, M.; Itoh, H.; Ikeda, T.; Tabata, Y.; Komeda, M., Simultaneous application of basic fibroblast growth factor and hepatocyte growth factor to enhance the blood vessels formation. *J Vasc Surg* 2005, 41, (1), 82–90.

54. Sobajima, S.; Shimer, A. L.; Chadderdon, R. C.; Kompel, J. F.; Kim, J. S.; Gilbertson, L. G.; Kang, J. D., Quantitative analysis of gene expression in a rabbit model of intervertebral disc degeneration by real-time polymerase chain reaction. *Spine J* 2005, 5, (1), 14–23.

55. Layman, H.; Sacasa, M.; Murphy, A. E.; Murphy, A. M.; Pham, S. M.; Andreopoulos, F. M., Co-delivery of FGF-2 and G-CSF from gelatin-based hydrogels as angiogenic therapy in a murine critical limb ischemic model. *Acta Biomater* 2009, 5, (1), 230–239.

56. Hill, E.; Boontheekul, T.; Mooney, D. J., Regulating activation of transplanted cells controls tissue regeneration. *Proc Natl Acad Sci USA* 2006, 103, (8), 2494–2499.

57. Hao, X.; Silva, E. A.; Mansson-Broberg, A.; Grinnemo, K. H.; Siddiqui, A. J.; Dellgren, G.; Wardell, E.; Brodin, L. A.; Mooney, D. J.; Sylven, C., Angiogenic effects of sequential release of VEGF-A165 and PDGF-BB with alginate hydrogels after myocardial infarction. *Cardiovasc Res* 2007, 75, (1), 178–185.

58. Kim, H. D.; Valentini, R. F., Human osteoblast response in vitro to platelet-derived growth factor and transforming growth factor-beta delivered from controlled-release polymer rods. *Biomaterials* 1997, 18, (17), 1175–1184.

59. Simmons, C. A.; Alsberg, E.; Hsiong, S.; Kim, W. J.; Mooney, D. J., Dual growth factor delivery and controlled scaffold degradation enhance in vivo bone formation by transplanted bone marrow stromal cells. *Bone*, 2004, 35, (2), 562–569.

60. Holland, T. A.; Tabata, Y.; Mikos, A. G., Dual growth factor delivery from degradable oligo(poly(ethylene glycol) fumarate) hydrogel scaffolds for cartilage tissue engineering. *J Control Release* 2005, 101, (1–3), 111–125.

61. Holland, T. A.; Bodde, E. W.; Cuijpers, V. M.; Baggett, L. S.; Tabata, Y.; Mikos, A. G.; Jansen, J. A., Degradable hydrogel scaffolds for in vivo delivery of single and dual growth factors in cartilage repair. *Osteoarthr Cartilage* 2007, 15, (2), 187–197.

62. Kost, J.; Langer, R., Responsive polymeric delivery systems. *Adv Drug Deliv Rev* 2001, 46, (1–3), 125–148.

63. King, W. J.; Mohammed, J. S.; Murphy, W. L., Modulating growth factor release from hydrogels via a protein conformational change. *Soft Matter* 2009, 5, (12), 2399–2406.

64. Mohammed, J. S.; Murphy, W. L., Bioinspired design of dynamic materials. *Adv Mater* 2009, 21, (23), 2361–2374.

65. Fan, V. H.; Tamama, K.; Au, A.; Littrell, R.; Richardson, L. B.; Wright, J. W.; Wells, A.; Griffith, L. G., Tethered epidermal growth factor provides a survival advantage to mesenchymal stem cells. *Stem Cells* 2007, 25, (5), 1241–1251.

66. Kuhl, P. R.; Griffith-Cima, L.G., Tethered epidermal growth factor as a paradigm for growth factor-induced stimulation from the solid phase. *Nat Med* 1996, 2, (9), 1022–1027.

67. Alberti, K.; Davey, R. E.; Onishi, K.; George, S.; Salchert, K.; Seib, F. P.; Bornhauser, M.; Pompe, T.; Nagy, A.; Werner, C.; Zandstra, P. W., Functional immobilization of signaling proteins enables control of stem cell fate. *Nat Methods* 2008, 5, (7), 645–650.

68. Gomez, N.; Schmidt, C. E., Nerve growth factor-immobilized polypyrrole: Bioactive electrically conducting polymer for enhanced neurite extension. *J Biomed Mater Res A* 2007, 81, (1), 135–149.

69. DeLong, S. A.; Moon, J. J.; West, J. L., Covalently immobilized gradients of bFGF on hydrogel scaffolds for directed cell migration. *Biomaterials* 2005, 26, (16), 3227–3234.

70. Nur, E. K. A.; Ahmed, I.; Kamal, J.; Babu, A. N.; Schindler, M.; Meiners, S., Covalently attached FGF-2 to three-dimensional polyamide nanofibrillar surfaces demonstrates enhanced biological stability and activity. *Mol Cell Biochem* 2008, 309, (1–2), 157–166.
71. Backer, M. V.; Patel, V.; Jehning, B. T.; Claffey, K. P.; Backer, J. M., Surface immobilization of active vascular endothelial growth factor via a cysteine-containing tag. *Biomaterials* 2006, 27, (31), 5452–5458.
72. Zisch, A. H.; Schenk, U.; Schense, J. C.; Sakiyama-Elbert, S. E.; Hubbell, J. A., Covalently conjugated VEGF–fibrin matrices for endothelialization. *J Control Release* 2001, 72, (1–3), 101–113.
73. Sakiyama-Elbert, S. E.; Panitch, A.; Hubbell, J. A., Development of growth factor fusion proteins for cell-triggered drug delivery. *FASEB J* 2001, 15, (7), 1300–1302.
74. Bentz, H.; Schroeder, J. A.; Estridge, T. D., Improved local delivery of TGF-beta2 by binding to injectable fibrillar collagen via difunctional polyethylene glycol. *J Biomed Mater Res* 1998, 39, (4), 539–548.
75. Shen, Y. H.; Shoichet, M. S.; Radisic, M., Vascular endothelial growth factor immobilized in collagen scaffold promotes penetration and proliferation of endothelial cells. *Acta Biomater* 2008, 4, (3), 477–489.
76. Zisch, A. H.; Lutolf, M. P.; Ehrbar, M.; Raeber, G. P.; Rizzi, S. C.; Davies, N.; Schmokel, H.; Bezuidenhout, D.; Djonov, V.; Zilla, P.; Hubbell, J. A., Cell-demanded release of VEGF from synthetic, biointeractive cell ingrowth matrices for vascularized tissue growth. *FASEB J* 2003, 17, (15), 2260–2262.
77. Ehrbar, M.; Djonov, V. G.; Schnell, C.; Tschanz, S. A.; Martiny-Baron, G.; Schenk, U.; Wood, J.; Burri, P. H.; Hubbell, J. A.; Zisch, A. H., Cell-demanded liberation of VEGF121 from fibrin implants induces local and controlled blood vessel growth. *Circ Res* 2004, 94, (8), 1124–1132.
78. Cetinkaya, G.; Turkoglu, H.; Arat, S.; Odaman, H.; Onur, M. A.; Gumusderelioglu, M.; Tumer, A., LIF-immobilized nonwoven polyester fabrics for cultivation of murine embryonic stem cells. *J Biomed Mater Res A* 2007, 81, (4), 911–919.
79. Kloxin, A. M.; Kasko, A. M.; Salinas, C. N.; Anseth, K. S., Photodegradable hydrogels for dynamic tuning of physical and chemical properties. *Science* 2009, 324, (5923), 59–63.
80. Kost, J.; Leong, K.; Langer, R., Ultrasound-enhanced polymer degradation and release of incorporated substances. *Proc Natl Acad Sci USA* 1989, 86, (20), 7663–7666.
81. Sakiyama-Elbert, S.; Hubbell, J., Functional biomaterials: Design of novel biomaterials. *Annu Rev Mater Res* 2001, 31, (1), 183–201.
82. Kato, K.; Sato, H.; Iwata, H., Ultrastructural study on the specific binding of genetically engineered epidermal growth factor to type I collagen fibrils. *Bioconjug Chem* 2007, 18, (6), 2137–2143.
83. Ishikawa, T.; Eguchi, M.; Wada, M.; Iwami, Y.; Tono, K.; Iwaguro, H.; Masuda, H.; Tamaki, T.; Asahara, T., Establishment of a functionally active collagen-binding vascular endothelial growth factor fusion protein in situ. *Arterioscler Thromb Vasc Biol* 2006, 26, (9), 1998–2004.
84. Kitajima, T.; Terai, H.; Ito, Y., A fusion protein of hepatocyte growth factor for immobilization to collagen. *Biomaterials* 2007, 28, (11), 1989–1997.
85. Lin, C. C.; Metters, A. T., Metal-chelating affinity hydrogels for sustained protein release. *J Biomed Mater Res A* 2007, 83, (4), 954–964.
86. Nakaji-Hirabayashi, T.; Kato, K.; Arima, Y.; Iwata, H., Oriented immobilization of epidermal growth factor onto culture substrates for the selective expansion of neural stem cells. *Biomaterials* 2007, 28, (24), 3517–3529.
87. Nakaji-Hirabayashi, T.; Kato, K.; Iwata, H., Surface-anchoring of spontaneously dimerized epidermal growth factor for highly selective expansion of neural stem cells. *Bioconjug Chem* 2009, 20, (1), 102–110.

88. Nakaji-Hirabayashi, T.; Kato, K.; Arima, Y.; Iwata, H., Multifunctional chimeric proteins for the sequential regulation of neural stem cell differentiation. *Bioconjug Chem* 2008, 19, (2), 516–524.

89. Cleland, J. L.; Duenas, E. T.; Park, A.; Daugherty, A.; Kahn, J.; Kowalski, J.; Cuthbertson, A., Development of poly-(D,L-lactide-coglycolide) microsphere formulations containing recombinant human vascular endothelial growth factor to promote local angiogenesis. *J Control Release* 2001, 72, (1–3), 13–24.

90. Tabata, Y.; Nagano, A.; Muniruzzaman, M.; Ikada, Y., In vitro sorption and desorption of basic fibroblast growth factor from biodegradable hydrogels. *Biomaterials* 1998, 19, (19), 1781–1789.

91. Tabata, Y.; Ikada, Y., Vascularization effect of basic fibroblast growth factor released from gelatin hydrogels with different biodegradabilities. *Biomaterials* 1999, 20, (22), 2169–2175.

92. Layman, H.; Spiga, M. G.; Brooks, T.; Pham, S.; Webster, K. A.; Andreopoulos, F. M., The effect of the controlled release of basic fibroblast growth factor from ionic gelatin-based hydrogels on angiogenesis in a murine critical limb ischemic model. *Biomaterials* 2007, 28, (16), 2646–2654.

93. Iwakura, A.; Fujita, M.; Kataoka, K.; Tambara, K.; Sakakibara, Y.; Komeda, M.; Tabata, Y., Intramyocardial sustained delivery of basic fibroblast growth factor improves angiogenesis and ventricular function in a rat infarct model. *Heart Vessels* 2003, 18, (2), 93–99.

94. Wissink, M. J.; Beernink, R.; Poot, A. A.; Engbers, G. H.; Beugeling, T.; van Aken, W. G.; Feijen, J., Improved endothelialization of vascular grafts by local release of growth factor from heparinized collagen matrices. *J Control Release* 2000, 64, (1–3), 103–114.

95. Pieper, J. S.; Hafmans, T.; van Wachem, P. B.; van Luyn, M. J.; Brouwer, L. A.; Veerkamp, J. H.; van Kuppevelt, T. H., Loading of collagen-heparan sulfate matrices with bFGF promotes angiogenesis and tissue generation in rats. *J Biomed Mater Res* 2002, 62, (2), 185–194.

96. Benoit, D. S.; Anseth, K. S., Heparin functionalized PEG gels that modulate protein adsorption for hMSC adhesion and differentiation. *Acta Biomater* 2005, 1, (4), 461–470.

97. Yamaguchi, N.; Kiick, K. L., Polysaccharide-poly(ethylene glycol) star copolymer as a scaffold for the production of bioactive hydrogels. *Biomacromolecules* 2005, 6, (4), 1921–1930.

98. Cai, S.; Liu, Y.; Zheng Shu, X.; Prestwich, G. D., Injectable glycosaminoglycan hydrogels for controlled release of human basic fibroblast growth factor. *Biomaterials* 2005, 26, (30), 6054–6067.

99. Pike, D. B.; Cai, S.; Pomraning, K. R.; Firpo, M. A.; Fisher, R. J.; Shu, X. Z.; Prestwich, G. D.; Peattie, R. A., Heparin-regulated release of growth factors in vitro and angiogenic response in vivo to implanted hyaluronan hydrogels containing VEGF and bFGF. *Biomaterials* 2006, 27, (30), 5242–5251.

100. Tanihara, M.; Suzuki, Y.; Yamamoto, E.; Noguchi, A.; Mizushima, Y., Sustained release of basic fibroblast growth factor and angiogenesis in a novel covalently crosslinked gel of heparin and alginate. *J Biomed Mater Res* 2001, 56, (2), 216–221.

101. Edelman, E. R.; Nugent, M. A.; Smith, L. T.; Karnovsky, M. J., Basic fibroblast growth factor enhances the coupling of intimal hyperplasia and proliferation of vasa vasorum in injured rat arteries. *J Clin Invest* 1992, 89, (2), 465–473.

102. Yamaguchi, N.; Zhang, L.; Chae, B. S.; Palla, C. S.; Furst, E. M.; Kiick, K. L., Growth factor mediated assembly of cell receptor-responsive hydrogels. *J Am Chem Soc* 2007, 129, (11), 3040–3041.

103. Sellke, F. W.; Laham, R. J.; Edelman, E. R.; Pearlman, J. D.; Simons, M., Therapeutic angiogenesis with basic fibroblast growth factor: Technique and early results. *Ann Thorac Surg* 1998, 65, (6), 1540–1544.

104. Laham, R. J.; Sellke, F. W.; Edelman, E. R.; Pearlman, J. D.; Ware, J. A.; Brown, D. L.; Gold, J. P.; Simons, M., Local perivascular delivery of basic fibroblast growth factor in patients undergoing coronary bypass surgery: Results of a phase I randomized, double-blind, placebo-controlled trial. *Circulation* 1999, 100, (18), 1865–1871.

105. Willerth, S. M.; Johnson, P. J.; Maxwell, D. J.; Parsons, S. R.; Doukas, M. E.; Sakiyama-Elbert, S. E., Rationally designed peptides for controlled release of nerve growth factor from fibrin matrices. *J Biomed Mater Res A* 2007, 80, (1), 13–23.

106. Lin, C. C.; Metters, A. T.; Anseth, K. S., Functional PEG-peptide hydrogels to modulate local inflammation induced by the pro-inflammatory cytokine TNFalpha. *Biomaterials* 2009, 30, (28), 4907–4914.

107. Willerth, S. M.; Rader, A.; Sakiyama-Elbert, S. E., The effect of controlled growth factor delivery on embryonic stem cell differentiation inside fibrin scaffolds. *Stem Cell Res* 2008, 1, (3), 205–218.

108. Sakiyama-Elbert, S. E.; Hubbell, J. A., Controlled release of nerve growth factor from a heparin-containing fibrin-based cell ingrowth matrix. *J Control Release* 2000, 69, (1), 149–158.

109. Lee, A. C.; Yu, V. M.; Lowe, J. B., 3rd; Brenner, M. J.; Hunter, D. A.; Mackinnon, S. E.; Sakiyama-Elbret, S. E., Controlled release of nerve growth factor enhances sciatic nerve regeneration. *Exp Neurol* 2003, 184, (1), 295–303.

110. Sakiyama-Elbert, S. E.; Hubbell, J. A., Development of fibrin derivatives for controlled release of heparin-binding growth factors. *J Control Release* 2000, 65, (3), 389–402.

111. Rajangam, K.; Behanna, H. A.; Hui, M. J.; Han, X.; Hulvat, J. F.; Lomasney, J. W.; Stupp, S. I., Heparin binding nanostructures to promote growth of blood vessels. *Nano Lett* 2006, 6, (9), 2086–2090.

112. Taylor, S. J.; McDonald, J. W., 3rd; Sakiyama-Elbert, S. E., Controlled release of neurotrophin-3 from fibrin gels for spinal cord injury. *J Control Release* 2004, 98, (2), 281–294.

113. Kanematsu, A.; Marui, A.; Yamamoto, S.; Ozeki, M.; Hirano, Y.; Yamamoto, M.; Ogawa, O.; Komeda, M.; Tabata, Y., Type I collagen can function as a reservoir of basic fibroblast growth factor. *J Control Release* 2004, 99, (2), 281–292.

114. Sahni, A.; Sporn, L. A.; Francis, C. W., Potentiation of endothelial cell proliferation by fibrin(ogen)-bound fibroblast growth factor-2. *J Biol Chem* 1999, 274, (21), 14936–14941.

115. Sahni, A.; Baker, C. A.; Sporn, L. A.; Francis, C. W., Fibrinogen and fibrin protect fibroblast growth factor-2 from proteolytic degradation. *Thromb Haemost* 2000, 83, (5), 736–741.

116. Sasisekharan, R.; Venkataraman, G., Heparin and heparan sulfate: Biosynthesis, structure and function. *Curr Opin Chem Biol* 2000, 4, (6), 626–631.

117. Kim, S. H.; Kiick, K. L., Heparin-mimetic sulfated peptides with modulated affinities for heparin-binding peptides and growth factors. *Peptides* 2007, 28, (11), 2125–2136.

118. Maynard, H. D.; Hubbell, J. A., Discovery of a sulfated tetrapeptide that binds to vascular endothelial growth factor. *Acta Biomater* 2005, 1, (4), 451–459.

119. Bottaro, D. P.; Fortney, E.; Rubin, J. S.; Aaronson, S. A., A keratinocyte growth factor receptor-derived peptide antagonist identifies part of the ligand binding site. *J Biol Chem* 1993, 268, (13), 9180–9183.

120. Pan, B.; Li, B.; Russell, S. J.; Tom, J. Y.; Cochran, A. G.; Fairbrother, W. J., Solution structure of a phage-derived peptide antagonist in complex with vascular endothelial growth factor. *J Mol Biol* 2002, 316, (3), 769–787.

121. Saito, H.; Kojima, T.; Takahashi, M.; Horne, W. C.; Baron, R.; Amagasa, T.; Ohya, K.; Aoki, K., A tumor necrosis factor receptor loop peptide mimic inhibits bone destruction to the same extent as anti-tumor necrosis factor monoclonal antibody in murine collagen-induced arthritis. *Arthritis Rheum* 2007, 56, (4), 1164–1174.

122. Takasaki, W.; Kajino, Y.; Kajino, K.; Murali, R.; Greene, M. I., Structure-based design and characterization of exocyclic peptidomimetics that inhibit TNF alpha binding to its receptor. *Nat Biotechnol* 1997, 15, (12), 1266–1270.

123. Chirinos-Rojas, C. L.; Steward, M. W.; Partidos, C. D., A peptidomimetic antagonist of TNF-alpha-mediated cytotoxicity identified from a phage-displayed random peptide library. *J Immunol* 1998, 161, (10), 5621–5626.

124. Ilkhanizadeh, S.; Teixeira, A. I.; Hermanson, O., Inkjet printing of macromolecules on hydrogels to steer neural stem cell differentiation. *Biomaterials* 2007, 28, (27), 3936–3943.

125. Phillippi, J. A.; Miller, E.; Weiss, L.; Huard, J.; Waggoner, A.; Campbell, P., Microenvironments engineered by inkjet bioprinting spatially direct adult stem cells toward muscle- and bone-like subpopulations. *Stem Cells* 2008, 26, (1), 127–134.

126. Chen, R. R.; Silva, E. A.; Yuen, W. W.; Mooney, D. J., Spatio-temporal VEGF and PDGF delivery patterns blood vessel formation and maturation. *Pharm Res* 2007, 24, (2), 258–264.

127. Mapili, G.; Lu, Y.; Chen, S.; Roy, K., Laser-layered microfabrication of spatially patterned functionalized tissue-engineering scaffolds. *J Biomed Mater Res B Appl Biomater* 2005, 75, (2), 414–424.

128. King, W. J.; Jongpaiboonkit, L.; Murphy, W. L., Influence of FGF2 and PEG hydrogel matrix properties on hMSC viability and spreading. *J Biomed Mater Res A* 2010, 93, (3), 1110–1123.

129. Koepsel, J. T.; Murphy, W. L., Patterning discrete stem cell culture environments via localized self-assembled monolayer replacement. *Langmuir* 2009, 25, (21), 12825–12834.

130. Derda, R.; Li, L.; Orner, B. P.; Lewis, R. L.; Thomson, J. A.; Kiessling, L. L., Defined substrates for human embryonic stem cell growth identified from surface arrays. *ACS Chem Biol* 2007, 2, (5), 347–355.

6 Regulation of Capillary Morphogenesis by the Adhesive and Mechanical Microenvironment

Colette J. Shen and Christopher S. Chen

CONTENTS

6.1　INTRODUCTION

Cardiovascular disease is the leading cause of death and disability in the United States and most Western countries. This mostly is due to the narrowing and eventual blockage of blood vessels to the normal flow of blood, causing tissue ischemia and damage. As such, therapeutic interventions to treat cardiovascular disease include strategies to revascularize these ischemic tissues. Among the two most common surgical procedures for coronary or peripheral vascular disease are the placement of stents (to reopen narrowing vessels physically) or introduction of a bypass graft to circumvent diseased vessels. While such approaches have proven to be widely successful in cases where vascular disease is restricted to large vessels, in many settings, the limitation for blood flow is in the microvasculature. In this setting, it is clear that the most promising approach is to promote the formation of additional microvessels in the ischemic tissue. In addition to what is now referred to as "therapeutic angiogenesis," our ability to promote controlled neovascularization is also critical to the success of tissue engineering as a strategy for organ replacement. Currently, investigators have demonstrated a number of promising avenues to engineer cells and biomaterials to form tissue-like structures in culture. Our ability to translate these *in vitro* tissues into transplantable replacement tissues is now limited by our ability to promote the successful vascularization of the engineered constructs. As most cells must reside within 200 μm of the nearest capillary for proper gas exchange, nutrient delivery, and waste removal,[1,2] a failure to vascularize an engineered implant would either result in massive failure of the implant or limit us to implant only tiny structures.

Historically, the strategies for promoting angiogenesis in these clinical settings have focused largely on gene- and cell-based therapies for delivering angiogenic cytokines and progenitor cells to promote neovascularization.[3] Studies of the fundamental process of angiogenesis over the past decade have revealed that in addition to the appropriate cells and soluble factors, the extracellular matrix environment is also a critical regulator that can either promote or prevent neovascularization of a tissue. In this chapter, we review the current understanding of how angiogenesis is regulated and then focus in particular on how adhesive and mechanical interactions with the extracellular matrix are being exploited to modulate the angiogenic behavior of endothelial cells.

6.2 REGULATION OF ANGIOGENESIS BY THE MICROENVIRONMENT

Angiogenesis is the sprouting of new capillaries from existing vessels and involves the activation of quiescent endothelium to degrade the surrounding ECM, proliferate and migrate away from existing vessels, and assemble to form hollow, elongated, branching tubes.[4] Angiogenesis requires the cooperative signaling of both soluble growth factor and matrix-mediated adhesive and mechanical cues. Several growth factors are known to regulate various aspects of angiogenesis, among them vascular endothelial growth factor (VEGF), basic fibroblast growth factor (bFGF), placental growth factor (PlGF), and platelet-derived growth factor (PDGF). Studies of angiogenesis in tumors have revealed that inhibition of VEGF, bFGF, PDGF, and, more recently, PlGF signaling blocks angiogenesis.[5–7] Clinically, these inhibitors potently inhibit angiogenesis in certain cancers and in diabetic retinopathy.[6,8–13] Likewise, agents that block endothelial cell adhesion to ECM profoundly inhibit angiogenesis in animal studies, and these approaches are being assessed as potential clinical cancer treatments.[14–17]

6.3 REGULATION OF ANGIOGENESIS BY CELL ADHESION TO EXTRACELLULAR MATRIX

While the mechanisms by which soluble factors promote angiogenesis are relatively straightforward, how cell adhesion to the ECM modulates endothelial cell behavior requires a more careful examination of the literature. Cell adhesion involves binding of integrins to ECM ligands as well as cell spreading against the substrate and subsequent generation of cytoskeletal tension. Each of these different aspects of adhesion (integrin binding, changes in cell shape, and alterations in cytoskeletal mechanics) appears to impact endothelial cell function.

The importance of integrin-mediated adhesion in angiogenesis has been well established: antagonists of $\alpha v \beta 3$, $\alpha v \beta 5$, $\alpha 5 \beta 1$, $\alpha 1 \beta 1$, and $\alpha 2 \beta 1$ have been shown to inhibit endothelial cell adhesion and migration *in vitro* and angiogenesis *in vivo* using the chick chorioallantoic membrane (CAM) model (Figure 6.1A) or vascularization of human skin transplanted onto SCID mice.[18–21] Interestingly, these integrins are not highly expressed on quiescent endothelium but are upregulated in tumors, wounds, and sites of inflammation and in response to angiogenic growth factors.[22] These integrins appear to regulate diverse angiogenic pathways, as blocking engagement of $\alpha_v \beta_3$ and $\alpha_5 \beta_1$ integrins inhibits bFGF-induced angiogenesis, while blocking $\alpha_v \beta_5$, $\alpha_1 \beta_1$, and $\alpha_2 \beta_1$ engagement inhibits VEGF-induced angiogenesis. Integrin knockout studies, on the other hand, suggest certain integrins—α_v, β_3, β_5, α_1, and α_2—are not always necessary for vascular development.[23–27] These discrepancies can be explained in part by compensatory upregulation of other angiogenesis signaling pathways, such as VEGF receptor 2 signaling in $\beta 3$-null mice[28] and VEGF receptor 1 signaling in $\alpha 2$-null mice,[27] though these studies still leave many questions unanswered.

The molecular intermediaries by which integrin engagement impacts endothelial cell behavior are largely found in focal adhesions. Binding and clustering of integrins to specific ECM ligands leads to the recruitment of numerous scaffolding and signaling proteins to the site of integrin ligation, forming these dynamic structures known as focal adhesions.[29–31] Signaling proteins (e.g., FAK, Src, ERK, RhoA) that are regulated by integrin activation also function in growth factor signaling, suggesting that focal adhesions act to coordinate integrin and growth factor signaling.[32,33] Indeed, growth factor receptors are concentrated within sites of focal adhesion formation,[31] and different integrins within focal adhesions interact with specific growth factor receptors.[34,35] For example, VEGF and bFGF signal synergistically with integrins $\alpha_v \beta_5$ and $\alpha_v \beta_3$, respectively, to activate the Ras–Erk pathway to modulate angiogenesis.[36] Focal adhesion kinase (FAK), a key signaling protein localized to focal adhesions, is known to transduce both soluble growth factor and integrin adhesion signals to regulate cell proliferation, migration, and survival.[37–40] Importantly, FAK is essential for vascular development *in vivo* as demonstrated by the embryonic lethality of endothelial-specific FAK knockout in mice.[41,42] Further downstream, sustained ERK activity in response to growth factor and integrin-mediated adhesive cues is required for angiogenesis *in vivo*.[43] Such studies provide a molecular context for how cell adhesion can modulate growth factor–mediated responses and emphasize the importance of integrin-mediated adhesion in transducing signals from the ECM.

While these studies clearly demonstrate the importance of integrin-mediated adhesion in regulating angiogenesis, several studies also suggest that the degree of cell adhesion and spreading can strongly impact endothelial cell behavior. Varying cell spreading by changing the density of the ECM protein, fibronectin, immobilized on culture surfaces results in increased proliferation at high densities and quiescence at low densities.[44,45] The use of micropatterned substrates with defined adhesive and surrounding nonadhesive regions extended these findings further. Such substrates dictate the area of cell–ECM contact and allow for more precise control of cell adhesion and shape.[46] Controlling cell spreading by culturing cells on different sizes of adhesive islands showed greater proliferation in endothelial cells cultured

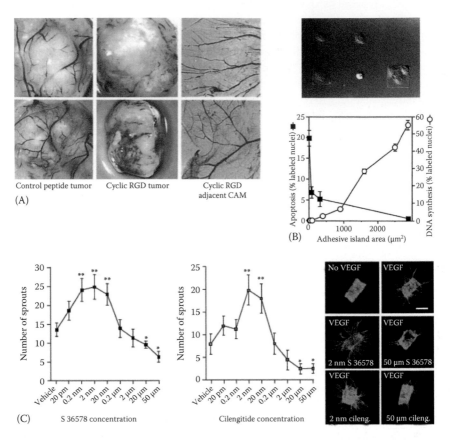

FIGURE 6.1 (See color insert.) Cell adhesion to the ECM regulates angiogenesis. (A) Tumor angiogenesis on the CAM is inhibited by cyclic RGD peptide, which antagonizes integrin αvβ3 binding to the ECM. (From *Cell*, 79(7), Brooks, P.C., Montgomery, A.M., Rosenfeld, M., Reisfeld, R.A., Hu, T., Klier, G., and Cheresh, D.A., Integrin alpha v beta 3 antagonists promote tumor regression by inducing apoptosis of angiogenic blood vessels, 1157–1164. Copyright 1994. With permission from Elsevier.) (B) The degree of cell spreading, controlled by varying sizes of micropatterned islands of fibronectin, regulates endothelial cell proliferation versus apoptosis, with greater proliferation in more highly spread cells. (From Chen, C.S., Mrksich, M., Huang, S., Whitesides, G.M., and Ingber, D.E., Geometric control of cell life and death, *Science*, 1997, 276(5317), 1425–1428. With permission of AAAS.) (C) Low concentrations of the αvβ3/αvβ5 inhibitors S 36578 and cilengitide promote VEGF-mediated angiogenesis from mouse aortic rings, compromising antiangiogenic effects. (By permission from Macmillan Publishers Ltd. *Nat Med*, Reynolds, A.R., Hart, I.R., Watson, A.R., Welti, J.C., Silva, R.G., Robinson, S.D., Da Violante, G., Gourlaouen, M., Salih, M., Jones, M.C., Jones, D.T., Saunders, G., Kostourou, V., Perron-Sierra, F., Norman, J.C., Tucker, G.C., and Hodivala-Dilke, K.M., Stimulation of tumor growth and angiogenesis by low concentrations of RGD-mimetic integrin inhibitors, 15(4) 392–400. Copyright 2009.)

on large adhesive islands and an increase in apoptosis on small adhesive islands (Figure 6.1B).[47] Interestingly, between these two extremes of high and low spreading, studies have shown enhanced cell–cell interaction and tubulogenesis at intermediate levels of adhesion. Intermediate cell–ECM adhesion, achieved either by introducing antibodies that block integrin binding or by coating substrates with intermediate amounts of ECM protein, resulted in increased tubulogenesis, decreased spreading, and decreased proliferation in vitro.[45,48] Similarly, confining endothelial cells on patterned adhesive lines of intermediate width promoted morphogenesis of cells into tubules with lumens.[49] A recent study also demonstrated an unexpected increase in tumor angiogenesis in vivo in response to low concentrations of RGD-mimetic integrin inhibitors, under development as antiangiogenic cancer therapeutics (Figure 6.1C).[50] A biphasic response was observed, such that no inhibitor had minimal effect on angiogenesis, high concentrations of inhibitor decreased angiogenesis, but intermediate concentrations actually increased angiogenesis above baseline levels. These studies all point to the possibility of enhanced capillary differentiation and angiogenesis at intermediate levels of adhesion.

Given the dynamic nature of cell invasion and sprouting during angiogenesis, the adhesive interaction between endothelial cells and the ECM is constantly changing. Matrix degradation leads to release of ECM fragments and exposure of cryptic binding sites, and new matrix deposition provides additional ECM ligands for integrin ligation. Specifically, denatured collagen is found around angiogenic, but not quiescent, vessels,[51] and collagen type IV cleavage reveals cryptic binding sites that interact preferentially with $\alpha_v\beta_3$ integrins upregulated in angiogenic vessels over $\alpha1\beta1$ integrins.[52] The importance of adhesive interactions with these cryptic sites is evidenced by the inhibition of angiogenesis in vivo with antibodies directed against these sites. Proteolytic activity also results in release of fragments of ECM proteins and proteases, all of which can ligate integrins and have been shown to block angiogenesis by altering integrin interactions with the ECM. These include endostatin, a fragment of collagen XVIII[53]; tumstatin, a fragment of collagen IV[54]; angiostatin, a fragment of plasminogen[55]; and PEX, a fragment of matrix metalloproteinase 2 (MMP-2).[56] Finally, new matrix deposition—either from the plasma as a result of increased endothelial permeability or directly produced by endothelial and supporting mesenchymal cells—provides additional ligands for adhesion. Fibronectin, vitronectin, and fibrinogen are among the provisional matrix proteins deposited from the plasma, and tenascin and thrombospondins are provisional ECM proteins produced by endothelial cells.[33,57] Returning full circle, endothelial–pericyte interactions appear to enhance production of basement membrane proteins such as laminins, collagen IV, and nidogens to stabilize newly formed vessels.[58] These examples underscore not only the importance of the adhesive interactions between endothelial cells and the ECM in modulating angiogenesis, but also the complex dynamic nature of the entire process.

6.4 MECHANICAL REGULATION OF ANGIOGENESIS

As cells attach and spread against an ECM substrate, RhoA and its effector ROCK are activated, leading to actin-myosin-generated contractility and further maturation of focal adhesions.[59–62] Studies from our group and others have demonstrated that progressively increasing cell spreading activates RhoA-ROCK- and myosin-mediated cytoskeletal tension, along with enhanced assembly of robust stress fibers and focal adhesions.[38,63–67] Using a microfabricated device containing arrays of microneedles that directly reports traction forces,[68] we demonstrated that the degree of cell spreading regulates the magnitude of cytoskeletal tension generated by cells cultured on the posts (Figure 6.2A). Together, these studies demonstrate a tight link between cellular mechanics and cell adhesion.

Importantly, these changes in cell contractility appear to be critical in modulating endothelial cell function. For example, RhoA-ROCK-mediated contractility is

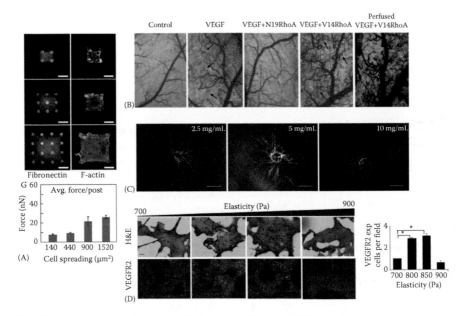

FIGURE 6.2 (See color insert.) Cell contractility and ECM mechanical properties regulate angiogenesis. (A) Increased cell spreading increases cell contractility, as measured by culture on arrays of flexible microposts. (From *Proceedings of the National Academy of the United States of America,* 100(4), Tan, J. L.; Tien, J.; Pirone, D. M.; Gray, D. S.; Bhadriraju, K.; Chen, C. S., Cells lying on a bed of microneedles: An approach to isolate mechanical force, 1484–1489. Copyright 2003. With permission from Elsevier.) (B) RhoA manipulations regulate VEGF-driven angiogenesis in mouse skin. (From Hoang, M.V., Whelan, M.C., and Senger, D.R., Rho activity critically and selectively regulates endothelial cell organization during angiogenesis. *Proc Natl Acad Sci USA* 2004, 101(7), 1874–1879. With permission.) (C) Fibrin density regulates endothelial cell sprouting from microcarrier beads, with higher densities inhibiting sprouting. (From *Biophys J*, 94(5), Ghajar, C.M., Chen, X., Harris, J.W., Suresh, V., Hughes, C.C., Jeon, N.L., Putnam, A.J., and George, S.C., The effect of matrix density on the regulation of 3-D capillary morphogenesis, 1930–1941. Copyright 2008. With permission from Elsevier.) (D) Intermediate stiffness of Matrigel promotes angiogenesis *in vivo* as measured by cellular infiltration and VEGFR2 expression. (By permission from Macmillan Publishers Ltd. *Nature*, Mammoto, A., Connor, K.M., Mammoto, T., Yung, C.W., Huh, D., Aderman, C.M., Mostoslavsky, G., Smith, L.E., and Ingber, D.E., A mechanosensitive transcriptional mechanism that controls angiogenesis, 457(7233), 1103–1138. Copyright 2009.)

key in regulating proliferation,[38] and Ingber and colleagues have shown that cell spreading–mediated changes in RhoA signaling regulate the G1/S transition in cell cycle progression by increasing expression of the transcriptional regulator Skp2.[69,70] RhoA-ROCK-mediated contractility has also been shown to be important for endothelial cell migration and capillary morphogenesis *in vitro* and angiogenesis *in vivo* (Figure 6.2B),[71–75] and directly inhibiting myosin signaling or disrupting the actin cytoskeleton decreases capillary sprouting and endothelial cell proliferation.[69,76]

Given that ECM stiffness has been reported to increase cellular contractility, might the mechanical properties of the ECM also regulate angiogenesis via alteration of cell-generated tension against the ECM? Increases in matrix stiffness have been shown in other cell types to activate FAK, RhoA-ROCK, and myosin activity.[77–79] Furthermore, the combination of increased contractility and increased rigidity of the substrate against which cells pull leads to higher stresses generated between the cell and substrate. Studies suggest that it is the tension generated when cells pull against a rigid substrate, rather than contractile activity within cells themselves, that drives cell function. In the absence of adhesion to a rigid substrate, activation of RhoA cannot generate cytoskeletal tension and fails to support focal adhesion formation or proliferative signaling.[63,67,80–82]

Early studies demonstrated changes in endothelial cell behavior when grown on malleable substrates versus on traditional rigid tissue culture surfaces—for example, the arrangement of endothelial cells into capillary tubes on soft gels and into monolayers on rigid surfaces[83,84]—and have suggested the possibility that mechanical cues from the ECM indeed play an important role in regulating endothelial cell behavior. However, only recently have the tools to study the role of matrix elasticity in angiogenesis well become available. Primarily in other biological contexts, matrix stiffness has been shown to regulate numerous cellular behaviors, including adhesion, contractility, spreading, motility, proliferation, and differentiation.[77–79,85–89] In studies with endothelial cells, it has been shown using natural ECMs such as fibrin and Matrigel that lower density and thus more compliant matrices tend to support capillary morphogenesis, while denser and thus more rigid matrices promote endothelial proliferation and migration (Figure 6.2C).[90–93] Matrix compliance is but one of many properties varied in these ECMs, however, and endothelial cell responses have been attributed to limitations in diffusion and changes in MMP production in addition to compliance.[90,94] To address this shortcoming, others have varied the stiffness of collagen gels by using glycation-induced cross-linking, while holding collagen density constant. Consistent with studies varying ECM density, increased ECM stiffness suppresses tubulogenesis while increasing proliferation.[95] Because cross-linking of the matrix itself can locally alter ligand density, hydration state, flexibility, or conformation (thereby affecting cell adhesion through changes in integrin binding rather than by mechanical effects), investigators have developed nonadhesive hydrogels, such as polyacrylamide or poly(ethylene glycol) (PEG), where cross-linking occurs on the nonadhesive backbone, and the ECM ligand is covalently attached in a separate chemical step to decouple ECM ligand density from stiffness manipulations.[86,96] Gelatin-coupled polyacrylamide gels of different stiffness exhibit enhanced tubulogenesis on more compliant matrices.[97] In addition, a recent study demonstrated enhanced angiogenesis at an intermediate stiffness of Matrigel *in vivo* (where stiffness was controlled by transglutaminase-mediated protein cross-linking) (Figure 6.2D) and showed that related endothelial cell responses are similarly biphasic on 2D polyacrylamide gels of varying stiffness.[98]

Although these studies are beginning to uncover a role for matrix mechanics in angiogenesis, it should be cautioned that mechanical properties of tissues could be dynamically changing over time due to degradation by cellular proteases and deposition of new matrix proteins, so studies of matrix stiffness currently can only accurately report effects of initial mechanical properties. In angiogenesis, membrane type 1 MMP (MT1-MMP, or MMP14) has been shown to be particularly important and is highly expressed at the tips of angiogenic sprouts.[90,99–104] It is possible, then, that a certain local mechanical compliance—different from initial conditions—must be reached before sprout invasion into a matrix can occur. However, the rate of sprouting into matrices of different initial stiffness should vary, as may the character of sprouting given the differences in angiogenesis in stiffer tumor tissues versus softer healthy tissues. Mechanical gradients along sprouts resulting from differential matrix degradation and deposition also point to the importance of studying relative differences in mechanical properties on angiogenesis, even if initial conditions are not maintained. Finally, studies have suggested that increased mechanical stiffness can increase matrix deposition,[105,106] resulting in a positive feedback loop that could significantly impact the dynamic properties of the ECM surrounding angiogenic vessels.

It can be appreciated from these studies that not only the extent but also the quality or character of angiogenesis could potentially be controlled simply by changing the adhesive or mechanical properties of the ECM. Shifts in cellular responses to these manipulations from proliferation to capillary morphogenesis and apoptosis, all in the presence of similar soluble factor conditions, could alter the overall architecture of vascular networks as they form. Such a perspective could potentially help explain why certain cells in an angiogenic sprout proliferate while others undergo capillary morphogenesis, as well as why pathological angiogenesis varies so much in character from normal angiogenesis. The similar trends in angiogenic responses to changes in the amount of integrin-mediated adhesion, cell spreading, and matrix compliance suggest that perhaps similar downstream signaling pathways are responding to control cellular responses. Importantly, we can also use these findings to inform our efforts to vascularize engineered tissues for tissue regeneration using rationally designed materials.

6.5   ENGINEERED MATERIALS TO PROMOTE VASCULARIZATION

Given current understanding of how adhesive and mechanical properties of the ECM regulate angiogenesis, a major goal is to employ this knowledge toward engineering materials to promote optimal vascularization for tissue engineering and therapeutic angiogenesis. While most current strategies focus on soluble factor delivery to induce neovascularization of ischemic tissue or material implants, it is clear that proper signals from the ECM are also required for optimal tissue vascularization. Soluble angiogenic factors, such as VEGF and bFGF, delivered by either controlled-release scaffolds or transfected cells cannot attract native microvessels into the implanted construct without the appropriate adhesive cues from the ECM.[107–110]

Conversely, ECMs designed specifically to promote vascularization of implanted tissues require soluble growth factors for efficient angiogenesis, Yoon et al., 2006,[111–114] highlighting the importance of both soluble and matrix cues in engineered systems for angiogenesis.

In engineering materials for angiogenesis, we can use principles guided by our studies of angiogenesis in natural ECMs. *In vivo*, sprouting angiogenic vessels first degrade and migrate through laminin- and collagen IV-rich basement membrane that surrounds quiescent vessels, then extend into an interstitial matrix containing collagen I and provisionally deposited fibrin and fibronectin matrices.[115] Many studies of tissue implant vascularization, as well as *in vitro* systems to investigate the mechanisms and processes involved in angiogenesis, use these natural ECMs. They provide the adhesive, mechanical, and degradable properties known to promote robust vascularization in tissues *in vivo*, as well as the ability to tether and retain soluble growth factors via heparin binding sites,[116] another important signal for vessel ingrowth. In a key study in efforts to engineer functional, stable vessels, Jain and colleagues implanted type I collagen scaffolds containing endothelial and mesenchymal precursor cells in mice and reported the assembly of vascular networks that anastomosed with the host vasculature and remained stable for several months (Figure 6.3A).[117] Similar results were reported with the implantation of endothelial progenitor cells and smooth muscle cells in laminin-rich Matrigel[118] and preformed endothelial cell spheroids in a fibrin–Matrigel matrix.[119] Several others have employed *in vitro* assays of capillary sprouting or morphogenesis in fibrin and type I collagen matrices to uncover the contribution of supporting mesenchymal cells and MMPs in these processes, as well as the signaling mechanisms regulating vessel invasion and lumen formation.[90,103,120–126] As noted previously, the adhesive and mechanical properties of these natural ECMs can be altered to some extent by varying matrix density, resulting in variations in angiogenic behavior; however, changes in density are accompanied by changes in adhesive ligand concentration, mechanical stiffness, degradability, and porosity, so the exact contribution of each parameter to angiogenic responses cannot easily be determined.

Because of such limitations, despite the availability and known vascularization potential of natural ECMs, investigators continue to create increasingly synthetic systems in attempts to control more precisely material parameters such as compliance, adhesive ligand density and type, and porosity.[109] Synthetic materials also allow for covalent incorporation of known proangiogenic factors such as soluble growth factors and cell–cell guidance molecules, as well as prepatterning of regions within a material to control vascularization spatially. Some natural ECMs have been "engineered" to incorporate additional functionalities to enhance vascularization. For instance, increased growth factor retention in fibrin and collagen matrices has been achieved through covalent coupling of VEGF itself,[127–129] heparin-binding peptides or heparin itself that promote binding of heparin-bound growth factors such as VEGF and bFGF,[130–132] and collagen-mimetic peptides that attract growth factors to collagen gels via charge–charge interactions.[133] Natural ECMs can also be chemically cross-linked to alter compliance independent of adhesive ligand density.[95,134]

Because many of these natural ECMs are derived from animal tissues and so present the potential risk of immunogenicity and infectious pathogens,[109,135] fully synthetic

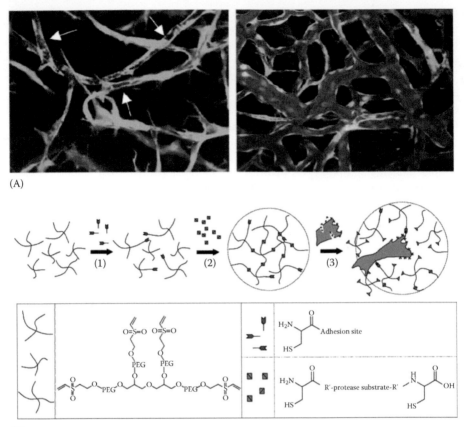

FIGURE 6.3 (See color insert.) Engineering materials to promote tissue vascularization. (A) Endothelial cells and pericytes co-seeded in collagen gel implants begin to form vascular networks at 4 days (left) and are stable and fully perfused at 4 months (right). (By permission from Macmillan Publishers Ltd. *Nature*, Koike, N., Fukumura, D., Gralla, O., Au, P., Schechner, J.S., and Jain, R.K., Tissue engineering: Creation of long-lasting blood vessels, 428(6979), 138–139. Copyright 2004.) (B) One synthetic scheme used to create biomimetic PEG materials. Monocysteine adhesive peptides and bis-cysteine MMP-sensitive peptides are coupled to multiarm PEGs via Michael-type addition reaction to allow cell migration through the matrix. (From Lutolf, M.P., Raeber, G.P., Zisch, A.H., Tirelli, N., and Hubbell, J.A.: Cell-responsive synthetic hydrogels. *Adv Mater.* 2003. 15(11). 888. Copyright Wiley-VCH Verlag GmbH & Co. KGaA. With permission.)

material systems are ultimately preferred for purposes of generating vascularized tissue constructs for implantation. Two primary material backbones, nontoxic and approved for *in vivo* clinical applications, have been used for this purpose: porous scaffolds formed from poly(lactic-*co*-glycolic acid) (PLGA) and PEG hydrogels functionalized with oligopeptides rendering them cell adhesive and degradable. Both materials satisfy the requirements of complete degradation and resorption *in vivo* after serving as temporary mechanical supports for cell ingrowth. PLGA scaffolds have long been used in medical implants such as sutures and prosthetic devices due to their biocompatibility and biodegradability.[136,137] Their mechanical and degradation properties can be altered by varying the ratio of lactic and glycolic acids. PLGA itself is not cell adhesive, but when implanted it nonspecifically adsorbs proteins from fluids in the body, rendering implanted PLGA scaffolds cell adhesive.[138,139] Mooney and colleagues have demonstrated vascular ingrowth in PLGA scaffolds containing VEGF alone[140,141] and VEGF with PDGF,[142] with the addition of PDGF promoting the formation of mature, stable vessels supported by smooth muscle cells. Langer and colleagues prevascularized skeletal muscle constructs before implantation by co-seeding endothelial cells, myoblasts, and embryonic fibroblasts on scaffolds composed of PLGA and poly(L-lactic acid) (PLLA).[143] They demonstrated that preformed vascular networks indeed improved the viability and vascularization of the skeletal muscle constructs after implantation.

While PLGA scaffolds have proven capable of supporting vascularization in implanted constructs, PEG hydrogels possess significant advantages over other material systems as biomimetic, tunable scaffolds for engineered tissue vascularization. Since natural ECMs themselves are hydrogels,[109] the basic structure of PEG hydrogels is ideal for serving as a native tissue replacement. Cross-linked polymers, like natural ECM protein fibrils, resist tensile stresses, while interstitial fluid resists compressive stresses. Given the importance of matrix rigidity in angiogenesis, the mechanical tunability of PEG, by varying either its cross-linking density or chain length between cross-links, is significant. The rigidity of PEG can be varied widely, with reported Young's modulus values between 300 Pa and 100 kPa,[144–146] spanning a significant range of biological tissues. While PEG itself is not adhesive to proteins or cells,[147] oligopeptides representing key integrin-binding regions within native ECM adhesion proteins such as fibronectin or laminin can be covalently cross-linked to the PEG network,[148] such that cells adhere to the PEG matrix only through those specific adhesive peptides, while the remaining polymer network is relatively inert. The most common adhesive ligand incorporated is the ubiquitous RGD ligand,[149] found in fibronectin, vitronectin, fibrinogen, and several other ECM adhesion proteins,[150] but other adhesive sequences such as IKVAV (from laminin) or REDV (from the fibronectin IIICS region) have also been incorporated so that the adhesive ligand can be tailored to the cell type of interest.[151] This synthetic scheme allows for variations of adhesive ligand density independent of other parameters such as stiffness and porosity, permitting investigation of the contributions of these material parameters to cell behavior independently. For instance, independent variation of adhesive ligand density has demonstrated a biphasic effect of ligand density on cell migration in 3D matrices[152–155] and may provide one explanation for the enhanced angiogenesis observed at intermediate degrees of adhesion.[45,50]

The ability to incorporate oligopeptides into a PEG backbone endows it with not only a cell adhesive property but also several other biological functionalities characteristic of natural ECMs. Proteolytically degradable sequences allow cells to degrade and migrate through the material via cell-secreted proteases, as the nanometer-scale pores in PEG hydrogels do not normally permit invasion and migration.[96,145,154] These sequences can be tailored to be sensitive to various proteases, including MMPs and plasmin, and thus specific to the biological process being studied. Similar to the "engineered" natural ECMs described earlier, growth factors can be incorporated into PEG hydrogels either directly or through binding to heparin.[110,156–158] The chemistry to polymerize and incorporate biological functionalities into PEG materials is now sufficiently well established and compatible with living cells and tissues such that the material can even be polymerized in situ after encapsulation of cells. West and colleagues have utilized a scheme in which acrylated PEG derivatives are photopolymerized with proteolytically degradable and cell-adhesive oligopeptides such that degradable sequences are incorporated in the PEG backbone, and adhesive sequences are attached as pendant chains.[96,153,159] Hubbell and colleagues employ Michael-type cross-linking reactions between end-functionalized PEG macromers and thiol-bearing compounds (such as the amino acid cysteine) to form similar degradable and adhesive matrices (Figure 6.3B).[144,145,154] Recently, Anseth and colleagues have developed chemistries involving thiol-ene photopolymerization[160] and sequential click reactions to pattern 3D PEG hydrogels.[161] The end result is a biocompatible, biomimetic material that can be customed tailored for the biological process or tissue type being studied via modular incorporation of bioactive oligopeptides.

In the context of angiogenesis, the potential for PEG hydrogels to serve as scaffolds for vascularization has only begun to be realized. Hubbell and colleagues demonstrated enhanced angiogenesis in response to VEGF covalently bound to and then released from PEG scaffolds by cell-mediated proteolytic degradation.[110] Vascularization on the chick CAM was improved with covalently bound VEGF over soluble VEGF, and scaffolds implanted subcutaneously in rats were replaced by vascularized tissue. West and colleagues have added several functionalities to PEG to improve endothelial cell function: covalent immobilization of VEGF enhanced endothelial cell tubulogenesis on the surface of PEG hydrogels and migration and cell–cell interaction of cells encapsulated within degradable hydrogels.[108] Endothelial cell adhesion and tubule formation were promoted by covalent incorporation of ephrin A1, a cell–cell adhesion molecule important in regulating vascular guidance and assembly.[162] By patterning lines of RGDS on the surface of PEG hydrogels, they demonstrated formation of endothelial cords only on lines of intermediate width, and only at intermediate RGDS concentrations,[163] consistent with previous results demonstrating enhanced tubulogenesis at intermediate densities of fibronectin and on fibronectin lines of intermediate width.[45,49] Further techniques developed to pattern PEG hydrogels in 2D and 3D[164,165] can potentially be used to control spatially the adhesive, mechanical, and degradable properties of the matrix to generate complex vascularized tissues. Taking further advantage of the mechanical, adhesive, and degradable tunability of PEG will provide further insight into ECM regulation of angiogenesis and aid in the rational design of materials for tissue repair.

6.6   MULTICELLULAR INTERACTIONS IN ANGIOGENESIS

In vivo, vascular cells exist in multicellular tubular structures and rarely as single cells. While studying ECM control of single endothelial cell behavior without the confounding contribution of cell–cell interaction is necessary, it is also crucial to develop a more focused understanding of the role of cell–cell adhesion itself in endothelial cell behavior, as well as how cell–cell interactions might affect ECM regulation of multicellular vascular structures. For instance, culture of single endothelial cells on adhesive islands of defined area results in a switch between proliferation with high cell spreading and apoptosis with low cell spreading.[47] However, it is only when endothelial cells are cultured on lines of fibronectin of intermediate width, permitting multicellular interactions, that cells differentiate into capillary tube-like structures.[49] While cell–cell contact is traditionally thought to inhibit endothelial cell proliferation,[166] our lab has observed proliferation within confluent sheets of endothelial cells at regions of higher stress as dictated by the surrounding ECM environment (Figure 6.4A), suggesting that cytoskeletal tension propagated through cell–cell contacts can be an important driver of proliferation in physiological contexts.[167] These observations provide one potential explanation for increased proliferation of stalk cells in angiogenic sprouts,[168] with migrating tip cells pulling and propagating stress toward stalk cells at the base of the sprouts. In addition to propagating tension, our lab has demonstrated that cell–cell contact itself can activate cytoskeletal tension signaling and subsequent proliferation. VE-cadherin-mediated contact normally reduced the degree of cell spreading and proliferation, but when cell spreading was kept constant via a micropatterning technique, the presence of cell–cell contact actually promoted cell proliferation via increased actomyosin-generated tension (Figure 6.4B).[169,170] The decrease in cell–ECM adhesion by VE-cadherin-mediated cell–cell adhesion could occur through several mechanisms: cell–cell contact-induced changes in the tension and structure of the actin cytoskeleton, Adams CL et al., 1998, cadherin-induced decrease in the expression of integrins,[171] and recruitment of vinculin to cell–cell contacts away from focal adhesions.[172] Conversely, integrin-mediated adhesion has been shown to disrupt VE-cadherin-containing cell–cell junctions,[173] potentially contributing to the initiation of angiogenic sprouting and subsequent vascular morphogenesis. These results highlight the complex interplay between cell–cell and cell–ECM adhesion and the importance of understanding how they affect each other in the multicellular processes of angiogenesis. In addition to the interaction between cell–cell and cell–ECM adhesion, Dejana and colleagues have importantly shown that endothelial cell–cell adhesion can regulate soluble VEGF signaling: engagement of VE-cadherin at cell–cell junctions can sequester VEGF receptor 2 at the cell surface, preventing its internalization and subsequent signaling from the receptor.[174,175] This reduction in VEGF signaling in confluent monolayers provides one explanation for contact-dependent inhibition of proliferation and quiescence in confluent endothelial cells.

In addition to the importance of homotypic endothelial cell interactions in angiogenesis, several recent studies have explored the role of mesenchymal support cells in inducing formation of and stabilizing vascular networks (reviewed in greater depth in[176,177]). Early work by D'Amore and colleagues demonstrated recruitment of 10T1/2 pericytes toward developing vessels and their differentiation to a smooth muscle fate by signals from endothelial cells,[178] as well as mutual inhibition of proliferation in pericytes and endothelial cells in contacting cocultures.[179,180] These responses resulted from both soluble PDGF and TGF-β signals[178] and direct N-cadherin contacts between endothelial cells and pericytes.[181] The importance of pericytes in stabilizing vessels is evidenced by leaky, heterogeneous tumor blood vessels, which have incomplete pericyte coverage,[182–184] as well as microvascular defects in PDGF-B or

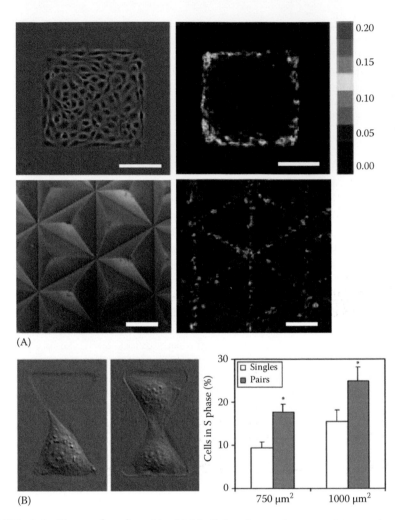

FIGURE 6.4 (See color insert.) Multicellular interactions regulate angiogenesis. (A) Confluent monolayers of endothelial cells proliferate at regions of higher stress: at monolayer edges (top) or in valleys of undulating monolayers without edges (bottom). (From Nelson, C.M. et al., *Proc Natl Acad Sci USA*, 102(33), 11594, 2005.) (B) Cell–cell contact increases proliferation when cell spreading is controlled. (From Nelson, C.M. and Chen, C.S., *J Cell Sci*, 116(Pt 17), 3571, 2003. With permission.)

PDGF beta-receptor knockout mice due to decreased pericyte recruitment.[185–187] As such, the majority of efforts to create stable vascular networks in engineered tissues now incorporate mural support cells in some form. Dermal fibroblasts seeded on the surface of fibrin gels are used to stimulate capillary sprouting from endothelial cell-coated beads into the gels,[124] extending previous work demonstrating stimulation of endothelial tubule formation by 3T3 fibroblasts.[188] The same endothelial sprouting assay can be carried out with mesenchymal stem cells (MSCs) distributed throughout the matrix in place of surface-seeded fibroblasts,[90] and these interactions are thought to occur primarily through secreted soluble factors. Others are employing direct cell–cell interactions between endothelial and mural cells to stabilize the vasculature in implanted tissues. A recent report demonstrated that MSCs can in fact serve as perivascular precursor cells by stabilizing endothelial cell networks implanted *in vivo* for several months,[189] producing similar results as earlier work by this group using 10T1/2 pericytes.[117] Others have demonstrated the ability of neural progenitor cells[190] and adipose stromal cells[191] to stabilize vessels *in vivo*. Perhaps the most intriguing finding relevant to vascularizing tissues, though, is the discovery that empty basement membrane "sleeves" and accompanying pericytes left behind after vessels regressed in response to inhibition of VEGF receptor—a potential antitumor treatment—quickly promoted revascularization upon withdrawal of the drug.[192] Endothelial sprouts grew into the basement membrane sleeves and fully revascularized the tumor by 7 days. While this study has obvious implications in cancer therapy, it also suggests that perhaps ECM secreted by endothelial cells and pericytes can be used to guide and vascularize engineered tissues efficiently. While cell-secreted ECM is one way endothelial cells might be regulated mechanically by pericytes, it remains to be seen whether other means of interaction, for example, pericyte-generated traction transmitted through the matrix or alignment of matrix fibers, also have a significant impact on endothelial cell behavior.

6.7 CONCLUSION

Multiple properties of the ECM—from mechanical stiffness to presentation of adhesive ligands—are crucial regulators of angiogenesis. While the effects of integrin-mediated adhesion, cell spreading, and matrix stiffness are being explored, many basic questions remain: how do these properties interact in regulating angiogenesis? Can a decrease in matrix stiffness be rescued by increased adhesive ligand presentation to promote a similar angiogenic response? Is there an "optimal" angiogenic response or are there merely variations in the quality of angiogenesis, and can these responses be traced directly back to specific signaling pathways responsive to changes in cell–ECM interaction? How do endothelial cell interactions with other endothelial cells and supporting mesenchymal cells affect ECM regulation of angiogenesis?

In order to answer such questions, knowledge and expertise in materials engineering and biology need to be tightly coupled. To engineer materials that optimize vascular ingrowth to support implanted tissues, we need to take advantage of our basic knowledge of how ECM ligand presentation, stiffness, and degradation, as well as soluble factor cues and cell–cell interactions, regulate angiogenesis. Similarly, we can utilize tools available to control material properties and cellular interactions precisely and systematically to further our understanding of the basic processes of angiogenesis. Significant progress has been made with the development of synthetic materials that are mechanically tunable and can be functionalized in a modular fashion with adhesive, degradable, and proangiogenic factor–presenting properties. Spatial patterning of cells, adhesive ligands, or mechanical gradients will also allow us to study and recapitulate some of the multicellular and microenvironmental complexity present in angiogenesis. We can continue to borrow proven design principles from nature—for example, the use of factor

XIII cross-linking in fibrin gels in the synthesis of synthetic materials[193] and heparin to sequester multiple growth factors[131]—that have been so effective in supporting vascularization in development and wound healing. As we integrate our basic biological understanding and engineering toolbox, we will make significant progress in designing solutions for therapeutic angiogenesis and tissue engineering, as well as in advancing our knowledge of the fundamental processes of angiogenesis.

ACKNOWLEDGMENTS

We apologize to the authors whose work could not be cited due to space limitations. We thank Jordan S. Miller for helpful discussions, as well as support by grants from the National Institutes of Health (EB00262, HL73305, GM74048, to C.S.C.), Ruth L. Kirschstein National Research Service Award (C.J.S.), and Paul and Daisy Soros Foundation (C.J.S.).

REFERENCES

1. Jain, R. K., Transport of molecules, particles, and cells in solid tumors. *Annu Rev Biomed Eng* 1999, 1, 241–263.
2. Jain, R. K.; Au, P.; Tam, J.; Duda, D. G.; Fukumura, D., Engineering vascularized tissue. *Nat Biotechnol* 2005, 23 (7), 821–823.
3. Renault, M. A.; Losordo, D. W., Therapeutic myocardial angiogenesis. *Microvasc Res* 2007, 74 (2–3), 159–171.
4. Adams, R. H.; Alitalo, K., Molecular regulation of angiogenesis and lymphangiogenesis. *Nat Rev Mol Cell Biol* 2007, 8 (6), 464–478.
5. Drevs, J.; Muller-Driver, R.; Wittig, C.; Fuxius, S.; Esser, N.; Hugenschmidt, H.; Konerding, M. A.; Allegrini, P. R.; Wood, J.; Hennig, J.; Unger, C.; Marme, D., PTK787/ZK 222584, a specific vascular endothelial growth factor-receptor tyrosine kinase inhibitor, affects the anatomy of the tumor vascular bed and the functional vascular properties as detected by dynamic enhanced magnetic resonance imaging. *Cancer Res* 2002, 62 (14), 4015–4022.
6. Ferrara, N., Vascular endothelial growth factor as a target for anticancer therapy. *Oncologist* 2004, 9, 2–10.
7. Levin, E. G.; Sikora, L.; Ding, L.; Rao, S. P.; Sriramarao, P., Suppression of tumor growth and angiogenesis in vivo by a truncated form of 24-kd fibroblast growth factor (FGF)-2. *Am J Pathol* 2004, 164 (4), 1183–1190.
8. Cabebe, E.; Fisher, G. A., Clinical trials of VEGF receptor tyrosine kinase inhibitors in pancreatic cancer. *Expert Opin Investig Drugs* 2007, 16 (4), 467–476.
9. Hanrahan, E. O.; Heymach, J. V., Vascular endothelial growth factor receptor tyrosine kinase inhibitors vandetanib (ZD6474) and AZD2171 in lung cancer. *Clin Cancer Res* 2007, 13 (15), 4617s–4622s.
10. Hilberg, F.; Roth, G. J.; Krssak, M.; Kautschitsch, S.; Sommergruber, W.; Tontsch-Grunt, U.; Garin-Chesa, P.; Bader, G.; Zoephel, A.; Quant, J.; Heckel, A.; Rettig, W. J., BIBF 1120: Triple angiokinase inhibitor with sustained receptor blockade and good antitumor efficacy. *Cancer Res* 2008, 68 (12), 4774–4782.
11. Manegold, C., Bevacizumab for the treatment of advanced non-small-cell lung cancer. *Expert Rev Anticancer Ther* 2008, 8 (5), 689–699.
12. Goodman, L., Persistence—Luck—Avastin. *J Clin Invest* 2004, 113 (7), 934.
13. Simo, R.; Hernandez, C., Intravitreous anti-VEGF for diabetic retinopathy: hopes and fears for a new therapeutic strategy. *Diabetologia* 2008, 51 (9), 1574–1580.

14. Drake, C. J.; Cheresh, D. A.; Little, C. D., An antagonist of integrin alpha v beta 3 prevents maturation of blood-vessels during embryonic neovascularization. *J Cell Sci* 1995, 108, 2655–2661.

15. Gutheil, J. C.; Campbell, T. N.; Pierce, P. R.; Watkins, J. D.; Huse, W. D.; Bodkin, D. J.; Cheresh, D. A., Targeted antiangiogenic therapy for cancer using Vitaxin: A humanized monoclonal antibody to the integrin alphavbeta3. *Clin Cancer Res* 2000, 6 (8), 3056–3061.

16. Hood, J. D.; Bednarski, M.; Frausto, R.; Guccione, S.; Reisfeld, R. A.; Xiang, R.; Cheresh, D. A., Tumor regression by targeted gene delivery to the neovasculature. *Science* 2002, 296 (5577), 2404–2407.

17. Trikha, M.; Zhou, Z.; Nemeth, J. A.; Chen, Q. M.; Sharp, C.; Emmell, E.; Giles-Komar, J.; Nakada, M. T., CNTO 95, a fully human monoclonal antibody that inhibits alpha v integrins, has antitumor and antiangiogenic activity in vivo. *Int J Cancer* 2004, 110 (3), 326–335.

18. Brooks, P. C.; Clark, R. A.; Cheresh, D. A., Requirement of vascular integrin alpha v beta 3 for angiogenesis. *Science* 1994, 264 (5158), 569–571.

19. Friedlander, M.; Brooks, P. C.; Shaffer, R. W.; Kincaid, C. M.; Varner, J. A.; Cheresh, D. A., Definition of two angiogenic pathways by distinct alpha v integrins. *Science* 1995, 270 (5241), 1500–1502.

20. Kim, S.; Bell, K.; Mousa, S. A.; Varner, J. A., Regulation of angiogenesis in vivo by ligation of integrin alpha5beta1 with the central cell-binding domain of fibronectin. *Am J Pathol* 2000, 156 (4), 1345–1362.

21. Senger, D. R.; Claffey, K. P.; Benes, J. E.; Perruzzi, C. A.; Sergiou, A. P.; Detmar, M., Angiogenesis promoted by vascular endothelial growth factor: Regulation through alpha-1beta1 and alpha2beta1 integrins. *Proc Natl Acad Sci USA* 1997, 94 (25), 13612–13617.

22. Avraamides, C. J.; Garmy-Susini, B.; Varner, J. A., Integrins in angiogenesis and lymphangiogenesis. *Nat Rev Cancer* 2008, 8 (8), 604–617.

23. Bader, B. L.; Rayburn, H.; Crowley, D.; Hynes, R. O., Extensive vasculogenesis, angiogenesis, and organogenesis precede lethality in mice lacking all alpha v integrins. *Cell* 1998, 95 (4), 507–519.

24. Hodivala-Dilke, K. M.; McHugh, K. P.; Tsakiris, D. A.; Rayburn, H.; Crowley, D.; Ullman-Cullere, M.; Ross, F. P.; Coller, B. S.; Teitelbaum, S.; Hynes, R. O., Beta3-integrin-deficient mice are a model for Glanzmann thrombasthenia showing placental defects and reduced survival. *J Clin Invest* 1999, 103 (2), 229–238.

25. Huang, X.; Griffiths, M.; Wu, J.; Farese, R. V., Jr.; Sheppard, D., Normal development, wound healing, and adenovirus susceptibility in beta5-deficient mice. *Mol Cell Biol* 2000, 20 (3), 755–759.

26. Pozzi, A.; Moberg, P. E.; Miles, L. A.; Wagner, S.; Soloway, P.; Gardner, H. A., Elevated matrix metalloprotease and angiostatin levels in integrin alpha 1 knockout mice cause reduced tumor vascularization. *Proc Natl Acad Sci USA* 2000, 97 (5), 2202–2207.

27. Zhang, Z.; Ramirez, N. E.; Yankeelov, T. E.; Li, Z.; Ford, L. E.; Qi, Y.; Pozzi, A.; Zutter, M. M., Alpha2beta1 integrin expression in the tumor microenvironment enhances tumor angiogenesis in a tumor cell-specific manner. *Blood* 2008, 111 (4), 1980–1988.

28. Reynolds, A. R.; Reynolds, L. E.; Nagel, T. E.; Lively, J. C.; Robinson, S. D.; Hicklin, D. J.; Bodary, S. C.; Hodivala-Dilke, K. M., Elevated Flk1 (vascular endothelial growth factor receptor 2) signaling mediates enhanced angiogenesis in beta3-integrin-deficient mice. *Cancer Res* 2004, 64 (23), 8643–8650.

29. Burridge, K.; Fath, K.; Kelly, T.; Nuckolls, G.; Turner, C., Focal adhesions—Transmembrane junctions between the extracellular-matrix and the cytoskeleton. *Annu Rev Cell Biol* 1988, 4, 487–525.

30. Miyamoto, S.; Teramoto, H.; Coso, O. A.; Gutkind, J. S.; Burbelo, P. D.; Akiyama, S. K.; Yamada, K. M., Integrin function—Molecular hierarchies of cytoskeletal and signaling molecules. *J Cell Biol* 1995, 131 (3), 791–805.

31. Plopper, G. E.; Mcnamee, H. P.; Dike, L. E.; Bojanowski, K.; Ingber, D. E., Convergence of integrin and growth-factor receptor signaling pathways within the focal adhesion complex. *Mol Biol Cell* 1995, 6 (10), 1349–1365.

32. Schwartz, M. A., Integrins, oncogenes, and anchorage independence. *Journal of Cell Biology* 1997, 139 (3), 575–578.

33. Stupack, D. G.; Cheresh, D. A., Integrins and angiogenesis. *Curr Top Dev Biol* 2004, 64, 207–238.

34. Mettouchi, A.; Klein, S.; Guo, W. J.; Lopez-Lago, M.; Lemichez, E.; Westwick, J. K.; Giancotti, F. G., Integrin-specific activation of Rac controls progression through the G(1) phase of the cell cycle. *Mol Cell* 2001, 8 (1), 115–127.

35. Schneller, M.; Vuori, K.; Ruoslahti, E., Alpha v beta 3 integrin associates with activated insulin and PDGF beta receptors and potentiates the biological activity of PDGF. *EMBO J* 1997, 16 (18), 5600–5607.

36. Hood, J. D.; Frausto, R.; Kiosses, W. B.; Schwartz, M. A.; Cheresh, D. A., Differential alphav integrin-mediated Ras-ERK signaling during two pathways of angiogenesis. *J Cell Biol* 2003, 162 (5), 933–943.

37. Parsons, J. T., Focal adhesion kinase: The first ten years. *J Cell Sci* 2003, 116, (Pt 8), 1409–1416.

38. Pirone, D. M.; Liu, W. F.; Ruiz, S. A.; Gao, L.; Raghavan, S.; Lemmon, C. A.; Romer, L. H.; Chen, C. S., An inhibitory role for FAK in regulating proliferation: A link between limited adhesion and RhoA-ROCK signaling. *J Cell Biol* 2006, 174 (2), 277–288.

39. Schober, M.; Raghavan, S.; Nikolova, M.; Polak, L.; Pasolli, H. A.; Beggs, H. E.; Reichardt, L. F.; Fuchs, E., Focal adhesion kinase modulates tension signaling to control actin and focal adhesion dynamics. *J Cell Biol* 2007, 176 (5), 667–680.

40. Vadali, K.; Cai, X.; Schaller, M. D., Focal adhesion kinase: An essential kinase in the regulation of cardiovascular functions. *IUBMB Life* 2007, 59 (11), 709–716.

41. Braren, R.; Hu, H.; Kim, Y. H.; Beggs, H. E.; Reichardt, L. F.; Wang, R., Endothelial FAK is essential for vascular network stability, cell survival, and lamellipodial formation. *J Cell Biol* 2006, 172 (1), 151–162.

42. Shen, T. L.; Park, A. Y.; Alcaraz, A.; Peng, X.; Jang, I.; Koni, P.; Flavell, R. A.; Gu, H.; Guan, J. L., Conditional knockout of focal adhesion kinase in endothelial cells reveals its role in angiogenesis and vascular development in late embryogenesis. *J Cell Biol* 2005, 169 (6), 941–952.

43. Eliceiri, B. P.; Klemke, R.; Stromblad, S.; Cheresh, D. A., Integrin alphavbeta3 requirement for sustained mitogen-activated protein kinase activity during angiogenesis. *J Cell Biol* 1998, 140 (5), 1255–1263.

44. Ingber, D. E., Fibronectin controls capillary endothelial cell growth by modulating cell shape. *Proc Natl Acad Sci USA* 1990, 87 (9), 3579–3583.

45. Ingber, D. E.; Folkman, J., Mechanochemical switching between growth and differentiation during fibroblast growth factor-stimulated angiogenesis in vitro: Role of extracellular matrix. *J Cell Biol* 1989, 109 (1), 317–330.

46. Singhvi, R.; Kumar, A.; Lopez, G. P.; Stephanopoulos, G. N.; Wang, D. I.; Whitesides, G. M.; Ingber, D. E., Engineering cell shape and function. *Science* 1994, 264 (5159), 696–698.

47. Chen, C. S.; Mrksich, M.; Huang, S.; Whitesides, G. M.; Ingber, D. E., Geometric control of cell life and death. *Science* 1997, 276 (5317), 1425–1428.

48. Gamble, J. R.; Matthias, L. J.; Meyer, G.; Kaur, P.; Russ, G.; Faull, R.; Berndt, M. C.; Vadas, M. A., Regulation of in vitro capillary tube formation by anti-integrin antibodies. *J Cell Biol* 1993, 121 (4), 931–943.

49. Dike, L. E.; Chen, C. S.; Mrksich, M.; Tien, J.; Whitesides, G. M.; Ingber, D. E., Geometric control of switching between growth, apoptosis, and differentiation during angiogenesis using micropatterned substrates. *In Vitro Cell Dev Biol Anim* 1999, 35 (8), 441–448.

50. Reynolds, A. R.; Hart, I. R.; Watson, A. R.; Welti, J. C.; Silva, R. G.; Robinson, S. D.; Da Violante, G.; Gourlaouen, M.; Salih, M.; Jones, M. C.; Jones, D. T.; Saunders, G.; Kostourou, V.; Perron-Sierra, F.; Norman, J. C.; Tucker, G. C.; Hodivala-Dilke, K. M., Stimulation of tumor growth and angiogenesis by low concentrations of RGD-mimetic integrin inhibitors. *Nat Med* 2009, 15 (4), 392–400.

51. Brooks, P. C.; Stromblad, S.; Sanders, L. C.; vonSchalscha, T. L.; Aimes, R. T.; StetlerStevenson, W. G.; Quigley, J. P.; Cheresh, D. A., Localization of matrix metalloproteinase MMP-2 to the surface of invasive cells by interaction with integrin alpha v beta 3. *Cell* 1996, 85 (5), 683–693.

52. Xu, J.; Rodriguez, D.; Petitclerc, E.; Kim, J. J.; Hangai, M.; Moon, Y. S.; Davis, G. E.; Brooks, P. C., Proteolytic exposure of a cryptic site within collagen type IV is required for angiogenesis and tumor growth in vivo. *J Cell Biol* 2001, 154 (5), 1069–1079.

53. O'Reilly, M. S.; Boehm, T.; Shing, Y.; Fukai, N.; Vasios, G.; Lane, W. S.; Flynn, E.; Birkhead, J. R.; Olsen, B. R.; Folkman, J., Endostatin: An endogenous inhibitor of angiogenesis and tumor growth. *Cell* 1997, 88 (2), 277–285.

54. Maeshima, Y.; Colorado, P. C.; Torre, A.; Holthaus, K. A.; Grunkemeyer, J. A.; Ericksen, M. B.; Hopfer, H.; Xiao, Y.; Stillman, I. E.; Kalluri, R., Distinct antitumor properties of a type IV collagen domain derived from basement membrane. *J Biol Chem* 2000, 275 (28), 21340–21348.

55. O'Reilly, M. S.; Holmgren, L.; Shing, Y.; Chen, C.; Rosenthal, R. A.; Moses, M.; Lane, W. S.; Cao, Y.; Sage, E. H.; Folkman, J., Angiostatin: A novel angiogenesis inhibitor that mediates the suppression of metastases by a Lewis lung carcinoma. *Cell* 1994, 79 (2), 315–328.

56. Brooks, P. C.; Silletti, S.; von Schalscha, T. L.; Friedlander, M.; Cheresh, D. A., Disruption of angiogenesis by PEX, a noncatalytic metalloproteinase fragment with integrin binding activity. *Cell* 1998, 92 (3), 391–400.

57. Davis, G. E.; Senger, D. R., Endothelial extracellular matrix: biosynthesis, remodeling, and functions during vascular morphogenesis and neovessel stabilization. *Circ Res* 2005, 97 (11), 1093–1107.

58. Stratman, A. N.; Malotte, K. M.; Mahan, R. D.; Davis, M. J.; Davis, G. E., Pericyte recruitment during vasculogenic tube assembly stimulates endothelial basement membrane matrix formation. *Blood* 2009, 114 (24), 5091–5101.

59. Amano, M.; Ito, M.; Kimura, K.; Fukata, Y.; Chihara, K.; Nakano, T.; Matsuura, Y.; Kaibuchi, K., Phosphorylation and activation of myosin by Rho-associated kinase (Rho-kinase). *J Biol Chem* 1996, 271 (34), 20246–20249.

60. Ishizaki, T.; Maekawa, M.; Fujisawa, K.; Okawa, K.; Iwamatsu, A.; Fujita, A.; Watanabe, N.; Saito, Y.; Kakizuka, A.; Morii, N.; Narumiya, S., The small GTP-binding protein Rho binds to and activates a 160 kDa Ser/Thr protein kinase homologous to myotonic dystrophy kinase. *EMBO J* 1996, 15 (8), 1885–1893.

61. Kimura, K.; Ito, M.; Amano, M.; Chihara, K.; Fukata, Y.; Nakafuku, M.; Yamamori, B.; Feng, J.; Nakano, T.; Okawa, K.; Iwamatsu, A.; Kaibuchi, K., Regulation of myosin phosphatase by Rho and Rho-associated kinase (Rho-kinase). *Science* 1996, 273 (5272), 245–248.

62. Chrzanowska-Wodnicka, M.; Burridge, K., Rho-stimulated contractility drives the formation of stress fibers and focal adhesions. *J Cell Biol* 1996, 133 (6), 1403–1415.

63. Bhadriraju, K.; Yang, M.; Ruiz, S. A.; Pirone, D.; Tan, J.; Chen, C. S., Activation of ROCK by RhoA is regulated by cell adhesion, shape, and cytoskeletal tension. *Exp Cell Res* 2007, 313 (16), 3616–3623.

64. Chen, C. S.; Alonso, J. L.; Ostuni, E.; Whitesides, G. M.; Ingber, D. E., Cell shape provides global control of focal adhesion assembly. *Biochem Biophys Res Commun* 2003, 307 (2), 355–361.

65. McBeath, R.; Pirone, D. M.; Nelson, C. M.; Bhadriraju, K.; Chen, C. S., Cell shape, cytoskeletal tension, and RhoA regulate stem cell lineage commitment. *Dev Cell* 2004, 6 (4), 483–495.

66. Polte, T. R.; Eichler, G. S.; Wang, N.; Ingber, D. E., Extracellular matrix controls myosin light chain phosphorylation and cell contractility through modulation of cell shape and cytoskeletal prestress. *Am J Physiol Cell Physiol* 2004, 286 (3), C518–C528.

67. Ren, X. D.; Wang, R. X.; Li, Q. Y.; Kahek, L. A. F.; Kaibuchi, K.; Clark, R. A. F., Disruption of Rho signal transduction upon cell detachment. *J Cell Sci* 2004, 117 (16), 3511–3518.

68. Tan, J. L.; Tien, J.; Pirone, D. M.; Gray, D. S.; Bhadriraju, K.; Chen, C. S., Cells lying on a bed of microneedles: An approach to isolate mechanical force. *Proceedings of the National Academy of Sciences of the United States of America* 2003, 100 (4), 1484–1489.

69. Huang, S.; Chen, C. S.; Ingber, D. E., Control of cyclin D1, p27(Kip1), and cell cycle progression in human capillary endothelial cells by cell shape and cytoskeletal tension. *Mol Biol Cell* 1998, 9 (11), 3179–3193.

70. Mammoto, A.; Huang, S.; Moore, K.; Oh, P.; Ingber, D. E., Role of RhoA, mDia, and ROCK in cell shape-dependent control of the Skp2-p27kip1 pathway and the G1/S transition. *J Biol Chem* 2004, 279 (25), 26323–26330.

71. Bryan, B. A.; D'Amore, P. A., What tangled webs they weave: Rho-GTPase control of angiogenesis. *Cell Mol Life Sci* 2007, 64 (16), 2053–2065.

72. Hoang, M. V.; Whelan, M. C.; Senger, D. R., Rho activity critically and selectively regulates endothelial cell organization during angiogenesis. *Proc Natl Acad Sci USA* 2004, 101 (7), 1874–1879.

73. Liu, Y.; Senger, D. R., Matrix-specific activation of Src and Rho initiates capillary morphogenesis of endothelial cells. *FASEB J* 2004, 18 (3), 457–468.

74. Uchida, S.; Watanabe, G.; Shimada, Y.; Maeda, M.; Kawabe, A.; Mori, A.; Arii, S.; Uehata, M.; Kishimoto, T.; Oikawa, T.; Imamura, M., The suppression of small GTPase rho signal transduction pathway inhibits angiogenesis in vitro and in vivo. *Biochem Biophys Res Commun* 2000, 269 (2), 633–640.

75. van Nieuw Amerongen, G. P.; Koolwijk, P.; Versteilen, A.; van Hinsbergh, V. W., Involvement of RhoA/Rho kinase signaling in VEGF-induced endothelial cell migration and angiogenesis in vitro. *Arterioscler Thromb Vasc Biol* 2003, 23 (2), 211–217.

76. Kniazeva, E.; Putnam, A. J., Endothelial cell traction and ECM density influence both capillary morphogenesis and maintenance in 3-D. *Am J Physiol Cell Physiol* 2009, 297 (1), C179–C187.

77. Engler, A. J.; Sen, S.; Sweeney, H. L.; Discher, D. E., Matrix elasticity directs stem cell lineage specification. *Cell* 2006, 126 (4), 677–689.

78. Paszek, M. J.; Zahir, N.; Johnson, K. R.; Lakins, J. N.; Rozenberg, G. I.; Gefen, A.; Reinhart-King, C. A.; Margulies, S. S.; Dembo, M.; Boettiger, D.; Hammer, D. A.; Weaver, V. M., Tensional homeostasis and the malignant phenotype. *Cancer Cell* 2005, 8 (3), 241–254.

79. Wozniak, M. A.; Desai, R.; Solski, P. A.; Der, C. J.; Keely, P. J., ROCK-generated contractility regulates breast epithelial cell differentiation in response to the physical properties of a three-dimensional collagen matrix. *J Cell Biol* 2003, 163 (3), 583–595.

80. Assoian, R. K.; Schwartz, M. A., Coordinate signaling by integrins and receptor tyrosine kinases in the regulation of G(1) phase cell-cycle progression. *Curr Opin Gene Dev* 2001, 11 (1), 48–53.

81. Ren, X. D.; Kiosses, W. B.; Schwartz, M. A., Regulation of the small GTP-binding protein Rho by cell adhesion and the cytoskeleton. *EMBO J* 1999, 18 (3), 578–585.

82. Renshaw, M. W.; Toksoz, D.; Schwartz, M. A., Involvement of the small GTPase Rho in integrin-mediated activation of mitogen-activated protein kinase. *J Biol Chem* 1996, 271 (36), 21691–21694.

83. Montesano, R.; Orci, L.; Vassalli, P., In vitro rapid organization of endothelial cells into capillary-like networks is promoted by collagen matrices. *J Cell Biol* 1983, 97 (5 Pt 1), 1648–1652.

84. Schor, A. M.; Schor, S. L.; Allen, T. D., Effects of culture conditions on the proliferation, morphology and migration of bovine aortic endothelial cells. *J Cell Sci* 1983, 62, 267–285.

85. Discher, D. E.; Janmey, P.; Wang, Y. L., Tissue cells feel and respond to the stiffness of their substrate. *Science* 2005, 310 (5751), 1139–1143.

86. Pelham, R. J., Jr.; Wang, Y., Cell locomotion and focal adhesions are regulated by substrate flexibility. *Proc Natl Acad Sci USA* 1997, 94 (25), 13661–13665.

87. Peyton, S. R.; Raub, C. B.; Keschrumrus, V. P.; Putnam, A. J., The use of poly(ethylene glycol) hydrogels to investigate the impact of ECM chemistry and mechanics on smooth muscle cells. *Biomaterials* 2006, 27 (28), 4881–4893.

88. Wang, H. B.; Dembo, M.; Wang, Y. L., Substrate flexibility regulates growth and apoptosis of normal but not transformed cells. *Am J Physiol Cell Physiol* 2000, 279 (5), C1345–C1350.

89. Yeung, T.; Georges, P. C.; Flanagan, L. A.; Marg, B.; Ortiz, M.; Funaki, M.; Zahir, N.; Ming, W.; Weaver, V.; Janmey, P. A., Effects of substrate stiffness on cell morphology, cytoskeletal structure, and adhesion. *Cell Motil Cytoskel* 2005, 60 (1), 24–34.

90. Ghajar, C. M.; Blevins, K. S.; Hughes, C. C.; George, S. C.; Putnam, A. J., Mesenchymal stem cells enhance angiogenesis in mechanically viable prevascularized tissues via early matrix metalloproteinase upregulation. *Tissue Eng* 2006, 12 (10), 2875–2888.

91. Nehls, V.; Herrmann, R., The configuration of fibrin clots determines capillary morphogenesis and endothelial cell migration. *Microvasc Res* 1996, 51 (3), 347–364.

92. Sieminski, A. L.; Hebbel, R. P.; Gooch, K. J., The relative magnitudes of endothelial force generation and matrix stiffness modulate capillary morphogenesis in vitro. *Exp Cell Res* 2004, 297 (2), 574–584.

93. Vailhe, B.; Ronot, X.; Tracqui, P.; Usson, Y.; Tranqui, L., In vitro angiogenesis is modulated by the mechanical properties of fibrin gels and is related to alpha(v)beta3 integrin localization. *In Vitro Cell Dev Biol Anim* 1997, 33 (10), 763–773.

94. Ghajar, C. M.; Chen, X.; Harris, J. W.; Suresh, V.; Hughes, C. C.; Jeon, N. L.; Putnam, A. J.; George, S. C., The effect of matrix density on the regulation of 3-D capillary morphogenesis. *Biophys J* 2008, 94 (5), 1930–1941.

95. Kuzuya, M.; Satake, S.; Ai, S.; Asai, T.; Kanda, S.; Ramos, M. A.; Miura, H.; Ueda, M.; Iguchi, A., Inhibition of angiogenesis on glycated collagen lattices. *Diabetologia* 1998, 41 (5), 491–499.

96. Mann, B. K.; Gobin, A. S.; Tsai, A. T.; Schmedlen, R. H.; West, J. L., Smooth muscle cell growth in photopolymerized hydrogels with cell adhesive and proteolytically degradable domains: Synthetic ECM analogs for tissue engineering. *Biomaterials* 2001, 22 (22), 3045–3051.

97. Deroanne, C. F.; Lapiere, C. M.; Nusgens, B. V., In vitro tubulogenesis of endothelial cells by relaxation of the coupling extracellular matrix-cytoskeleton. *Cardiovasc Res* 2001, 49 (3), 647–658.

98. Mammoto, A.; Connor, K. M.; Mammoto, T.; Yung, C. W.; Huh, D.; Aderman, C. M.; Mostoslavsky, G.; Smith, L. E.; Ingber, D. E., A mechanosensitive transcriptional mechanism that controls angiogenesis. *Nature* 2009, 457 (7233), 1103–1108.

99. Chun, T. H.; Sabeh, F.; Ota, I.; Murphy, H.; McDonagh, K. T.; Holmbeck, K.; Birkedal-Hansen, H.; Allen, E. D.; Weiss, S. J., MT1-MMP-dependent neovessel formation within the confines of the three-dimensional extracellular matrix. *J Cell Biol* 2004, 167 (4), 757–767.

100. Collen, A.; Hanemaaijer, R.; Lupu, F.; Quax, P. H.; van Lent, N.; Grimbergen, J.; Peters, E.; Koolwijk, P.; van Hinsbergh, V. W., Membrane-type matrix metalloproteinase-mediated angiogenesis in a fibrin-collagen matrix. *Blood* 2003, 101 (5), 1810–1817.

101. Hiraoka, N.; Allen, E.; Apel, I. J.; Gyetko, M. R.; Weiss, S. J., Matrix metalloproteinases regulate neovascularization by acting as pericellular fibrinolysins. *Cell* 1998, 95 (3), 365–377.

102. Lafleur, M. A.; Handsley, M. M.; Knauper, V.; Murphy, G.; Edwards, D. R., Endothelial tubulogenesis within fibrin gels specifically requires the activity of membrane-type-matrix metalloproteinases (MT-MMPs). *J Cell Sci* 2002, 115, (Pt 17), 3427–3438.

103. Stratman, A. N.; Saunders, W. B.; Sacharidou, A.; Koh, W.; Fisher, K. E.; Zawieja, D. C.; Davis, M. J.; Davis, G. E., Endothelial cell lumen and vascular guidance tunnel formation requires MT1-MMP-dependent proteolysis in 3-dimensional collagen matrices. *Blood* 2009, 114 (2), 237–247.

104. Yana, I.; Sagara, H.; Takaki, S.; Takatsu, K.; Nakamura, K.; Nakao, K.; Katsuki, M.; Taniguchi, S.; Aoki, T.; Sato, H.; Weiss, S. J.; Seiki, M., Crosstalk between neovessels and mural cells directs the site-specific expression of MT1-MMP to endothelial tip cells. *J Cell Sci* 2007, 120, (Pt 9), 1607–1614.

105. Khatiwala, C. B.; Peyton, S. R.; Putnam, A. J., Intrinsic mechanical properties of the extracellular matrix affect the behavior of pre-osteoblastic MC3T3-E1 cells. *Am J Physiol Cell Physiol* 2006, 290 (6), C1640–C1650.

106. Li, Z.; Dranoff, J. A.; Chan, E. P.; Uemura, M.; Sevigny, J.; Wells, R. G., Transforming growth factor-beta and substrate stiffness regulate portal fibroblast activation in culture. *Hepatology* 2007, 46 (4), 1246–1256.

107. Hall, H., Modified fibrin hydrogel matrices: both, 3D-scaffolds and local and controlled release systems to stimulate angiogenesis. *Curr Pharm Des* 2007, 13 (35), 3597–3607.

108. Leslie-Barbick, J. E.; Moon, J. J.; West, J. L., Covalently-immobilized vascular endothelial growth factor promotes endothelial cell tubulogenesis in poly(ethylene glycol) diacrylate hydrogels. *J Biomater Sci Polym Ed* 2009, 20 (12), 1763–1779.

109. Lutolf, M. P.; Hubbell, J. A., Synthetic biomaterials as instructive extracellular microenvironments for morphogenesis in tissue engineering. *Nat Biotechnol* 2005, 23 (1), 47–55.

110. Zisch, A. H.; Lutolf, M. P.; Ehrbar, M.; Raeber, G. P.; Rizzi, S. C.; Davies, N.; Schmokel, H.; Bezuidenhout, D.; Djonov, V.; Zilla, P.; Hubbell, J. A., Cell-demanded release of VEGF from synthetic, biointeractive cell ingrowth matrices for vascularized tissue growth. *FASEB J* 2003, 17 (15), 2260–2262.

111. Cai, S.; Liu, Y.; Zheng Shu, X.; Prestwich, G. D., Injectable glycosaminoglycan hydrogels for controlled release of human basic fibroblast growth factor. *Biomaterials* 2005, 26 (30), 6054–6067.

112. Perets, A.; Baruch, Y.; Weisbuch, F.; Shoshany, G.; Neufeld, G.; Cohen, S., Enhancing the vascularization of three-dimensional porous alginate scaffolds by incorporating controlled release basic fibroblast growth factor microspheres. *J Biomed Mater Res A* 2003, 65 (4), 489–497.

113. Smith, M. K.; Peters, M. C.; Richardson, T. P.; Garbern, J. C.; Mooney, D. J., Locally enhanced angiogenesis promotes transplanted cell survival. *Tissue Eng* 2004, 10 (1–2), 63–71.

114. Tanihara, M.; Suzuki, Y.; Yamamoto, E.; Noguchi, A.; Mizushima, Y., Sustained release of basic fibroblast growth factor and angiogenesis in a novel covalently crosslinked gel of heparin and alginate. *J Biomed Mater Res* 2001, 56 (2), 216–221.

115. Carmeliet, P., Angiogenesis in health and disease. *Nat Med* 2003, 9 (6), 653–660.

116. Ramirez, F.; Rifkin, D. B., Cell signaling events: A view from the matrix. *Matrix Biol* 2003, 22 (2), 101–107.

117. Koike, N.; Fukumura, D.; Gralla, O.; Au, P.; Schechner, J. S.; Jain, R. K., Tissue engineering: Creation of long-lasting blood vessels. *Nature* 2004, 428 (6979), 138–139.

118. Melero-Martin, J. M.; Khan, Z. A.; Picard, A.; Wu, X.; Paruchuri, S.; Bischoff, J., In vivo vasculogenic potential of human blood-derived endothelial progenitor cells. *Blood* 2007, 109 (11), 4761–4768.

119. Alajati, A.; Laib, A. M.; Weber, H.; Boos, A. M.; Bartol, A.; Ikenberg, K.; Korff, T.; Zentgraf, H.; Obodozie, C.; Graeser, R.; Christian, S.; Finkenzeller, G.; Stark, G. B.; Heroult, M.; Augustin, H. G., Spheroid-based engineering of a human vasculature in mice. *Nat Methods* 2008, 5 (5), 439–445.

120. Bayless, K. J.; Davis, G. E., The Cdc42 and Rac1 GTPases are required for capillary lumen formation in three-dimensional extracellular matrices. *J Cell Sci* 2002, 115, (Pt 6), 1123–1136.

121. Bayless, K. J.; Davis, G. E., Sphingosine-1-phosphate markedly induces matrix metallo-proteinase and integrin-dependent human endothelial cell invasion and lumen formation in three-dimensional collagen and fibrin matrices. *Biochem Biophys Res Commun* 2003, 312 (4), 903–913.

122. Griffith, C. K.; Miller, C.; Sainson, R. C.; Calvert, J. W.; Jeon, N. L.; Hughes, C. C.; George, S. C., Diffusion limits of an in vitro thick prevascularized tissue. *Tissue Eng* 2005, 11 (1–2), 257–266.

123. Koh, W.; Stratman, A. N.; Sacharidou, A.; Davis, G. E., In vitro three dimensional collagen matrix models of endothelial lumen formation during vasculogenesis and angiogenesis. *Methods Enzymol* 2008, 443, 83–101.

124. Nakatsu, M. N.; Sainson, R. C.; Aoto, J. N.; Taylor, K. L.; Aitkenhead, M.; Perez-del-Pulgar, S.; Carpenter, P. M.; Hughes, C. C., Angiogenic sprouting and capillary lumen formation modeled by human umbilical vein endothelial cells (HUVEC) in fibrin gels: The role of fibroblasts and Angiopoietin-1. *Microvasc Res* 2003, 66 (2), 102–112.

125. Sainson, R. C.; Aoto, J.; Nakatsu, M. N.; Holderfield, M.; Conn, E.; Koller, E.; Hughes, C. C., Cell-autonomous notch signaling regulates endothelial cell branching and proliferation during vascular tubulogenesis. *FASEB J* 2005, 19 (8), 1027–1029.

126. Saunders, W. B.; Bohnsack, B. L.; Faske, J. B.; Anthis, N. J.; Bayless, K. J.; Hirschi, K. K.; Davis, G. E., Coregulation of vascular tube stabilization by endothelial cell TIMP-2 and pericyte TIMP-3. *J Cell Biol* 2006, 175 (1), 179–191.

127. Ishikawa, T.; Eguchi, M.; Wada, M.; Iwami, Y.; Tono, K.; Iwaguro, H.; Masuda, H.; Tamaki, T.; Asahara, T., Establishment of a functionally active collagen-binding vascular endothelial growth factor fusion protein in situ. *Arterioscler Thromb Vasc Biol* 2006, 26 (9), 1998–2004.

128. Koch, S.; Yao, C.; Grieb, G.; Prevel, P.; Noah, E. M.; Steffens, G. C., Enhancing angiogenesis in collagen matrices by covalent incorporation of VEGF. *J Mater Sci Mater Med* 2006, 17 (8), 735–741.

129. Zisch, A. H.; Schenk, U.; Schense, J. C.; Sakiyama-Elbert, S. E.; Hubbell, J. A., Covalently conjugated VEGF–fibrin matrices for endothelialization. *J Control Release* 2001, 72 (1–3), 101–113.

130. Pieper, J. S.; Hafmans, T.; van Wachem, P. B.; van Luyn, M. J.; Brouwer, L. A.; Veerkamp, J. H.; van Kuppevelt, T. H., Loading of collagen-heparan sulfate matrices with bFGF promotes angiogenesis and tissue generation in rats. *J Biomed Mater Res* 2002, 62 (2), 185–194.

131. Sakiyama-Elbert, S. E.; Hubbell, J. A., Development of fibrin derivatives for controlled release of heparin-binding growth factors. *J Control Release* 2000, 65 (3), 389–402.

132. Steffens, G. C.; Yao, C.; Prevel, P.; Markowicz, M.; Schenck, P.; Noah, E. M.; Pallua, N., Modulation of angiogenic potential of collagen matrices by covalent incorporation of heparin and loading with vascular endothelial growth factor. *Tissue Eng* 2004, 10 (9–10), 1502–1509.

133. Wang, A. Y.; Leong, S.; Liang, Y. C.; Huang, R. C.; Chen, C. S.; Yu, S. M., Immobilization of growth factors on collagen scaffolds mediated by polyanionic collagen mimetic peptides and its effect on endothelial cell morphogenesis. *Biomacromolecules* 2008, 9 (10), 2929–2936.

134. Standeven, K. F.; Carter, A. M.; Grant, P. J.; Weisel, J. W.; Chernysh, I.; Masova, L.; Lord, S. T.; Ariens, R. A., Functional analysis of fibrin {gamma}-chain cross-linking by activated factor XIII: Determination of a cross-linking pattern that maximizes clot stiffness. *Blood* 2007, 110 (3), 902–907.

135. Brown, R. A.; Phillips, J. B., Cell responses to biomimetic protein scaffolds used in tissue repair and engineering. *Int Rev Cytol* 2007, 262, 75–150.

136. Gunatillake, P. A.; Adhikari, R., Biodegradable synthetic polymers for tissue engineering. *Eur Cell Mater* 2003, 5, 1–16; discussion 16.

137. Athanasiou, K. A.; Niederauer, G. G.; Agrawal, C. M., Sterilization, toxicity, biocompatibility and clinical applications of polylactic acid/polyglycolic acid copolymers. *Biomaterials* 1996, 17 (2), 93–102.

138. Miller, D. C.; Haberstroh, K. M.; Webster, T. J., Mechanism(s) of increased vascular cell adhesion on nanostructured poly(lactic-co-glycolic acid) films. *J Biomed Mater Res A* 2005, 73 (4), 476–484.

139. Tjia, J. S.; Aneskievich, B. J.; Moghe, P. V., Substrate-adsorbed collagen and cell secreted fibronectin concertedly induce cell migration on poly(lactide-glycolide) substrates. *Biomaterials* 1999, 20 (23–24), 2223–2233.

140. Peters, M. C.; Polverini, P. J.; Mooney, D. J., Engineering vascular networks in porous polymer matrices. *J Biomed Mater Res* 2002, 60 (4), 668–678.

141. Sheridan, M. H.; Shea, L. D.; Peters, M. C.; Mooney, D. J., Bioabsorbable polymer scaffolds for tissue engineering capable of sustained growth factor delivery. *J Control Release* 2000, 64 (1–3), 91–102.

142. Richardson, T. P.; Peters, M. C.; Ennett, A. B.; Mooney, D. J., Polymeric system for dual growth factor delivery. *Nat Biotechnol* 2001, 19 (11), 1029–1034.

143. Levenberg, S.; Rouwkema, J.; Macdonald, M.; Garfein, E. S.; Kohane, D. S.; Darland, D. C.; Marini, R.; van Blitterswijk, C. A.; Mulligan, R. C.; D'Amore, P. A.; Langer, R., Engineering vascularized skeletal muscle tissue. *Nat Biotechnol* 2005, 23 (7), 879–884.

144. Elbert, D. L.; Hubbell, J. A., Conjugate addition reactions combined with free-radical cross-linking for the design of materials for tissue engineering. *Biomacromolecules* 2001, 2 (2), 430–441.

145. Lutolf, M. P.; Raeber, G. P.; Zisch, A. H.; Tirelli, N.; Hubbell, J. A., Cell-responsive synthetic hydrogels. *Adv Mater* 2003, 15 (11), 888–892.

146. Raeber, G. P.; Lutolf, M. P.; Hubbell, J. A., Molecularly engineered PEG hydrogels: A novel model system for proteolytically mediated cell migration. *Biophys J* 2005, 89 (2), 1374–1388.

147. Gombotz, W. R.; Wang, G. H.; Horbett, T. A.; Hoffman, A. S., Protein adsorption to poly(ethylene oxide) surfaces. *J Biomed Mater Res* 1991, 25 (12), 1547–1562.

148. Hern, D. L.; Hubbell, J. A., Incorporation of adhesion peptides into nonadhesive hydrogels useful for tissue resurfacing. *J Biomed Mater Res* 1998, 39 (2), 266–276.

149. Hersel, U.; Dahmen, C.; Kessler, H., RGD modified polymers: Biomaterials for stimulated cell adhesion and beyond. *Biomaterials* 2003, 24 (24), 4385–4415.

150. Ruoslahti, E.; Pierschbacher, M. D., New perspectives in cell adhesion: RGD and integrins. *Science* 1987, 238 (4826), 491–497.

151. Shin, H.; Jo, S.; Mikos, A. G., Biomimetic materials for tissue engineering. *Biomaterials* 2003, 24 (24), 4353–4364.

152. Burgess, B. T.; Myles, J. L.; Dickinson, R. B., Quantitative analysis of adhesion-mediated cell migration in three-dimensional gels of RGD-grafted collagen. *Ann Biomed Eng* 2000, 28 (1), 110–118.

153. Gobin, A. S.; West, J. L., Cell migration through defined, synthetic ECM analogs. *FASEB J* 2002, 16 (7), 751–753.
154. Lutolf, M. P.; Lauer-Fields, J. L.; Schmoekel, H. G.; Metters, A. T.; Weber, F. E.; Fields, G. B.; Hubbell, J. A., Synthetic matrix metalloproteinase-sensitive hydrogels for the conduction of tissue regeneration: Engineering cell-invasion characteristics. *Proc Natl Acad Sci USA* 2003, 100 (9), 5413–5418.
155. Schense, J. C.; Hubbell, J. A., Three-dimensional migration of neurites is mediated by adhesion site density and affinity. *J Biol Chem* 2000, 275 (10), 6813–6818.
156. Benoit, D. S.; Durney, A. R.; Anseth, K. S., The effect of heparin-functionalized PEG hydrogels on three-dimensional human mesenchymal stem cell osteogenic differentiation. *Biomaterials* 2007, 28 (1), 66–77.
157. Gobin, A. S.; West, J. L., Effects of epidermal growth factor on fibroblast migration through biomimetic hydrogels. *Biotechnol Prog* 2003, 19 (6), 1781–1785.
158. Mann, B. K.; Schmedlen, R. H.; West, J. L., Tethered-TGF-beta increases extracellular matrix production of vascular smooth muscle cells. *Biomaterials* 2001, 22 (5), 439–444.
159. West, J. L.; Hubbell, J. A., Polymeric biomaterials with degradation sites for proteases involved in cell migration. *Macromolecules* 1999, 32 (1), 241–244.
160. Aimetti, A. A.; Machen, A. J.; Anseth, K. S., Poly(ethylene glycol) hydrogels formed by thiol-ene photopolymerization for enzyme-responsive protein delivery. *Biomaterials* 2009, 30 (30), 6048–6054.
161. DeForest, C. A.; Polizzotti, B. D.; Anseth, K. S., Sequential click reactions for synthesizing and patterning three-dimensional cell microenvironments. *Nat Mater* 2009, 8 (8), 659–664.
162. Moon, J. J.; Lee, S. H.; West, J. L., Synthetic biomimetic hydrogels incorporated with ephrin-A1 for therapeutic angiogenesis. *Biomacromolecules* 2007, 8 (1), 42–49.
163. Moon, J. J.; Hahn, M. S.; Kim, I.; Nsiah, B. A.; West, J. L., Micropatterning of poly(ethylene glycol) diacrylate hydrogels with biomolecules to regulate and guide endothelial morphogenesis. *Tissue Eng Part A* 2009, 15 (3), 579–585.
164. Hahn, M. S.; Miller, J. S.; West, J. L., Three-dimensional biochemical and biomechanical patterning of hydrogels for guiding cell behavior. *Adv Mater* 2006, 18 (20),2679–2684.
165. Hahn, M. S.; Taite, L. J.; Moon, J. J.; Rowland, M. C.; Ruffino, K. A.; West, J. L., Photolithographic patterning of polyethylene glycol hydrogels. *Biomaterials* 2006, 27 (12), 2519–2524.
166. Caveda, L.; Martin-Padura, I.; Navarro, P.; Breviario, F.; Corada, M.; Gulino, D.; Lampugnani, M. G.; Dejana, E., Inhibition of cultured cell growth by vascular endothelial cadherin (cadherin-5/VE-cadherin). *J Clin Invest* 1996, 98 (4), 886–893.
167. Nelson, C. M.; Jean, R. P.; Tan, J. L.; Liu, W. F.; Sniadecki, N. J.; Spector, A. A.; Chen, C. S., Emergent patterns of growth controlled by multicellular form and mechanics. *Proc Natl Acad Sci USA* 2005, 102 (33), 11594–11599.
168. Gerhardt, H.; Golding, M.; Fruttiger, M.; Ruhrberg, C.; Lundkvist, A.; Abramsson, A.; Jeltsch, M.; Mitchell, C.; Alitalo, K.; Shima, D.; Betsholtz, C., VEGF guides angiogenic sprouting utilizing endothelial tip cell filopodia. *J Cell Biol* 2003, 161 (6), 1163–1177.
169. Nelson, C. M.; Chen, C. S., Cell–cell signaling by direct contact increases cell proliferation via a PI3K-dependent signal. *FEBS Lett* 2002, 514 (2–3), 238–242.
170. Nelson, C. M.; Chen, C. S., VE-cadherin simultaneously stimulates and inhibits cell proliferation by altering cytoskeletal structure and tension. *J Cell Sci* 2003, 116, (Pt 17), 3571–3581.
171. Zhu, A. J.; Watt, F. M., Expression of a dominant negative cadherin mutant inhibits proliferation and stimulates terminal differentiation of human epidermal keratinocytes. *J Cell Sci* 1996, 109 (Pt 13), 3013–3023.
172. Levenberg, S.; Katz, B. Z.; Yamada, K. M.; Geiger, B., Long-range and selective autoregulation of cell–cell or cell-matrix adhesions by cadherin or integrin ligands. *J Cell Sci* 1998, 111 (Pt 3), 347–357.

173. Wang, Y.; Jin, G.; Miao, H.; Li, J. Y.; Usami, S.; Chien, S., Integrins regulate VE-cadherin and catenins: Dependence of this regulation on Src, but not on Ras. *Proc Natl Acad Sci USA* 2006, 103 (6), 1774–1779.

174. Grazia Lampugnani, M.; Zanetti, A.; Corada, M.; Takahashi, T.; Balconi, G.; Breviario, F.; Orsenigo, F.; Cattelino, A.; Kemler, R.; Daniel, T. O.; Dejana, E., Contact inhibition of VEGF-induced proliferation requires vascular endothelial cadherin, beta-catenin, and the phosphatase DEP-1/CD148. *J Cell Biol* 2003, 161 (4), 793–804.

175. Lampugnani, M. G.; Orsenigo, F.; Gagliani, M. C.; Tacchetti, C.; Dejana, E., Vascular endothelial cadherin controls VEGFR-2 internalization and signaling from intracellular compartments. *J Cell Biol* 2006, 174 (4), 593–604.

176. Gerhardt, H.; Betsholtz, C., Endothelial-pericyte interactions in angiogenesis. *Cell Tissue Res* 2003, 314 (1), 15–23.

177. Hughes, C. C., Endothelial-stromal interactions in angiogenesis. *Curr Opin Hematol* 2008, 15 (3), 204–209.

178. Hirschi, K. K.; Rohovsky, S. A.; D'Amore, P. A., PDGF, TGF-beta, and heterotypic cell-cell interactions mediate endothelial cell-induced recruitment of 10T1/2 cells and their differentiation to a smooth muscle fate. *J Cell Biol* 1998, 141 (3), 805–814.

179. Hirschi, K. K.; Rohovsky, S. A.; Beck, L. H.; Smith, S. R.; D'Amore, P. A., Endothelial cells modulate the proliferation of mural cell precursors via platelet-derived growth factor-BB and heterotypic cell contact. *Circ Res* 1999, 84 (3), 298–305.

180. Orlidge, A.; D'Amore, P. A., Inhibition of capillary endothelial cell growth by pericytes and smooth muscle cells. *J Cell Biol* 1987, 105 (3), 1455–1462.

181. Gerhardt, H.; Wolburg, H.; Redies, C., N-cadherin mediates pericytic-endothelial interaction during brain angiogenesis in the chicken. *Dev Dyn* 2000, 218 (3), 472–479.

182. Abramsson, A.; Berlin, O.; Papayan, H.; Paulin, D.; Shani, M.; Betsholtz, C., Analysis of mural cell recruitment to tumor vessels. *Circulation* 2002, 105 (1), 112–117.

183. Eberhard, A.; Kahlert, S.; Goede, V.; Hemmerlein, B.; Plate, K. H.; Augustin, H. G., Heterogeneity of angiogenesis and blood vessel maturation in human tumors: Implications for antiangiogenic tumor therapies. *Cancer Res* 2000, 60 (5), 1388–1393.

184. Yonenaga, Y.; Mori, A.; Onodera, H.; Yasuda, S.; Oe, H.; Fujimoto, A.; Tachibana, T.; Imamura, M., Absence of smooth muscle actin-positive pericyte coverage of tumor vessels correlates with hematogenous metastasis and prognosis of colorectal cancer patients. *Oncology* 2005, 69 (2), 159–166.

185. Enge, M.; Bjarnegard, M.; Gerhardt, H.; Gustafsson, E.; Kalen, M.; Asker, N.; Hammes, H. P.; Shani, M.; Fassler, R.; Betsholtz, C., Endothelium-specific platelet-derived growth factor-B ablation mimics diabetic retinopathy. *EMBO J* 2002, 21 (16), 4307–4316.

186. Leveen, P.; Pekny, M.; Gebre-Medhin, S.; Swolin, B.; Larsson, E.; Betsholtz, C., Mice deficient for PDGF B show renal, cardiovascular, and hematological abnormalities. *Genes Dev* 1994, 8 (16), 1875–1887.

187. Soriano, P., Abnormal kidney development and hematological disorders in PDGF beta-receptor mutant mice. *Genes Dev* 1994, 8 (16), 1888–1896.

188. Montesano, R.; Pepper, M. S.; Orci, L., Paracrine induction of angiogenesis in vitro by Swiss 3T3 fibroblasts. *J Cell Sci* 1993, 105 (Pt 4), 1013–1024.

189. Au, P.; Tam, J.; Fukumura, D.; Jain, R. K., Bone marrow-derived mesenchymal stem cells facilitate engineering of long-lasting functional vasculature. *Blood* 2008, 111 (9), 4551–4558.

190. Ford, M. C.; Bertram, J. P.; Hynes, S. R.; Michaud, M.; Li, Q.; Young, M.; Segal, S. S.; Madri, J. A.; Lavik, E. B., A macroporous hydrogel for the coculture of neural progenitor and endothelial cells to form functional vascular networks in vivo. *Proc Natl Acad Sci USA* 2006, 103 (8), 2512–2517.

191. Traktuev, D. O.; Prater, D. N.; Merfeld-Clauss, S.; Sanjeevaiah, A. R.; Saadatzadeh, M. R.; Murphy, M.; Johnstone, B. H.; Ingram, D. A.; March, K. L., Robust functional vascular network formation in vivo by cooperation of adipose progenitor and endothelial cells. *Circ Res* 2009, 104 (12), 1410–1420.
192. Mancuso, M. R.; Davis, R.; Norberg, S. M.; O'Brien, S.; Sennino, B.; Nakahara, T.; Yao, V. J.; Inai, T.; Brooks, P.; Freimark, B.; Shalinsky, D. R.; Hu-Lowe, D. D.; McDonald, D. M., Rapid vascular regrowth in tumors after reversal of VEGF inhibition. *J Clin Invest* 2006, 116 (10), 2610–2621.
193. Ehrbar, M.; Rizzi, S. C.; Hlushchuk, R.; Djonov, V.; Zisch, A. H.; Hubbell, J. A.; Weber, F. E.; Lutolf, M. P., Enzymatic formation of modular cell-instructive fibrin analogs for tissue engineering. *Biomaterials* 2007, 28 (26), 3856–3866.
194. Brooks, P. C.; Montgomery, A. M.; Rosenfeld, M.; Reisfeld, R. A.; Hu, T.; Klier, G.; Cheresh, D. A., Integrin alpha v beta 3 antagonists promote tumor regression by inducing apoptosis of angiogenic blood vessels. *Cell* 1994, 79 (7), 1157–1164.
195. Yoon, J. J.; Chung, H. J.; Lee, H. J.; Park, T. G., Heparin-immobilized biodegradable scaffolds for local and sustained release of angiogenic growth factor. *J Biomed Mater Res A* 2006, 79 (4), 934–942.
196. Adams, C. L.; Chen, Y. T.; Smith, S. J.; Nelson, W. J., Mechanisms of epithelial cell-cell adhesion and cell compaction revealed by high-resolution tracking of E-cadherin-green fluorescent protein. *J Cell Biol* 1998, 142 (4), 1105–1119.

7 Treating Cardiovascular Diseases by Enhancing Endogenous Stem Cell Mobilization

Liang Youyun, Ross J. DeVolder,
and Hyunjoon Kong

CONTENTS

7.1 INTRODUCTION

Recently, there has been increased implementation of bone marrow cell (BMC) transplantations in clinical settings to treat leukemia, lymphoma, and primary immunodeficiency disease.[1] BMCs consist of various cell populations including hematopoietic stem cells (HSCs), mesenchymal stem cells, also known as multipotent stromal cells (MSCs), and endothelial progenitor cells (EPCs).[2] Human HSCs are characterized by the expression of CD34, CD45, CD117, and CD133 markers; EPCs are identified through the expression of endothelial markers such as vascular endothelial growth factor (VEGF) receptor 2/KDR, along with the aforementioned HSC markers; and MSCs are commonly identified by markers such as Stro-1, CD73, CD106, CD105, CD73, and CD90.[3,4]

BMCs present multipotency and the ability to repair and regenerate various tissues. They are able to differentiate into an array of cell types including smooth muscular cells, cardiomyocytes, and endothelial cells.[5-7] BMCs also secret multiple cytokine and growth factors that stimulate host cells involved with the healing of wounds and tissue defects. Therefore, there have been extensive studies investigating the use of BMCs for treating various cardiovascular diseases such as acute and chronic myocardial ischemia, and peripheral vascular disease using BMCs.[6,8-15] In healthy individuals, very small populations of BMCs consisting of HSCs, MSCs, and EPCs circulate in the blood stream. The number of BMCs mobilized is usually too small to yield significant therapeutic effects. Interestingly, the population of BMCs in the circulating blood significantly increases with cardiovascular injury and trauma.[16] This increase is attributed to the upregulation of certain key factors that increases the number of BMCs. These mobilized cells stimulate the repair of damaged vessels and the formation of new capillary networks, following transmigration into ischemic tissue.[17] However, the efficacy of spontaneous BMC mobilization varies with the severity of vascular injury or trauma, and the patient's age and overall health.[18]

In the 1970s, certain studies reported that the administration of chemotherapeutic agents for cancer treatment significantly increased HSC populations in the circulating blood stream.[19] Reinfusion of cell mobilized into the peripheral blood presented excellent bone marrow rescue after chemotherapy in several preclinical trials.[20-22] These findings prompted extensive studies to discover and synthesize bioactive molecules that stimulate endogenous BMC mobilization and subsequently improve tissue repair and regeneration.

This chapter will describe several preclinical and clinical approaches to treat ischemic tissues by promoting endogenous BMC mobilization. First, we will review the mechanism of spontaneous BMC mobilization and ischemic tissue repair (Section 7.2). Next, we will describe various strategies to stimulate endogenous stem cell mobilization using several signaling, bioactive molecules (Section 7.3) and also guide the mobilized BMCs into target ischemic tissue (Section 7.4). Finally, we will discuss the mechanisms by which the mobilized cells treat ischemic tissues (Section 7.5).

7.2   ISCHEMIA-INDUCED SPONTANEOUS BMC MOBILIZATION

In the typical progression of myocardial infarction, thromboembolism and atherosclerosis lead to the occlusion of coronary arteries and ultimately result in the disruption of blood supply to the heart. These events lead to cardiomyocyte death in the ischemic tissues, which is remodeled through fibrous tissue deposition. The presence of fibrotic scar tissue significantly deteriorates the pumping function of the heart.[23] To restore normal heart function, it is desirable to recreate the myocardium and blood vessels within the infarcted region. Another important issue is ensuring the structural and functional integrity of newly regenerated tissues while maintaining normal functions of the surrounding heart tissue.[15]

Following cardiac infarction, the population of BMCs in circulation is spontaneously upregulated.[24,25] In a study conducted by Massa et al., CD34+ BMCs were found to reach maximal quantities (5.8-fold) several hours after the onset of acute myocardial infarction (AMI).[18] In another study conducted by *Leone* et al., CD34+ cells were also found to peak days after AMI. However, there was significant variance in the maximum density of CD34+ cells ranging from 0.84 to 33.4 cells per microliter of peripheral blood. Around 95% of the CD34+ cells collected in this study expressed the CD45 leukocyte common antigen, which indicates that the majority of the CD34+ cells were mobilized from the bone marrow.[26]

It has been proposed that BMCs, spontaneously mobilized into the circulation, are recruited to infarcted heart and subsequently involved in tissue repair as depicted in Figure 7.1. Specifically, several studies have demonstrated a positive correlation between the number of CD34+ cells in circulation and improvements in left ventricular function. For example, the left ventricular ejection fraction, which is a function of the end-diastolic and end-systolic volumes, was increased with an increasing density of CD34+ cells in the peripheral blood.[25] Another study demonstrated that the mobilization of BMCs led to an increase in myocardial blood flow, which indicates recovery of blood perfusion through the infarcted myocardium.[27]

BMC mobilization has also been observed with peripheral tissue ischemia. In a study by Takahashi et al., it was observed that the circulating population of endothelial progenitor cells (EPCs) increased by more than four times after the onset of ischemia in the hindlimb of mice, reaching a peak 7 days later. In the same study, cornea micropocket surgery was conducted in both mice with hindlimb ischemia and control mice. Enhanced ocular neovascularization was observed in mice with hindlimb ischemia as compared to controls. This difference in neovascularization demonstrates the spontaneous mobilization of BMCs in response to ischemia and the subsequent recruitment of the mobilized cells to sites of injury.[20]

Spontaneous BMC mobilization in response to ischemia has been related to the upregulation of several mobilizing factors in the circulating blood. Examples of such factors are granulocyte colony-stimulating factor (G-CSF) and stem cell factor (SCF).[10,28] These mobilizing factors disrupt the interactions between BMCs and the bone marrow niche, thus allowing the release of BMCs into peripheral blood. At the same time, an upregulation of various factors recruiting BMCs to ischemic and infarcted tissue have also been discovered. These recruiting factors include stromal-derived cell factor-1 (SDF-1), vascular endothelial growth factor (VEGF), and hepatocyte growth factor (HGF).[28] They are known to recruit endogenous circulating progenitor and stem cells by interacting with specific receptors upregulated on BMCs.

Following cell recruitment to ischemic tissue, it is suggested that BMCs perform various tissue repairing functions. These repair modes include the transdifferentiation of BMCs into cardiomyocytes and endothelial cells; the fusion of undifferentiated BMCs with existing cardiomyocytes; and the release of angiogenic and

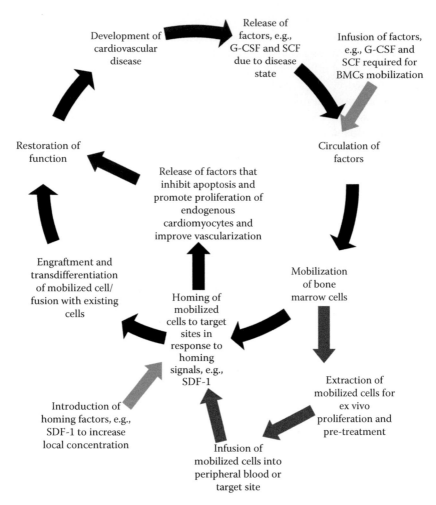

FIGURE 7.1 (See color insert.) Overview of bone marrow cell (BMC) mobilization for treatment of cardiovascular diseases. Blue arrows represent naturally occurring processes, red arrows represent conventional treatment involving extraction and reimplantation of cells, and green arrows represent strategies to directly recruit mobilized cells endogenously.

anti-apoptotic factors, for example, VEGF.[6] Each of these suggested repair modes will be discussed in more detail in later sections.

Even though BMCs spontaneously mobilize from bone marrow after myocardial infarction and peripheral tissue ischemia, the number of cells recruited to the ischemic tissue is usually insufficient to fully restore normal cardiovascular function.[29] In addition, preclinical outcomes of spontaneous BMC mobilization have been varied with several extrinsic variables including a patient's age and overall health. Therefore, extensive efforts are being made to develop various therapies that enhance both BMC mobilization and recruitment into ischemic tissues—as will be discussed in following sections.

7.3 MOLECULAR THERAPIES TO STIMULATE ENDOGENOUS BMC MOBILIZATION

Molecular therapies for enhancing endogenous BMC mobilization are largely categorized into three strategies: (1) administration of BMC inhibitors of receptor–ligand bonding, for example, specific antibodies; (2) administration of chemokines to regulate BMC phenotypic activity; and (3) administration of chemokines to stimulate matrix metalloproteinase (MMP) activity. It is common to administer these molecules subcutaneously, but limited efforts are being made to deliver the molecules in a sustained manner, such as methods using drug-encapsulating biomaterial.[30]

7.3.1 ADMINISTRATION OF ANTIBODIES OR INHIBITORS OF RECEPTOR–LIGAND BONDS

As mentioned in the introduction, BMCs in a quiescent stage mostly reside in bone marrow. The release of BMCs into the circulating blood is inhibited by BMCs' adhesion to extracellular matrix and neighboring cells. A number of integrins and adhesion molecules are involved in the retention of BMCs in the bone marrow. These include the very late antigen-4 (VLA-4) integrins, SDF-1, lymphocyte function-associated antigen-1 (LFA-1), very late antigen-5 (VLA-5), and lymphocyte function-associated antigen-3 (LFA-3).[31–36] These integrin receptors mediate cell attachment to the extracellular matrix or to other neighboring cells (Figure 7.2).[37] There is evidence that suggests significant changes of cell–cell and cell–matrix interactions occur in bone marrow before the spontaneous mobilization of BMCs following ischemia.[31] Therefore, various antibodies that bind to BMC receptors and are involved with bone marrow adhesion were examined to induce BMC mobilization. For example, the administration of antibodies to VLA-4 resulted in a significant increase of BMCs in the circulating blood.[32] The administration of antibodies to vascular cell adhesion molecule 1 (VCAM-1), which is a receptor of VLA-4, was also effective at inducing cell mobilization.[38] Several chemotherapeutic drugs such as cyclophosphamide and paclitaxel also stimulated BMC mobilization by disrupting VLA-4.[39]

Inhibitors of cellular receptor expression also effectively stimulate BMC mobilization. One other type of adhesion molecule that has been shown to play a key role in BMC anchorage and trafficking is SDF-1, also known as CXC ligand 12. Interaction between SDF-1 and its receptor, chemokine (C-X-C motif) receptor 4 (CXCR-4), is important for cell anchorage and retention. CXCR-4 is highly expressed within cells residing in the bone marrow, thus accounting for the accumulation of BMCs within the bone marrow.[31] The administration of CXCR-4 inhibitor, AMD3000, was found to enhance the mobilization of HSCs by blocking the binding of CXCR-4 with SDF-1.[33]

Other adhesion molecules that have been suggested to participate in the retention of BMCs in bone marrow include LFA-1, VLA-5, and LFA-3. It was found that these molecules are expressed at lower levels within mobilized HSCs compared to HSCs retained in the bone marrow. Also, it has been proposed that the lower expression of these adhesion molecules allow HSCs to leave the bone marrow niche (Figure 7.2).[34–36] Therefore, antibodies to these adhesion molecules and inhibitors blocking

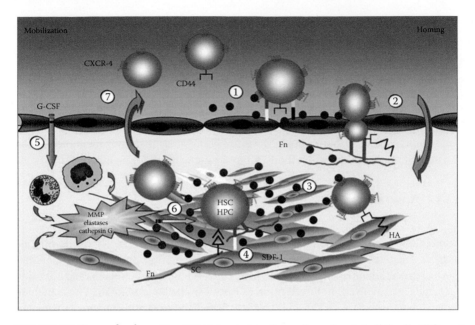

FIGURE 7.2 (See color insert.) SDF-1 located on the surface of endothelial cells (EC) and bound to heparan sulfates can optimally interact with its counter-receptor CXCR-4 expressed on the rolling HSC hematopoietic progenitor cells (HPC). CXCR-4 engagement induces an activation of integrins VLA-4 and LFA-1 by inside-out signaling, which converts their interactions with VCAM-1 and ICAM-1 constitutively expressed on EC to firm adhesion. This activation induces the arrest of HSC\HPC and stimulates actin polymerization, which results in transendothelial migration mediated by VLA-4 and VLA-5 in the presence of fibronectin (Fn), these interactions being enhanced by SDF-1. HSC\HPC then polarizes, migrates toward a local gradient of SDF-1 continuously produced by stromal cells (SC), and reaches the hematopoietic niche. SDF-1-induced migration is related to the presence of CD44, the receptor for hyaluronic acid (HA). The final anchoring of HSC\HPC within the niche depends mainly on interactions with SC and extracellular matrix (Fn, HA). Anchoring is maintained by the continuous production of SDF-1 by SC. *In vivo* administration of G-CSF induces local production of proteases (MMP, elastases, and cathepsin G) from leucocytes or SC. These molecules are able to degrade the ECM and to disrupt VLA-4\VCAM-1, c-Kit\SCF, and also CXCR-4\SDF-1 interactions by degrading both CXCR-4 and SDF-1. The loss of attachment to stromal cells and to the ECM, together with the loss of SDF-1 activity, favors the release of HSC\HPC into the peripheral blood. (From Lataillade, J.J. et al., *Eur. Cytokine Netw.*, 15(3), 177, 2004. With permission.)

their expression would serve as possible candidates for stimulating BMC mobilization as demonstrated in animal studies.[40,41]

7.3.2   ADMINISTRATION OF CHEMOKINES TO REGULATE PHENOTYPIC ACTIVITY OF BMCs

7.3.2.1   Action of Chemokines

Aside from the direct use of antibodies and inhibitors to influence receptor–ligand bonds, other strategies such as administration of chemokines to directly influence cellular phenotypic activity have also been explored.[42] These molecules include cytokines such as G-CSF, granulocyte-macrophage colony-stimulating factor (GM-CSF), stem cell factor (SCF), and fms-like tyrosine kinase receptor-3 ligand (Flt3-ligand), to name a few.[43–45] It has been suggested that cellular mobilization by these molecules involves pathways that downregulate integrin receptors and/or adhesion molecules.[42]

G-CSF is the most commonly used factor to increase BMC mobilization and has been shown to increase the level of peripheral blood progenitor cells by a factor of 80 following 4–5 days of treatment. The proposed mechanism in which G-CSF mobilizes BMCs is shown in Figure 7.2. In the study put forth by *Petit* et al., G-CSF administration resulted in the upregulation and accumulation of bone marrow neutrophils elastase and capthesin G, both of which cleave SDF-1 within the bone marrow niche. The elastase is also responsible for the cleavage of other adhesion molecules and extracellular matrix components of bone marrow, thus explaining the high mobilization efficiency exhibited by G-CSF.[46]

Subcutaneous injections of G-CSF have been regarded as a safe method for BMC mobilization through numerous preclinical and clinical trials, and have also been approved by the FDA.[14,41,44–49]

Flt-3 ligand (FL) is also an alternative medicine to mobilize BMCs. The FL binds to Flt-3 type III tyrosine kinase receptor, which is mainly expressed on hematopoietic cells. Daily subcutaneous injections of FL in mice induced endogenous BMC mobilization, specifically HSCs and EPCs, into the peripheral blood.[47] Additionally, elevated serum levels of FL were observed during G-CSF administrations, indicating FLs involved in G-CSF-induced mobilization.[48]

The efficacy of G-CSF in the treatment of both peripheral ischemia and myocardial infarction has been demonstrated in numerous studies. Daily intramuscular injections of G-CSF into mouse hindlimb ischemia models, over a 5-day period, resulted in a significant increase in capillary density and blood perfusion determined through histological analysis and laser Doppler perfusion imaging, respectively.[49] Similarly, subcutaneous administration of 100 µg G-SCF/kg/day for 5 days using murine myocardial infarction models resulted in a reduction of infarct area and an improvement in postinfarction survival.[9]

While many animal studies demonstrate promising outcomes, critics argue that the results from some of these studies show little clinical relevance because the cytokines administration took place before the coronary ligation. Also, the mouse models used were typically splenectomized to prevent immunological reactions. Furthermore, permanent coronary ligation was usually carried out in these experiments, whereas, patients with acute myocardial infarction generally received coronary reperfusion treatment.[13]

On the other hand, there are several animal studies with promising results that have prompted several clinical trials for treating ischemic diseases with G-CSF, as summarized in Table 7.1. In a clinical trial conducted by Ince et al., a group of 30 patients were randomly assigned to receiving 10 µg G-CSF/kg/day and a placebo treatment for 6 days after percutaneous coronary intervention. In this study, G-CSF administration resulted in left ventricular function improvement after 1 year with left ventricular ejection fraction of 56 ± 9% compared to 48 ± 4% prior to treatment, whereas no significant improvement was detected in the placebo group.[12]

TABLE 7.1

Summary of Recent Clinical Trials Involving Endogenous Mobilization of BMCs as Clinical Treatment for Myocardial Infarction

Study	No. of Subjects per Group	Settings and Patient Type	Treatment	Duration of Observation (Months)	No. of Cells Mobilized	Outcome
Ince et al.[12]	15	Randomized, placebo-controlled trial STEMI PCI (timing not stated)	Subcutaneous injection of 10 µg/kg/day of G-CSF for 6 days	12	$65 \pm 54 \times 10^3$ CD34+ cells/mL peripheral blood	LV function significantly improved only in G-CSF group No increase in restenosis, atherosclerosis, or other adverse effects
Kuethe et al.[77]	14	Non-randomized, placebo-controlled STEMI PCI (timing not stated)	Subcutaneous injection of 5 µg/kg of G-CSF twice daily till peripheral leukocyte count was above 50×10^9/L or when the CD34+ cell count decreased under continued administration of G-CSF stimulation	12	$76 \pm 45 \times 10^3$ CD34+ cells/mL peripheral blood	Significant differences between improvement in LV function for G-CSF group and control group No increased risk of restenosis
Zohlnhöfer et al.[78]	56	Randomized, double-blind, placebo-controlled trial STEMI PCI (>12 h after onset)	Subcutaneous injection of 10 µg/kg/day of G-CSF for 5 days	4 or 6	59 CD34+ cells/µL G-CSF	No significant difference in LV function when compared to control group No increase in risk of restenosis or other adverse effects

Study	Number	Trial design	Treatment		Cells	Results
Ripa et al.[79]	39	Randomized, double-blind, placebo-controlled trial STEMI PCI (>12 hours after onset)	Subcutaneous injection of 10 μg/kg/day of G-CSF for 6 days	6	$42 \pm 45 \times 10^3$ CD45-/CD34-/CRCX4+ cells/mL peripheral blood	No significant difference in LV function when compared to control group. No increase in atherosclerosis or other adverse effects
Ellis et al.[80]	6	Randomized, double-blind, controlled trial STEMI PCI (>4h after onset)	Subcutaneous injection of 10 μg/kg/day of G-CSF for 5 days	1	$29 \pm 14 \times 10^3$ CD34+ cells/mL peripheral blood	No significant difference in LV function when compared to control group. 1 patient with reinfarction. 1 patient with congestive heart failure
Ellis et al.[80]	6	Randomized, double-blind, placebo-controlled trial STEMI PCI (>4h after onset)	Subcutaneous injection of 5 μg/kg/day of G-CSF for 5 days	1	$37 \pm 30 \times 10^3$ CD34+ cells/mL peripheral blood	No significant difference in LV function when compared to control group. 1 patient with congestive heart failure
Engelmann et al.[50]	22	Randomized, double-blind, placebo-controlled trial STEMI PCI (>6h after and <7 days)	Subcutaneous injection of 10 μg/kg/day of G-CSF for 5 days	12	$41 \pm 27 \times 10^3$ CD34+ /c-kit+ cells/mL peripheral blood	No significant difference in LV function when compared to control group. 1 patient with in-stent thrombosis. 1 patient with sudden cardiac arrest, resulting in death

STEMI, ST-segment elevation myocardial infarction; PCI, percutaneous coronary intervention.

However, a clinical trial conducted by Engelmann et al. contends the effectiveness of G-CSF for treating patients who received late coronary reperfusion, etc. In this trial, after 3 months, patients who were given 10 μg G-CSF/kg/day subcutaneously for 5 days did not show significant improvements of global myocardial function as compared to the placebo group.[50]

The results from ten similar clinical trials were compared and analyzed to examine the effect of G-CSF administration after acute myocardial infarction. A total of 445 patients were included in this comparison. The G-CSF administration led to BMC mobilization in a dose-dependent manner, however, with no overall improvement in mean left ventricular function and no reduction in mean infarct size.[51] Certain studies have also reported that administration of G-CSF may promote restenosis and atherosclerosis, possibly due to passive entrapment of BMCs in the coronary arteries.[11,52]

7.3.2.2 Coadministration of Chemokines

Several studies have shown that combined delivery of cytokines often improved the efficacy of cell mobilization. For example, subcutaneous administration of FL and G-CSF resulted in an increased BMC population in peripheral blood, by a factor greater than 1000.[53] Furthermore, a proper combination of cytokines could control the specific population of BMCs. Pretreatment of mice with C-GSF before CXCR-4 inhibitor, AMD3000, administration increased the mobilization of HSCs but not EPCs. In contrast, administration of VEGF followed by AMD3000 reduced the mobilization of HSCs but increased the population of EPCs in blood.[33] The combined delivery of cytokines further improved the ischemic tissue repair. For example, the left ventricular function was improved with the coadministration of either G-CSF and FL or G-CSF and SCF.[13] The coadministration of G-CSF and AMD3100 also enhanced neovascularization of ischemic mouse hindlimb as compared to injection of G-CSF only (Figure 7.3).[37]

7.3.2.3 Sustained Delivery of Chemokines

Various strategies have also been explored to either improve the stability and function of G-CSF or serve as alternatives to daily G-CSF administration. The incorporation of G-CSF into PEGylated liposomes led to an extended circulation time. Subsequently, it doubled the number of HSCs mobilized compared to an equivalent bolus dosage of G-CSF.[54] Another study encapsulated G-CSF and fibroblast growth factor-2 (FGF-2) in gelatin hydrogels for sustained drug delivery. The injection of cytokine-encapsulating hydrogels into ischemic mouse hindlimbs increased the blood vessel density after a period of 8 weeks as compared to bolus injections of G-CSF and FGF-2.[30]

7.3.3 Administration of Chemokines to Stimulate MMP Activity

Chemokines such as interleukin-8 (IL-8) have also been demonstrated to effectively induce BMC mobilization by promoting the release of metalloproteinases (MMPs). It is well known that MMPs mediate degradation of the basement membrane and extracellular matrix. In fact, the administration of a single dose of IL-8

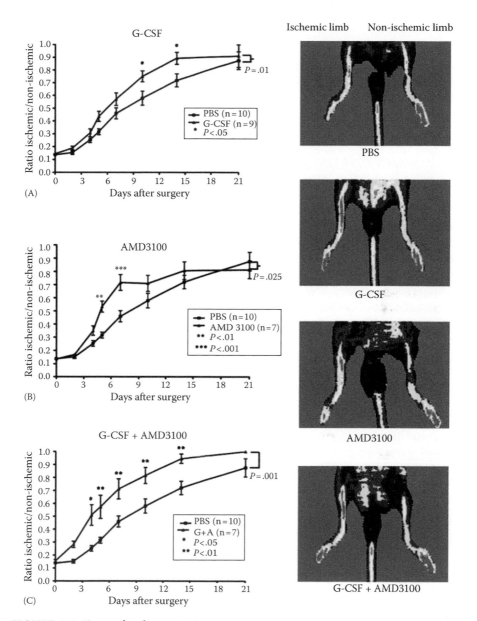

FIGURE 7.3 (See color insert.) Cytokine treatment and neovascularization. C57BL/6 mice were treated with saline alone (PBS), G-CSF, AMD3100, or G-CSF in combination with AMD3100 (G + A), and perfusion of the ischemic hindlimb relative to the nonischemic hindlimb was measured. Representative laser Doppler images taken 14 days after surgery are shown. Low perfusion is displayed as blue, while the highest level of perfusion is displayed as red. (A) G-CSF versus PBS. (B) AMD-3100 versus PBS. (C) G-CSF + AMD-3100 versus PBS. Data represent the mean ± standard deviation. (From Capoccia, B.J. et al., *Blood*, 108(7), 2438, 2006. With permission.)

into primates stimulated the rapid release of BMCs into peripheral blood, while G-CSF administration typically required several days of treatment. However, the number of BMCs in peripheral blood is much lower for mobilization using IL-8 as compared to G-CSF.[55] In a similar experiment, administration of MMP antibody completely inhibited mobilization, thus demonstrating the role of MMPs in IL-8-induced mobilization.[56]

7.4 HOMING OF CELLS TO TARGET ISCHEMIC TISSUE

In order to exert a therapeutic effect, mobilized cells must first be recruited to ischemic tissue. Such cell migration followed by tissue repair and regeneration is known as "homing." Many studies report an active recruitment process in which mobilized BMCs migrate to inflamed and injured tissues in the body. It is postulated that certain cytokines, upregulated in these ischemic regions, serve to attract mobilized cells.[57] However, without external intervention, the number of cells that home to the ischemic tissue is usually insufficient for significant tissue repair and regeneration.[58] One particular experiment conducted by *Kang* et al. showed that very low numbers of intravenously injected BMCs home to infarcted myocardium without the administration of external cues to promote cell recruitment.[58] Overall, increasing homing efficiency of BMCs mobilized in the circulating blood is a key strategy for stem cell mobilization in clinical ischemia treatments.

In order to increase the efficiency of stem cell homing, two strategies have been investigated: (1) the administration of supplemental drugs in conjunction with mobilizing factors, and (2) the modification of mobilized cells ex vivo to improve their affinity with the target site. As the second approach requires extraction of mobilized cells and reimplantation, this chapter will focus on describing the first approach.

Several bioactive molecules, upregulated during ischemia, were tested for improving BMC homing.[29,59] These molecules include VEGF, SDF-1, hypoxia inducible factor-1α (HIF-1α), VCAM, and hepatocyte growth factor (HGF). In one investigation, a transgenic mouse model overexpressing VEGF in the liver and myocardium, showed enhanced BMC infiltration in both the liver and myocardium compared to wild type controls. The infiltration of BMCs began shortly after VEGF induction, and an accumulation of BMCs was observed in the two organs over the first 4 days.[60] Schrepfer et al. also demonstrated that both VEGF and HGF promote the migration of MSCs *in vitro*. Homing of intravenously injected MSCs was significantly improved when VEGF or HGF was locally administered into the infracted region through intramyocardial injection.[52] It is hypothesized that stem cell recruitment, in response to VEGF, is mediated through VEGF receptor 1 (FLT-1), since the antibody to the receptor diminishes EPCs homing.[61]

Additionally, cell adhesion molecule SDF-1 improved stem cell homing. SDF-1 associates with CXCR-4, which is a receptor usually upregulated on CD34+ cells and is shown to strongly promote chemotaxis of CD34+ cells *in vitro*.[62] In a study implemented by Yamaguchi et al., 1 μg of SDF-1 was injected at the ischemic hindlimb of a mouse immediately after intravenous injection of fluorescently labeled EPCs. The SDF-1 injection resulted in an increased EPC localization at the ischemic site and improved ischemic neovascularization demonstrated by an increased capillary

density. This study also demonstrated that exposure of human umbilical vein endothelial cells (HUVEC) to SDF-1 *in vitro* increased their VEGF excretion.[63]

A limitation of using SDF-1A in clinical treatments is its short lifetime; it
is readily cleaved by several proteases including CD26/dipeptidylpeptidase IV
(DPP-IV).[64] Various strategies are being explored to extend the bioactivity of
SDF-1 under physiological conditions. These strategies, which include the use of
protease-resistant forms of SDF-1 and DPP IV inhibitors called Diprotin A, have
been shown to increase cell recruitment.[65,66] Alternatively, SDF-1 was modified to
impart resistance to MMP-2 and exopeptidase cleavage while retaining chemotactic bioactivity as demonstrated by Vincent et al. The resulting S-SDF-1(S4V) is
less neurotoxic compared to the native SDF-1. Direct tethering of S-SDF-1(S4V)
to self-assembling peptide nanofibers, followed by intramyocardial delivery,
resulted in a significant cardiac ejection function improvement: an increase from
$34.0 \pm 2.5\%$ to $50.7 \pm 3.1\%$.[66]

A combination of homing factors with G-CSF significantly improved neovascularization in ischemic tissues. Ieda et al. examined the effect of G-CSF, HGF, and
G-CSF + HGF administration on a murine hindlimb ischemia model. The G-CSF
group was injected with 300 µg/kg/day of G-CSF for 10 days, the HGF group was
injected with 500 µg of plasmid in the ischemic abductor muscles on the first day,
and the G CSF + HGF group received both of the preceding treatments. At the end
of 4 weeks, approximately 50% of the G-CSF + HGF group and 30% of both the
G-CSF and HGF individual groups had no necrosis in the foot and toes compared
to 27% for the control group.[67] These results support the idea that both mobilization and recruitment with G-CSF and HGF, respectively, are required for greater
therapeutic effects.

7.5   MECHANISM OF TISSUE REPAIR BY MOBILIZED BMCs

It has been suggested that mobilized BMCs repair ischemic tissues in two ways:
(1) differentiation into cardiomyocytes or fusion with neighboring cardiomyocytes,
and (2) release of multiple signaling factors such as anti-apoptotic factors, angiogenic factors, and stem cell homing factors.

7.5.1   TISSUE REPAIR THROUGH CELLULAR DIFFERENTIATION OR INFUSION

There is evidence suggesting BMCs recruited into ischemic tissue engraft infarcted
myocardium and ischemic peripheral tissues. However, it is still unclear whether the
engrafted BMCs differentiate into cardiomyocytes and vascular cells, or fuse with
local cells in the myocardium and vasculature, to restore function.

One common hypothesis for the improvement of neovascularization in ischemic
tissue, following BMC mobilization, is the differentiation of BMCs into endothelial
cells. This differentiation has been demonstrated with *in vitro* and *in vitro* studies.[68] One particular study demonstrated that the implantation of Lac-Z+ MSCs into
hindlimb ischemia mouse models resulted in improved vascular density compared
to controls. It was also found that the same populations of Lac-Z+ cells, identified
by positive staining with X-gal, expressed endothelial markers VIII while other

populations expressed α-smooth muscle actin. This result indicates the differentiation into endothelial and vascular smooth muscle cells, respectively.[69]

Orlic et al. transfected murine BMCs with the green fluorescent protein (GFP) gene to study the fate of circulating BMCs during myocardial infarction treatment. Cross-gender cell transplantation was then performed to identify implanted cells through the expression of GFP and the presence of the Y chromosome. The injection of the GFP-expressing donor cells in the peri-infarcted left ventricle of female mice resulted in myocardial regeneration in 12 out of the 30 mice investigated. Markers specific to myocytes, endothelial cells, and smooth muscle cells were found on the GFP-expressing BMCs. The downregulation of c-kit receptors, characteristic of stem cells, and the enhanced myocardial regeneration were attributed to the differentiation of engrafted BMCs into cardiomyocytes and endothelial cells.[15]

In contrast, Alvarez-Dolado et al. suggested that BMCs undergo fusion with existing cardiomyocytes instead of differentiation. In this study, two genetically modified mouse lines were used: a Cre reporter line, in which β-galactosidase (β-gal) was expressed in the presence of Cre recombinase, and an engineered line to constitutively express Cre recombinase and GFP. The first mouse line was irradiated to upregulate circulating homing signals and implanted with BMCs isolated from the second mouse line. Fusion between donor and host cells was indicated by the expression of β-gal. β-gal+ cells were detected in the myocardial wall, cerebellum, and Purkinje fibers, indicating the recruitment of BMCs and their fusion to host cells (Figure 7.4). The β-gal+ cells in the myocardial wall had similar morphology and alignment compared to other cardiomyocytes.[70] In another study, BMCs expressing GFP were transplanted into mice carrying the lacZ-gene, resulting in the detection of X-gal precipitate in the GFP+ cardiomyocyte cytoplasm.[70]

7.5.2 Paracrine Signaling of Mobilized BMCs

Another tissue repair mechanism of recruited BMCs is the expression of various signaling factors to stimulate cellular activities. In a study conducted by Timmers et al., MSC-conditioned medium was obtained by culturing MSCs for 3 days in a serum-free, chemically defined medium. The medium was then concentrated and subsequently injected into a pig with a myocardial infarction. Interestingly, infarction size was reduced with the administration of BMC-conditioned medium compared to nonconditioned medium and saline perfusion.[71] Other studies also found that BMCs upregulated the expression of VEGF, basic fibroblast growth factor (bFGF), insulin-like growth factor (IGF), and SDF-1 under ischemic conditions.[72] As addressed in the previous sections, these factors serve to prevent apoptosis of cells in the ischemic area, recruit endogenous cardiac stem cells, and further enhance neovascularization.

Numerous studies have also demonstrated that EPCs in peripheral blood exert proangiogenic effects on surrounding tissues due to their secretion of numerous angiogenic factors including VEGF, HGF, platelet-derived growth factor (PDGF), angiopoietin-1 (Ang-1), G-CSF, and GM-CSF. These factors help to recruit more stem and progenitor cells and mediate blood vessel formation.[73,74]

Apart from inducing BMC mobilization for myocardial repair, it is also suggested that G-CSF act directly on cardiomyocytes to promote their survival after myocardial

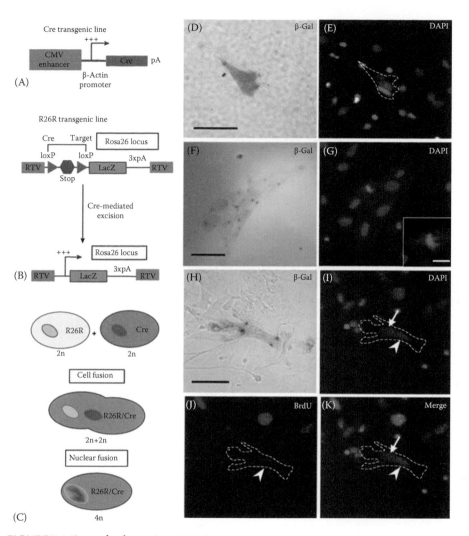

FIGURE 7.4 (See color insert.) (A, B) Schematic representation of the transgenes expressed by the mouse lines used. (C) When a cell expressing Cre recombinase (A) fuses with a cell bearing the LacZ reporter transgene (B), the floxed stop cassette is excised and the LacZ reporter is expressed in the fused cell. LacZ expression can be detected by the generation of a blue precipitate after X-gal staining. RTV, Integration retroviral sequence (LTR). (D–G) Cocultures of R26R BMSCs with Cre+ neurospheres stained for β-gal and the nuclear dye DAPI (4,6-diamidino-2-phenylindole). (D, E) Multinucleated β-gal+ fused cell (4 DIV). (F, G) Colony of β-gal+ (15 DIV). These cells were mononucleated and mitotically active, as demonstrated by a cell in metaphase (inset in (G)). (H–K) Coculture of R26R BMCs with BrdU-labeled Cre+ neurospheres. Binucleated β-gal+ fused cells (H) with one nucleus immunopositive for BrdU (J, K, arrowhead) and the second nucleus negative for BrdU (I, K, arrow). Scale bars: (D), (F), (H) 20 µm; (inset in (G)) 5 µm. (By permission from Macmillan Publishers Ltd., *Nature*, Alvarez-Dolado, M., Pardal, R., Garcia-Verdugo, J.M., Fike, J.R., Lee, H.O., Pfeffer, K., Lois, C., Morrison, S.J., Alvarez-Buylla, A., Fusion of bone-marrow-derived cells with Purkinje neurons, cardiomyocytes and hepatocytes, 425(6961), 968–973. Copyright 2003.)

infarction. Harada et al. demonstrated that the G-CSF receptor is expressed on cardiomyocytes. The *in vitro* binding of G-CSF to cardiomyocytes activated the Jak-Stat pathway, which minimized the death of cardiomyocytes induced by hydrogen peroxide. Expression levels of anti-apoptotic proteins, such as Bcl-2, were also found to be higher in cardiomyocytes preincubated with G-CSF before exposure to hydrogen peroxide. This anti-apoptotic effect was abolished in the presence of Jak2 inhibitor.[8]

7.6 DISCUSSION

Overall, studies aiming to enhance the treatment of cardiovascular diseases with endogenous stem cell mobilization have yielded mixed results. While there have been some favorable outcomes in animal studies, there are also modest or no improvements observed in clinical trials. However, there is still room to improve the efficacy of endogenous stem cell mobilization through a more robust and systematic translation of preclinical animal studies into clinical trials. For example, as described in previous sections, coadministration of cytokines to enhance both stem cell mobilization and homing has significant potentials to improve clinical outcomes, but these effects were not yet tested. It is also important to adopt proper cytokine administration strategies to improve the efficacy of cytokines.

In addition, the use of proper imaging tools to monitor the transmigration of mobilized stem cells through the endothelium would greatly improve our current understanding of BMC mobilization and the homing process. This understanding of cell transport would greatly contribute to the development of advanced strategies for endogenous stem cell mobilization. Furthermore, a better understanding of the signals involved in BMC mobilization, homing, and further phenotypic activities would also allow one to significantly improve the therapeutic efficacy of BMCs.

In parallel, more attention should be paid to the adverse effects of BMC mobilization. Some studies have reported restenosis and atherosclerosis stimulated by the administration of G-CSF. It was proposed that these adverse outcomes are related to accumulation of smooth muscular cell precursors that might be present in the mobilized BMCs population, and /or an increased local inflammation of plaque caused by G-CSF.[75] Since not all of the risks involved with G-CSF treatments have been clearly established, more extensive and thorough investigations should be conducted to determine whether the reported adverse effects were linked to G-CSF administration or purely coincidental.

REFERENCES

1. Terry Wikle Shapiro, *Clinical Guide to Stem Cell and Bone Marrow Transplantation.* Jones & Bartlett Publishers: Sudbury, MA, 1997; p. 454.
2. Kassis, I.; Zangi, L.; Rivkin, R.; Levdansky, L.; Samuel, S.; Marx, G.; Gorodetsky, R., Isolation of mesenchymal stem cells from G-CSF-mobilized human peripheral blood using fibrin microbeads. *Bone Marrow Transplant* 2006, *37* (10), 967–976.
3. Bengel, F. M.; Schachinger, V.; Dimmeler, S., Cell-based therapies and imaging in cardiology. *European Journal of Nuclear Medicine and Molecular Imaging* 2005, *32* (Suppl 2), S404–S416.

4. Kolf, C. M.; Cho, E.; Tuan, R. S., Mesenchymal stromal cells. Biology of adult mesenchymal stem cells: Regulation of niche, self-renewal and differentiation. *Arthritis Research & Therapy* 2007, *9* (1), 204.

5. Boheler, K. R.; Czyz, J.; Tweedie, D.; Yang, H. T.; Anisimov, S. V.; Wobus, A. M., Differentiation of pluripotent embryonic stem cells into cardiomyocytes. *Circulation Research* 2002, *91* (3), 189–201.

6. Wollert, K. C.; Drexler, H., Clinical applications of stem cells for the heart. *Circulation Research* 2005, *96* (2), 151–163.

7. Martin-Rendon, E.; Snowden, J. A.; Watt, S. M., Stem cell-related therapies for vascular diseases. *Transfusion Medicine* 2009, *19* (4), 159–171.

8. Harada, M.; Qin, Y.; Takano, H.; Minamino, T.; Zou, Y.; Toko, H.; Ohtsuka, M.; Matsuura, K.; Sano, M.; Nishi, J.-i.; Iwanaga, K.; Akazawa, H.; Kunieda, T.; Zhu, W.; Hasegawa, H.; Kunisada, K.; Nagai, T.; Nakaya, H.; Yamauchi-Takihara, K.; Komuro, I., G-CSF prevents cardiac remodeling after myocardial infarction by activating the Jak-Stat pathway in cardiomyocytes. *Nature Medicine* 2005, *11* (3), 305–311.

9. Deindl, E.; Zaruba, M.-M.; Brunner, S.; Huber, B.; Mehl, U.; Assmann, G.; Hoefer, I. E.; Mueller-Hoecker, J.; Franz, W.-M., G-CSF administration after myocardial infarction in mice attenuates late ischemic cardiomyopathy by enhanced arteriogenesis. *The FASEB Journal: Official Publication of the Federation of American Societies for Experimental Biology* 2006, *20* (7), 956–958.

10. Fazel, S.; Chen, L.; Weisel, R. D.; Angoulvant, D.; Seneviratne, C.; Fazel, A.; Cheung, P.; Lam, J.; Fedak, P. W. M.; Yau, T. M.; Li, R.-K., Cell transplantation preserves cardiac function after infarction by infarct stabilization: Augmentation by stem cell factor. *The Journal of Thoracic and Cardiovascular Surgery* 2005, *130* (5), 1310.

11. Hill, J. M.; Syed, M. A.; Arai, A. E.; Powell, T. M.; Paul, J. D.; Zalos, G.; Read, E. J.; Khuu, H. M.; Leitman, S. F.; Horne, M.; Csako, G.; Dunbar, C. E.; Waclawiw, M. A.; Cannon, R. O., Outcomes and risks of granulocyte colony-stimulating factor in patients with coronary artery disease. *Journal of the American College of Cardiology* 2005, *46* (9), 1643–1648.

12. Ince, H.; Petzsch, M.; Kleine, H. D.; Eckard, H.; Rehders, T.; Burska, D.; Kische, S.; Freund, M.; Nienaber, C. A., Prevention of left ventricular remodeling with granulocyte colony-stimulating factor after acute myocardial infarction: Final 1-year results of the front-integrated revascularization and stem cell liberation in evolving acute myocardial infarction by granulocyte colony-stimulating factor (FIRSTLINE-AMI) trial. *Circulation* 2005, *112* (9 Suppl), I73–180.

13. Dawn, B.; Guo, Y.; Rezazadeh, A.; Huang, Y.; Stein, A. B.; Hunt, G.; Tiwari, S.; Varma, J.; Gu, Y.; Prabhu, S. D.; Kajstura, J.; Anversa, P.; Ildstad, S. T.; Bolli, R., Postinfarct cytokine therapy regenerates cardiac tissue and improves left ventricular function. *Circulation Research* 2006, *98* (8), 1098–1105.

14. Kang, H.; Kim, H., Safety and efficacy of intracoronary infusion of mobilized peripheral blood stem cell in patients with myocardial infarction: MAGIC Cell-1 and MAGIC Cell-3-DES-trials. *European Heart Journal Supplements* 2008, *10* (Supplement K), 39–43.

15. Orlic, D.; Kajstura, J.; Chimenti, S.; Jakoniuk, I.; Anderson, S. M.; Li, B.; Pickel, J.; McKay, R.; Nadal-Ginard, B.; Bodine, D. M.; Leri, A.; Anversa, P., Bone marrow cells regenerate infarcted myocardium. *Nature* 2001, *410* (6829), 701–705.

16. Wang, C.-H.; Cherng, W.-J.; Yang, N.-I.; Kuo, L.-T.; Hsu, C.-M.; Yeh, H.-I.; Lan, Y.-J.; Yeh, C.-H.; Stanford, W. L., Late-outgrowth endothelial cells attenuate intimal hyperplasia contributed by mesenchymal stem cells after vascular injury. *Arteriosclerosis, Thrombosis and Vascular Biology* 2008, *28* (1), 54–60.

17. Zhang, Q. H.; She, M. P., Biological behaviour and role of endothelial progenitor cells in vascular diseases. *Chinese Medical Journal* 2007, *120* (24), 2297–2303.

18. Massa, M.; Rosti, V.; Ferrario, M.; Campanelli, R.; Ramajoli, I.; Rosso, R.; De, F.; Ferlini, M.; Goffredo, L.; Bertoletti, A.; Klersy, C.; Pecci, A.; Moratti, R.; Tavazzi, L., Increased circulating hematopoietic and endothelial progenitor cells in the early phase of acute myocardial infarction. *Blood* 2005, *105* (1), 199–206.

19. Richman, C. M.; Weiner, R. S.; Yankee, R. A., Increase in circulating stem cells following chemotherapy in man. *Blood* 1976, *47* (6), 1031–1039.

20. Cantin, G.; Marchand-Laroche, D.; Bouchard, M. M.; Demers, C.; Leblond, P. F.; Lyonnais, J.; Petitclerc, C.; Delage, R., Hematopoietic engraftment from a minimal number of apheresis procedures after mobilization of peripheral blood stem cells with chemotherapy and rhG-CSF. *Transfusion Science* 1995, *16* (2), 145–154.

21. Kolbe, K.; Peschel, C.; Rupilius, B.; Després, D.; Burger, K.; Sklenar, I.; Färber, L.; Huber, C.; Derigs, H. G., Peripheral blood stem cell (PBSC) mobilization with chemotherapy followed by sequential IL-3 and G-CSF administration in extensively pretreated patients. *Bone Marrow Transplant* 1997, *20* (12), 1027–1032.

22. Bertolini, F.; de, V.; Lanata, L.; Lemoli, R. M.; Maccario, R.; Majolino, I.; Ponchio, L.; Rondelli, D.; Tabilio, A.; Zanon, P.; Tura, S., Allogeneic hematopoietic stem cells from sources other than bone marrow: Biological and technical aspects. *Haematologica* 1997, *82* (2), 220–238.

23. Agocha, A.; Lee, H. W.; Eghbali-Webb, M., Hypoxia regulates basal and induced DNA synthesis and collagen type I production in human cardiac fibroblasts: Effects of transforming growth factor-beta1, thyroid hormone, angiotensin II and basic fibroblast growth factor. *Journal of Molecular and Cellular Cardiology* 1997, *29* (8), 2233–2244.

24. Shintani, S.; Murohara, T.; Ikeda, H.; Ueno, T.; Honma, T.; Katoh, A.; Sasaki, K.; Shimada, T.; Oike, Y.; Imaizumi, T., Mobilization of endothelial progenitor cells in patients with acute myocardial infarction. *Circulation* 2001, *103* (23), 2776–2779.

25. Leone, A. M.; Rutella, S.; Bonanno, G.; Contemi, A. M.; de Ritis, D. G.; Giannico, M. B.; Rebuzzi, A. G.; Leone, G.; Crea, F., Endogenous G-CSF and CD34+ cell mobilization after acute myocardial infarction. *International Journal of Cardiology* 2006, *111* (2), 202–208.

26. Leone, A. M.; Rutella, S.; Bonanno, G.; Abbate, A.; Rebuzzi, A. G.; Giovannini, S.; Lombardi, M.; Galiuto, L.; Liuzzo, G.; Andreotti, F.; Lanza, G. A.; Contemi, A. M.; Leone, G.; Crea, F., Mobilization of bone marrow-derived stem cells after myocardial infarction and left ventricular function. *European Heart Journal* 2005, *26* (12), 1196–1204.

27. Norol, F.; Merlet, P.; Isnard, R.; Sebillon, P.; Bonnet, N.; Cailliot, C.; Carrion, C.; Ribeiro, M.; Charlotte, F.; Pradeau, P.; Mayol, J.-F.; Peinnequin, A.; Drouet, M.; Safsafi, K.; Vernant, J.-P.; Herodin, F., Influence of mobilized stem cells on myocardial infarct repair in a nonhuman primate model. *Blood* 2003, *102* (13), 4361–4368.

28. Wojakowski, W.; Tendera, M.; Zebzda, A.; Michalowska, A.; Majka, M.; Kucia, M.; Maslankiewicz, K.; Wyderka, R.; Król, M.; Ochala, A.; Kozakiewicz, K.; Ratajczak, M. Z., Mobilization of CD34(+), CD117(+), CXCR4(+), c-met(+) stem cells is correlated with left ventricular ejection fraction and plasma NT-proBNP levels in patients with acute myocardial infarction. *European Heart Journal* 2006, *27* (3), 283–289.

29. Seeger, F. H.; Zeiher, A. M.; Dimmeler, S., Cell-enhancement strategies for the treatment of ischemic heart disease. *Nature Clinical Practice Cardiovascular Medicine* 2007, S110–S113.

30. Layman, H.; Sacasa, M.; Murphy, A. E.; Murphy, A. M.; Pham, S. M.; Andreopoulos, F. M., Co-delivery of FGF-2 and G-CSF from gelatin-based hydrogels as angiogenic therapy in a murine critical limb ischemic model. *Acta Biomaterialia* 2009, *5* (1), 230–239.

31. Fu, S.; Liesveld, J., Mobilization of hematopoietic stem cells. *Blood Reviews* 2000, *14* (4), 205–218.
32. Craddock, C. F.; Nakamoto, B.; Andrews, R. G.; Priestley, G. V.; Papayannopoulou, T., Antibodies to VLA4 integrin mobilize long-term repopulating cells and augment cytokine-induced mobilization in primates and mice. *Blood* 1997, *90* (12), 4779–4788.
33. Pitchford, S. C.; Furze, R. C.; Jones, C. P.; Wengner, A. M.; Rankin, S. M., Differential mobilization of subsets of progenitor cells from the bone marrow. *Cell Stem Cell* 2009, *4* (1), 62–72.
34. Turner, M. L.; McIlwaine, K.; Anthony, R. S.; Parker, A. C., Differential expression of cell adhesion molecules by human hematopoietic progenitor cells from bone marrow and mobilized adult peripheral blood. *Stem Cells* 1995, *13* (3), 311–316.
35. Dercksen, M. W.; Gerritsen, W. R.; Rodenhuis, S.; Dirkson, M. K.; Slaper-Cortenbach, I. C.; Schaasberg, W. P.; Pinedo, H. M.; von, d.; van, d., Expression of adhesion molecules on CD34+ cells: CD34+ L-selectin+ cells predict a rapid platelet recovery after peripheral blood stem cell transplantation. *Blood* 1995, *85* (11), 3313–3319.
36. Möhle, R.; Murea, S.; Kirsch, M.; Haas, R., Differential expression of L-selectin, VLA-4, and LFA-1 on CD34+ progenitor cells from bone marrow and peripheral blood during G-CSF-enhanced recovery. *Experimental Hematology* 1995, *23* (14), 1535–1542.
37. Capoccia, B. J.; Shepherd, R. M.; Link, D. C., G-CSF and AMD3100 mobilize monocytes into the blood that stimulate angiogenesis in vivo through a paracrine mechanism. *Blood* 2006, *108* (7), 2438–2445.
38. Papayannopoulou, T.; Craddock, C.; Nakamoto, B.; Priestley, G. V.; Wolf, N. S., The VLA4/VCAM-1 adhesion pathway defines contrasting mechanisms of lodgement of transplanted murine hemopoietic progenitors between bone marrow and spleen. *Proceedings of the National Academy of Sciences of United States of America* 1995, *92* (21), 9647–9651.
39. Papayannopoulou, T., Mechanisms of stem-/progenitor-cell mobilization: The anti-VLA-4 paradigm. *Seminars in Hematology* 2000, *37* (1 Suppl 2), 11–18.
40. Papayannopoulou, T.; Priestley, G. V.; Nakamoto, B.; Zafiropoulos, V.; Scott, L. M.; Harlan, J. M., Synergistic mobilization of hemopoietic progenitor cells using concurrent beta1 and beta2 integrin blockade or beta2-deficient mice. *Blood* 2001, *97* (5), 1282–1288.
41. Velders, G. A.; Pruijt, J. F. M.; Verzaal, P.; van Os, R.; van Kooyk, Y.; Figdor, C. G.; de Kruijf, E.-J. F. M.; Willemze, R.; Fibbe, W. E., Enhancement of G-CSF-induced stem cell mobilization by antibodies against the beta 2 integrins LFA-1 and Mac-1. *Blood* 2002, *100* (1), 327–333.
42. Lapidot, T.; Petit, I., Current understanding of stem cell mobilization: The roles of chemokines, proteolytic enzymes, adhesion molecules, cytokines, and stromal cells. *Experimental Hematology* 2002, *30* (9), 973–981.
43. Drize, N.; Chertkov, J.; Samoilina, N.; Zander, A., Effect of cytokine treatment (granulocyte colony-stimulating factor and stem cell factor) on hematopoiesis and the circulating pool of hematopoietic stem cells in mice. *Experimental Hematology* 1996, *24* (7), 816–822.
44. Mielcarek, M.; Torok-Storb, B., Phenotype and engraftment potential of cytokine-mobilized peripheral blood mononuclear cells. *Current Opinion in Hematology* 1997, *4* (3), 176–182.
45. Heil, M.; Clauss, M.; Suzuki, K.; Buschmann, I. R.; Willuweit, A.; Fischer, S.; Schaper, W., Vascular endothelial growth factor (VEGF) stimulates monocyte migration through endothelial monolayers via increased integrin expression. *European Journal Cell Biology* 2000, *79* (11), 850–857.

46. Petit, I.; Szyper-Kravitz, M.; Nagler, A.; Lahav, M.; Peled, A.; Habler, L.; Ponomaryov, T.; Taichman, R. S.; Arenzana-Seisdedos, F.; Fujii, N.; Sandbank, J.; Zipori, D.; Lapidot, T., G-CSF induces stem cell mobilization by decreasing bone marrow SDF-1 and up-regulating CXCR4. *Nature Immunology* 2002, *3* (7), 687–694.

47. Brasel, K.; McKenna, H. J.; Morrissey, P. J.; Charrier, K.; Morris, A. E.; Lee, C. C.; Williams, D. E.; Lyman, S. D., Hematologic effects of flt3 ligand in vivo in mice. *Blood* 1996, *88* (6), 2004–2012.

48. Dettke, M.; Jurko, S.; Rüger, B. M.; Leitner, G.; Greinix, H. T.; Kalhs, P.; Fischer, M. B.; Höcker, P., Increased serum flt3-ligand in healthy donors undergoing granulo-cyte colony-stimulating factor-induced peripheral stem cell mobilization. *Journal of Hematotherapy and Stem Cell Research* 2001, *10* (2), 317–320.

49. Ohki, Y.; Heissig, B.; Sato, Y.; Akiyama, H.; Zhu, Z.; Hicklin, D. J.; Shimada, K.; Ogawa, H.; Daida, H.; Hattori, K.; Ohsaka, A., Granulocyte colony-stimulating fac-tor promotes neovascularization by releasing vascular endothelial growth factor from neutrophils. *The FASEB Journal: Official Publication of the Federation of American Societies for Experimental Biology* 2005, *19* (14), 2005–2007.

50. Engelmann, M. G.; Theiss, H. D.; Hennig-Theiss, C.; Huber, A.; Wintersperger, B. J.; Werle-Ruedinger, A.-E.; Schoenberg, S. O.; Steinbeck, G.; Franz, W.-M., Autologous bone marrow stem cell mobilization induced by granulocyte colony-stimulating factor after subacute ST-segment elevation myocardial infarction undergoing late revascular-ization: Final results from the G-CSF-STEMI (Granulocyte Colony-Stimulating Factor ST-Segment Elevation Myocardial Infarction) trial. *Journal of the American College of Cardiology* 2006, *48* (8), 1712–1721.

51. Zohlnhöfer, D.; Dibra, A.; Koppara, T.; de Waha, A.; Ripa, R. S.; Kastrup, J.; Valgimigli, M.; Schömig, A.; Kastrati, A., Stem cell mobilization by granulocyte colony-stimulating factor for myocardial recovery after acute myocardial infarc-tion: A meta-analysis. *Journal of the American College of Cardiology* 2008, *51* (15), 1429–1437.

52. Kang, H.-J.; Kim, H.-S.; Zhang, S.-Y.; Park, K.-W.; Cho, H.-J.; Koo, B.-K.; Kim, Y.-J.; Soo Lee, D.; Sohn, D.-W.; Han, K.-S.; Oh, B.-H.; Lee, M.-M.; Park, Y.-B., Effects of intracoronary infusion of peripheral blood stem-cells mobilised with granulocyte-colony stimulating factor on left ventricular systolic function and restenosis after coronary stenting in myocardial infarction: The MAGIC cell randomised clinical trial. *Lancet* 2004, *363* (9411), 751–756.

53. Brasel, K.; McKenna, H. J.; Charrier, K.; Morrissey, P. J.; Williams, D. E.; Lyman, S. D., Flt3 ligand synergizes with granulocyte-macrophage colony-stimulating factor or granulocyte colony-stimulating factor to mobilize hematopoietic progenitor cells into the peripheral blood of mice. *Blood* 1997, *90* (9), 3781–3788.

54. Yatuv, R.; Carmel-Goren, L.; Dayan, I.; Robinson, M.; Baru, M., Binding of proteins to PEGylated liposomes and improvement of G-CSF efficacy in mobilization of hema-topoietic stem cells. *Journal of Controlled Release: Official Journal of the Controlled Release Society* 2009, *135* (1), 44–50.

55. Laterveer, L.; Lindley, I. J.; Hamilton, M. S.; Willemze, R.; Fibbe, W. E., Interleukin-8 induces rapid mobilization of hematopoietic stem cells with radioprotective capacity and long-term myelolymphoid repopulating ability. *Blood* 1995, *85* (8), 2269–2275.

56. Fibbe, W. E.; Pruijt, J. F.; Velders, G. A.; Opdenakker, G.; van, K.; Figdor, C. G.; Willemze, R., Biology of IL-8-induced stem cell mobilization. *Annals of the New York Academy of Sciences* 1999, 872, 71–82.

57. Karp, J. M.; Leng Teo, G. S., Mesenchymal stem cell homing: The devil is in the details. *Cell Stem Cell* 2009, *4* (3), 206–216.

58. Kang, W. J.; Kang, H. J.; Kim, H. S.; Chung, J. K.; Lee, M. C.; Lee, D. S., Tissue distribution of 18F-FDG-labeled peripheral hematopoietic stem cells after intracoronary administration in patients with myocardial infarction. *Journal of Nuclear Medicine* 2006, *47* (8), 1295–1301.

59. Theiss, H. D.; David, R.; Engelmann, M. G.; Barth, A.; Schotten, K.; Naebauer, M.; Reichart, B.; Steinbeck, G.; Franz, W.-M., Circulation of CD34+ progenitor cell populations in patients with idiopathic dilated and ischaemic cardiomyopathy (DCM and ICM). *European Heart Journal* 2007, *28* (10), 1258–1264.

60. Grunewald, M.; Avraham, I.; Dor, Y.; Bachar-Lustig, E.; Itin, A.; Jung, S.; Yung, S.; Chimenti, S.; Landsman, L.; Abramovitch, R.; Keshet, E., VEGF-induced adult neovascularization: Recruitment, retention, and role of accessory cells. *Cell* 2006, *124* (1), 175–189.

61. Li, B.; Sharpe, E. E.; Maupin, A. B.; Teleron, A. A.; Pyle, A. L.; Carmeliet, P.; Young, P. P., VEGF and PlGF promote adult vasculogenesis by enhancing EPC recruitment and vessel formation at the site of tumor neovascularization. *The FASEB Journal: Official Publication of the Federation of American Societies for Experimental Biology* 2006, *20* (9), 1495–1497.

62. Faber, A.; Roderburg, C.; Wein, F.; Saffrich, R.; Seckinger, A.; Horsch, K.; Diehlmann, A.; Wong, D.; Bridger, G.; Eckstein, V.; Ho, A. D.; Wagner, W., The many facets of SDF-1alpha, CXCR4 agonists and antagonists on hematopoietic progenitor cells. *Journal of Biomedicine and Biotechnology* 2007, (3), 26065.

63. Yamaguchi, J.; Kusano, K. F.; Masuo, O.; Kawamoto, A.; Silver, M.; Murasawa, S.; Bosch-Marce, M.; Masuda, H.; Losordo, D. W.; Isner, J. M.; Asahara, T., Stromal cell-derived factor-1 effects on ex vivo expanded endothelial progenitor cell recruitment for ischemic neovascularization. *Circulation* 2003, *107* (9), 1322–1328.

64. De, L.; Yang, F.; Narazaki, M.; Salvucci, O.; Davis, D.; Yarchoan, R.; Zhang, H. H.; Fales, H.; Tosato, G., Differential processing of stromal-derived factor-1alpha and stromal-derived factor-1beta explains functional diversity. *Blood* 2004, *103* (7), 2452–2459.

65. Zaruba, M.-M.; Theiss, H. D.; Vallaster, M.; Mehl, U.; Brunner, S.; David, R.; Fischer, R.; Krieg, L.; Hirsch, E.; Huber, B.; Nathan, P.; Israel, L.; Imhof, A.; Herbach, N.; Assmann, G.; Wanke, R.; Mueller-Hoecker, J.; Steinbeck, G.; Franz, W.-M., Synergy between CD26/DPP-IV inhibition and G-CSF improves cardiac function after acute myocardial infarction. *Cell Stem Cell* 2009, *4* (4), 313–323.

66. Segers, V. F. M.; Tokunou, T.; Higgins, L. J.; MacGillivray, C.; Gannon, J.; Lee, R. T., Local delivery of protease-resistant stromal cell derived factor-1 for stem cell recruitment after myocardial infarction. *Circulation* 2007, *116* (15), 1683–1692.

67. Ieda, Y.; Fujita, J.; Ieda, M.; Yagi, T.; Kawada, H.; Ando, K.; Fukuda, K., G-CSF and HGF: Combination of vasculogenesis and angiogenesis synergistically improves recovery in murine hind limb ischemia. *Journal of Molecular and Cellular Cardiology* 2007, *42* (3), 540–548.

68. Hirata, K.; Li, T.-S.; Nishida, M.; Ito, H.; Matsuzaki, M.; Kasaoka, S.; Hamano, K., Autologous bone marrow cell implantation as therapeutic angiogenesis for ischemic hindlimb in diabetic rat model. *American Journal of Physiology. Heart and Circulatory Physiology* 2003, *284* (1), H66–H70.

69. Al-Khaldi, A.; Al-Sabti, H.; Galipeau, J.; Lachapelle, K., Therapeutic angiogenesis using autologous bone marrow stromal cells: Improved blood flow in a chronic limb ischemia model. *The Annals of Thoracic Surgery* 2003, *75* (1), 204–209.

70. Alvarez-Dolado, M.; Pardal, R.; Garcia-Verdugo, J. M.; Fike, J. R.; Lee, H. O.; Pfeffer, K.; Lois, C.; Morrison, S. J.; Alvarez-Buylla, A., Fusion of bone-marrow-derived cells with Purkinje neurons, cardiomyocytes and hepatocytes. *Nature* 2003, *425* (6961), 968–973.

71. Timmers, L.; Lim, S.; Lee, C.; Choo, A.; Pasterkamp, G.; de Kleijn, D., Clinically compliant mesenchymal stem cell conditioned medium reduces myocardial infarct size in a pig model of ischemia and reperfusion injury. *Circulation* 2007, *116* (16 Meeting Abstracts), II 132b.

72. Brunner, S.; Engelmann, M. G.; Franz, W. M., Stem cell mobilisation for myocardial repair. *Expert Opinion on Biological Therapy* 2008, *8* (11), 1675–1690.

73. Rehman, J.; Li, J.; Orschell, C. M.; March, K. L., Peripheral blood "endothelial progenitor cells" are derived from monocyte/macrophages and secrete angiogenic growth factors. *Circulation* 2003, *107* (8), 1164–1169.

74. Miyamoto, Y.; Suyama, T.; Yashita, T.; Akimaru, H.; Kurata, H., Bone marrow subpopulations contain distinct types of endothelial progenitor cells and angiogenic cytokine-producing cells. *Journal of Molecular and Cellular Cardiology* 2007, *43* (5), 627–635.

75. Kang, H. J.; Kim, H. S.; Koo, B. K.; Kim, Y. J.; Lee, D.; Sohn, D. W.; Oh, B. H.; Park, Y. B., Intracoronary infusion of the mobilized peripheral blood stem cell by G-CSF is better than mobilization alone by G-CSF for improvement of cardiac function and remodeling: 2-year follow-up results of the myocardial regeneration and angiogenesis in myocardial infarction with G-CSF and intra-coronary stem cell infusion (MAGIC Cell) 1 trial. *American Heart Journal* 2007, *153* (2), 237.

76. Lataillade, J. J.; Domenech, J.; Le Bousse-Kerdilès, M. C., Stromal cell-derived factor-1 (SDF-1)/CXCR4 couple plays multiple roles on haematopoietic progenitors at the border between the old cytokine and new chemokine worlds: Survival, cell cycling and trafficking. *European Cytokine Network* 2004, *15* (3), 177–188.

77. Kuethe, F.; Figulla, H. R.; Herzau, M.; Voth, M.; Fritzenwanger, M.; Opfermann, T.; Pachmann, K.; Krack, A.; Sayer, H. G.; Gottschild, D.; Werner, G. S., Treatment with granulocyte colony-stimulating factor for mobilization of bone marrow cells in patients with acute myocardial infarction. *American Heart Journal* 2005, *150* (1), 115.

78. Zohlnhöfer, D.; Ott, I.; Mehilli, J.; Schömig, K.; Michalk, F.; Ibrahim, T.; Meisetschläger, G.; von Wedel, J.; Bollwein, H.; Seyfarth, M.; Dirschinger, J.; Schmitt, C.; Schwaiger, M.; Kastrati, A.; Schömig, A.; for the REVIVAL-2 Investigators, Stem cell mobilization by granulocyte colony-stimulating factor in patients with acute myocardial infarction: A randomized controlled trial. *JAMA: The Journal of the American Medical Association* 2006, *295* (9), 1003–1010.

79. Ripa, R. S.; Jørgensen, E.; Wang, Y.; Thune, J. J.; Nilsson, J. C.; Søndergaard, L.; Johnsen, H. E.; Køber, L.; Grande, P.; Kastrup, J., Stem cell mobilization induced by subcutaneous granulocyte-colony stimulating factor to improve cardiac regeneration after acute ST-elevation myocardial infarction: Result of the double-blind, randomized, placebo-controlled stem cells in myocardial infarction (STEMMI) trial. *Circulation* 2006, *113* (16), 1983–1992.

80. Ellis, S. G.; Penn, M. S.; Bolwell, B.; Garcia, M.; Chacko, M.; Wang, T.; Brezina, K. J.; McConnell, G.; Topol, E. J., Granulocyte colony stimulating factor in patients with large acute myocardial infarction: Results of a pilot dose-escalation randomized trial. *American Heart Journal* 2006, *152* (6), 1051.e9–1051.e14.

Part III

Stem Cell Mobilization Strategies

8 Stem Cell Homing to Sites of Injury and Inflammation

Weian Zhao, James Ankrum, Debanjan Sarkar,
Namit Kumar, Wei Suong Teo, and Jeffrey M. Karp

CONTENTS

ABBREVIATIONS

HSC, hematopoietic stem cells; MSC, mesenchymal stem cells; GVHD, graft versus host disease; MI, myocardial infarction; ESC, embryonic stem cells; iPSC, induced pluripotent stem cells; PSGL-1, P-selectin glycoprotein ligand 1; VLA, very late antigen; LFA, lymphocyte function-associated antigen; ICAM-1, intercellular adhesion molecule 1; VCAM-1, vascular cell-adhesion molecule 1; MadCAM-1, mucosal vascular addressin cell adhesion molecule 1; PI3K, phosphoinositide 3-kinase; MAC-1, macrophage receptor 1; PECAM-1, platelet/endothelial-cell adhesion molecule 1; JAM, junction adhesion molecule; ESAM, endothelial cell-selective adhesion molecule; PNAd, peripheral lymph node addressin; SLeX, sialyl-Lewis X tetrasaccharide; ESL, E selectin ligand; HEV, high endothelial venule; TNF, tumor-necrosis factor; ECM, extracellular matrix; GPCR, G-protein-coupled receptors; CXCR, chemokine.

(C-X-C motif) receptor; CCR, chemokine (C-C motif) receptor; CRTH2, Chemoattractant receptor-homologous molecule expressed on Th2 lymphocytes; PGD2, Prostaglandin D2; PAFR, platelet-activating factor receptor; mAbs, monoclonal antibodies; EPCs, endothelial progenitor cells; PCR, polymerase chain reaction; ELISA, enzyme-linked immunosorbent assay; MRI, magnetic resonance imaging; CT, computed tomography; HCELL, hematopoietic cell E-/L-selectin ligand; NOD/SCID, nonobese diabetic/severe combined immunodeficient; HCAM, homing-associated cell adhesion molecule; SDF-1, stromal cell-derived factor-1; IL, interleukin; SCF, stem cell factor; HIF, hypoxia-inducible factor; G-CSF, granulocyte colony-stimulating factor; GROβ, growth related protein beta; VEGF, Vascular endothelial growth factor; cysLT1, cysteinyl leukotriene receptor 1; S1P lyase, sphingosine-1-phosphate Lyase; SH3, Src homology-3; HUVEC, human umbilical vein endothelial cell; CD, cluster of differentiation; ALCAM, activated leukocyte cell adhesion molecule; CMECs, cardiac microvascular endothelial cells; bFGF, basic fibroblast growth factor; MMP, matrix metalloproteinase; MT1-MMP, membrane type 1 matrix metalloproteinase; BST-1, bone marrow stromal antigen-1; TLR, toll-like receptor; PDGF, platelet derived growth factor; EGFR, hepatocyte growth factor receptor; HGFR, hepatocyte growth factor receptor; IGF, insulin-like growth factor; NGFR, nerve growth factor receptor; TNFRSF, tumor necrosis factor receptor super family; MCP, monocyte-chemoattractant protein; MDC, macrophage-derived cytokine; MIP, macrophage inflammatory proteins; vWF, von Willebrand factor; eNOS, endothelial nitric oxide synthase; CSC, cancer stem cell; IFN, interferon.

8.1   INTRODUCTION

8.1.1   STEM CELL THERAPY

Stem cells are unspecialized precursor cells that are capable of both self-renewal (to sustain the stem cell pool) and differentiation (to produce mature daughter cells with tissue specific functions).[1] In adulthood, stem cells play important roles in continuously replacing short-lived mature differentiated cells as well as regeneration of diseased/damaged tissue.[1] More recently, it has been recognized that stem cells such as hematopoietic stem cells (HSCs) and mesenchymal stem cells (MSCs) can participate directly in immune surveillance and defense against pathogens[1] or have immunomodulatory effects and are capable of regulating tissue regeneration/repair via tropic/paracrine mechanisms.[2,3] Therefore, the use of stem cells for regenerative therapy holds great promise for the treatment of a variety of diseases. In fact, HSCs, or bone marrow transplantation, has been used in the clinic for several decades to treat leukemia, sickle cell anemia, and other bone/blood cancers and diseases.[4] Other adult stem cells including MSCs have recently been tested in clinical trials for treatment of numerous diseases including graft versus host disease (GVHD), myocardial infarction (MI), multiple sclerosis, and skeletal tissue repair, among others.[2] In addition to multipotent adult stem cells, pluripotent embryonic stem cells (ESCs)[5] and induced pluripotent stem cells (iPSCs),[6] although still in their infancy, hold great potential for drug screening and possibly treatment of practically any disease.

8.1.2 DELIVERY ROUTES IN STEM CELL THERAPY

The concept of stem cell therapy, and cell therapy in general, is to (1) stimulate endogenous cells, or (2) transplant exogenous cells, often culture expanded, to repair or replace damaged/injured tissues or cells.[7] Cells can be delivered locally to a tissue of interest, often seeded within a scaffold, or infused systemically. For some conditions, such as bone and cartilage regeneration, it is appropriate to locally administer stem cells to a defect (see Chapter 11 for details); however, the systemic infusion route is frequently used in the clinic because it is convenient, minimally invasive, and often desirable for systemic conditions.[2,8] Indeed, systemic infusion can address some of the challenges faced by local transplantation including potential invasiveness (e.g., into the heart or brain) and massive cell death due to limited diffusion of nutrients and oxygen.[2,8]

8.1.3 STEM CELL TRAFFICKING AND HOMING

The key to the success of systemic stem cell therapy is whether the cells are able to traffic or "home" to a target niche where they can survive and exert their therapeutic function locally through cell–cell contacts, paracrine signaling, and differentiation, or systemically through production and secretion of soluble factors into the circulation that can impact distant tissues. Natural endogenous stem cells (i.e., HSCs) and leukocytes have efficient homing pathways that guide them to sites of injury and inflammation. Some tumors are capable of shedding cells from the primary tumor site, intravasating into the blood stream, and eventually homing to an appropriate niche to seed secondary tumors.[1] While the important physiological roles (i.e., tissue regeneration, immune surveillance) of stem cell homing has been recognized, many of the processes that guide stem cell homing are poorly understood. From a therapeutic perspective, a major challenge is that systemically administrated *in vitro* expanded stem cells often have extremely poor homing efficiency (typically <1%) to the targeted tissue, which is generally thought to be due to the stem cell's lack of key homing receptors.[2] Therefore, understanding stem cell homing is not only crucial from a fundamental biology standpoint, but may help us consider new approaches for enhancing cell targeting.

8.1.4 Scope of This Chapter

We attempt in this chapter to summarize the current knowledge of stem cell homing to sites of injury and inflammation with a focus on systemically administrated stem cells, with particular emphasis on those that are currently being clinically applied (i.e., HSCs) or being explored within clinical trials (i.e., MSCs). Specifically, we will discuss the stem cell homing mechanisms and molecules (i.e., adhesion molecules, chemokines, and protease enzymes) that are involved in the stepwise homing process including cell tethering, rolling, firm adhesion, transendothelial migration, and invasion into underlying extracellular matrix. As stem cell homing is believed to share some common characteristics to the well-established leukocyte homing, we will begin with a brief introduction of the leukocyte homing cascade. We will then discuss the current knowledge of how stem cells home to specific tissues. Subsequently, we will summarize the approaches one can employ to engineer stem cells to improve the homing efficiency and how targeted stem cell homing can be used for therapeutic applications. Finally, we will discuss the challenges, unaddressed questions, and emerging opportunities in this field.

Note that the mobilization of endogenous stem cells from their niche to blood circulation will be the topic of Chapter 10. Also, stem cell trafficking during development (i.e., at fetal stage) and interstitial migration (i.e., skeletal muscle satellite cell migration) that is independent of systemic trafficking are not the focus of this chapter. Furthermore, the use of pluripotent stem cells such as ESCs and iPSCs for practical therapy, a field still in its infancy, will not be included herein, and we recommend some excellent reviews[9,10] on this topic for interested readers.

8.2 LEUKOCYTE HOMING CASCADE

8.2.1 Definition and Characteristics of "Homing"

Classically, homing has been used to describe the process by which circulating leukocytes are recruited to lymphoid organs and peripheral sites of injury and inflammation. This homing process is schematically illustrated in Figure 8.1 and discussed specifically in the following.[1,11]

8.2.2 Leukocyte Tethering and Rolling

To migrate to tissue(s), circulating leukocytes must first marginate in the blood stream and interact with the vascular endothelium.[12] This first essential "tethering/capture" step is mediated by a cell "rolling" process (Figure 8.1) where leukocytes roll along endothelium with a typical velocity of ~2–100 μm/s for anywhere between a few seconds to several minutes.[13] This cell rolling process is regulated by a family of adhesion molecules called selectins together with a subset of integrins.[14]

The selectin family[13] (Table 8.1) includes three members of highly conserved C-type lectins, which bind to sialyl-Lewis X tetrasaccharide (SLeX)–like carbohydrate ligands presented by sialomucin-like surface molecules such as P-selectin glycoprotein ligand 1 (PSGL-1). Essentially, L-selectin constitutively expressed on most circulating leukocytes binds to PSGL-1 and others ligands, which regulates

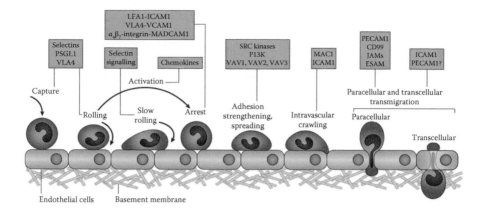

FIGURE 8.1 (See color insert.) General scheme of the leukocyte homing cascade, which includes rolling (mediated by selectins), activation (mediated by chemokines), arrest (mediated by integrins), intravascular crawling, and paracellular and transcellular transmigration. Key molecules involved in each step are indicated in boxes. (Adapted by permission from Macmillan Publishers Ltd. *Nat. Rev. Immunol.*, Ley, K., Laudanna, C., Cybulsky, M.I., Nourshargh, S., Getting to the site of inflammation: The leukocyte adhesion cascade updated, 7(9), 678–689. Copyright 2007.)

TABLE 8.1

Selectins and Their Ligands

Selectins	Selectin Ligands
L-selectin	PSGL-1, CD34, MadCAM-1, GlyCAM-1
P-selectin	PSGL-1
E-selectin	PSGL-1, HCELL, CD44, ESL-1, CD18, CD66, L-selectin, and CD43

Source: Data from Kansas, G., *Blood*, 88(9), 3259, 1996.

---

leukocyte capture/rolling events particularly in high endothelial venules (HEVs) of secondary lymphoid tissues. Note that L-selectin and its ligand(s) coexpressed on leukocytes can initiate a secondary binding event where leukocytes arrested on endothelium further capture newly incoming leukocytes, which might be responsible for leukocyte accumulation during inflammation.[15] P- and E-selectins are inducibly expressed on inflamed endothelium on which leukocytes are captured. PSGL-1 expressed on leukocytes is the major ligand for P-selectin and E-selectin, although E-selectin also binds to other molecules including glycosylated CD44 and E-selectin ligand (ESL).[16]

Selectin-mediated adhesion bonds are characterized by their exceptionally high "on and off rates," which lasts only a few seconds or less.[14] A rapid association rate facilitates the initial tethering in flow. A rapid dissociation rate ensures that even with multiple selectin–ligand bonds, it will not take long before the bond that is most upstream randomly dissociates, allowing the cell to roll forward until it is held by the next bonds.[14] Interestingly, the off rate is shear dependent. In fact, L-selectin and P-selectin require shear stress to support adhesion.[12,17,18] Note that the topographical distribution of selectins and their ligands are crucial to initiate an effective rolling process. For instance, L-selectin and PSGL-1 are selectively located at the tip of cell membrane protrusion structures called microvilli, which facilitates their binding with their counterparts. Moreover, actin cytoskeleton arrangement, selectin shedding, and soluble selectins are additional factors that collectively regulate the leukocyte rolling process.[1,18]

While selectins are known as the most effective rolling receptors, a subset of integrins can also participate in particularly slow rolling interactions, which often coordinate with and stabilize the selectin-mediated rolling. For instance, $\alpha_4\beta_7$-integrin-expressing cells roll on immobilized mucosal vascular addressin cell adhesion molecule 1 (MadCAM-1). Very late antigen 4 (VLA-4 or $\alpha_4\beta_1$-integrin)-expressing lymphocytes can roll on immobilized vascular cell adhesion molecule 1 (VCAM-1).[11] It has also been shown that neutrophils roll on a substrate of co-immobilized E-selectin and intercellular adhesion molecule 1 (ICAM-1). The engagement of E-selectin upregulates lymphocyte function-associated antigen 1 (LFA1; $\alpha_L\beta_2$-integrin), which transiently binds to ICAM-1.[19] In another study, lymphocyte rolling was shown to be slower when ICAM-1 was coexpressed with L-selectin ligands on human vascular endothelial cells.[20] In an *in vivo* study, neutrophils rolled very slowly (<5 µm/s) on E-selectin and ICAM-1 coexpressing endothelium in tumor-necrosis factor (TNF)-stimulated mice.[21] Specifically, it was demonstrated that such slow rolling was mediated not only by E-selectin, but also by the engagement of β_2-integrins (LFA-1 or macrophage receptor 1 [MAC-1; CD11b/CD18 or $\alpha_M\beta_2$-integrin]) with ICAM-1. These studies confirm the potential involvement and stabilizing effect of integrins in leukocyte cell rolling.

8.2.3 Leukocyte Activation and Firm Adhesion

The rolling of leukocytes along the endothelium allows cells to sample the local milieu.[12] If a desirable combination of chemokines is present, certain integrins will be activated to facilitate cell arrest on endothelium. Integrins include a family of two dozen heterodimers whose ligand-binding activity can be rapidly regulated by induced conformational changes and cytoskeleton rearrangement.[11,18] As summarized in Table 8.2, the most relevant integrins for leukocyte adhesion and migration include β_2 integrins (particularly LFA-1 and MAC-1), VLA-4 ($\alpha_4\beta_1$ integrin), and $\alpha_4\beta_7$ integrin. Integrins bind endothelial cell surface receptors belonging to the immunoglobulin "superfamily," such as ICAMs (ligands for LFA-1 and MAC-1), VCAM-1 (ligand for VLA-4), and mucosal addressin cell adhesion molecule 1 (MAdCAM-1) (ligand for $\alpha_4\beta_7$ integrin) as well as certain extracellular matrix (ECM) molecules.

The activation of integrins on endothelium is known to be mediated mainly by chemokines. Chemokines are a superfamily of small proteins,[22] produced by endothelial cells, leukocytes or stromal cells, and function as potent chemotactic agents to regulate cell trafficking and migration to injured/inflamed

TABLE 8.2

Integrins and Endothelial Ligands Involved in Leukocyte Rolling and Arrest

Leukocyte Integrin	Endothelial Ligands	Functions
LFA-1 ($\alpha_L\beta_2$)	ICAM-1 ICAM-2 (?)	Nearly all arrest processes of all leukocyte types
VLA-4 ($\alpha_4\beta_1$)	VCAM-1	Rolling and arrest in chronic inflammation; arrest on inflamed arteries
Mac-1 ($\alpha_M\beta_2$)	ICAM-1	Rolling stabilization and arrest; crawling; extravasation
$\alpha_4\beta_7$	MadCAM-1	Rolling on resting and inflamed gut venules

Source: Adapted by permission from Macmillan Publishers Ltd. Luster, A.D., Alon, R., von Andrian, U.H., Immune cell migration in inflammation: Present and future therapeutic targets, *Nat. Immunol.*, 6, 1182–1190. Copyright 2005.

sites. Table 8.3 summarizes the major chemokines and their ligands involved in leukocyte homing. In addition to their potent chemotactic effects, endothelium bound or soluble chemokines activate integrins on trafficking leukocytes through binding to G-protein-coupled receptors (GPCRs) on the leukocyte membrane. The chemokine/GPCR complex initiates a multipart intracellular signaling network within milliseconds that activates integrins by triggering a conformational change that leads to a higher binding affinity and further lateral mobility to form polarized clusters.[11,18] Note that some membrane partners (i.e., tetraspanins, CD47, CD98, and CD44) also bind to integrins and participate in the integrin activation process by regulating integrin localization, mobility, and signaling pathways.[11] The activation of integrins leads to firm adhesion of leukocytes to the vascular endothelium and the induction of cytoskeleton-driven leukocyte migration, which is responsible for lateral cell migration and paracellular and transcellular migration. Readers are referred to Refs. [11,23] on the detailed mechanism of integrin activation by chemokine/GPCR signaling.

8.2.4 Transmigration/Crossing Vascular and Tissue Barriers

Upon chemokine activation, the arrested leukocytes use the functionally polarized cell-migration machinery (i.e., integrin/ligands interactions and subsequently activated intracellular signaling pathways) to travel along (crawl) the luminal side of vascular endothelial cells and then transmigrate through the endothelial cell barrier and, subsequently, its associated basement membrane and the pericyte sheath.[1,17,18] Intraluminal crawling is known to be primarily mediated by MAC-1/ICAM-1. Regarding transmigration, leukocytes can migrate through endothelial cells via endothelial junctions (paracellular route) and, in few cases, through the body of endothelial cells (transcellular route). This transmigration or "diapedesis" process is known to be regulated by binding of specific integrins and other adhesion

TABLE 8.3

Chemoattractant Receptors and Their Chemokine and Lipid Ligands Implicated in Adhesion and Chemotaxis

Leukocyte Receptor	Ligand	Chemotaxis and Adhesion In Vitro and In Vivo
CXCR1, CXCR2	CXCL1-4, CXCL-8	Neutrophils
CXCR3	CXCL-9-11	Th1 cells, CD8 effectors, NK, NKT cells
CXCR5	CXCL13	B cell and follicular helper CD4 T cells
CXCR4	CXCL12 (SDF-1)	HSCs, T and B cells, platelets
CXCR6	CXCL16	Th1 cells, CD8 effectors, NK, NKT cells
CCR2	CCL2, CCL7-8, CCL12-13	Inflammatory monocytes
CCR3	CCL5, CCL11, CCL13, CCL24, CCL26	Eosinophils and mast cells
CCR4	CCL17, CCL21	Th2 cell, Tregs, skin homing T cell
CCR6	CCL20	Immature DCs, Memory T and B cells
CCR7	CCL19, CCL21	Naïve and effector-memory T cells; mature DCs
CCR8	CCL1	Th2 and Treg cells
CCR9	CCL28	Gut homing T cells
CCR10	CCL27	Skin homing T cells
CX3CR1	CX3CL1	Monocytes, DCs, microglial, NK cells
BLT1	LTB4	Neutrophils, monocytes, eosinophils, mast cells, effector T cells
SIP1	SIP	Naïve T cells and B cells
CRTH2	PGD2	Th2 cells
PAFR	PAF	Neutrophils, eosinophils, monocytes
C5aR	C5a	Neutrophils, eosinophils, monocytes

Source: Adapted by permission from Macmillan Publishers Ltd. Luster, A.D., Alon, R., von Andrian, U.H., Immune cell migration in inflammation: Present and future therapeutic targets, *Nat. Immunol.*, 6, 1182–1190. Copyright 2005.

molecules expressed on leukocytes with receptors on endothelial cells and cell junctions (Table 8.4). These receptors include platelet/endothelial cell adhesion molecule 1 (PECAM-1), ICAM-1, VCAM-1, junction adhesion molecule A (JAM-A), JAM-B, JAM-C, and CD99. Essentially, inflammation-activated endothelial cells, particularly when engaged with (chemokine) activated leukocytes, rearrange their junctional molecules and therefore disassemble the connections (i.e., endothelial-expressed vascular endothelial cadherin [VE-cadherin]) between endothelial cells, which allows leukocytes to migrate through endothelium (paracellular route). While the paracellular route is thought to be the predominant pathway for leukocyte transmigration both *in vitro* and *in vivo*, there is evidence indicating a transcellular route may occur under certain circumstances particularly *in vitro*.[11] It has been suggested

TABLE 8.4
Key Transmigration and ECM Counter Receptors

Leukocyte Receptor	Endothelial/Matrix Ligands	Evidence for Function
LFA-1	ICAM-1, ICAM-2 (?), JAM-A	Blocking mAbs *in vitro*; Blocking mAbs and soluble JAMs *in vitro*
JAM-A	JAM-A	*In vitro*
PECAM-1	PECAM-1	Blocking mAbs *in vitro* and *in vivo* inhibit emigration in a stimulus specific manner. Knockout mice: a defect in crossing of basement membranes
$\alpha_v\beta_3$	PECAM-1	*In vitro*
None	VE-Cadherin	Blocking mAbs *in vivo* accelerates neutrophil transmigration
CD99	CD99	Blocking mAbs *in vitro* memory T cell CD99
VLA-4, JAM-C	JAM-B	Biochemical
Mac-1	Numerous matrix and stromal ligands including ICAM-1, glycosaminoglycans(GAGs) and JAM-C	Neutrophil and monocyte migration; platelet-neutrophil aggregates *in vitro*
VLA-6	Laminin	Neutrophil transmigration *in vivo*
VLA-4,5	Fibronectin	Neutrophil recruitment to inflamed lung
VLA-1,2	Collagens	Lymphocyte retention in inflamed tissues; neutrophil locomotion *in vivo*

Source: Adapted by permission from Macmillan Publishers Ltd. Luster, A.D., Alon, R., von Andrian, U.H., Immune cell migration in inflammation: Present and future therapeutic targets, *Nat. Immunol.*, 6, 1182–1190. Copyright 2005.

that the ligation between ICAM-1 and leukocyte integrins trigger intracellular signaling pathways that result in the rearrangement of cytoskeleton and membrane, which collectively form channels facilitating leukocyte passage through the endothelial cell.[11]

Having passed through the endothelial barrier, leukocytes then penetrate the endothelial basement membrane (comprising networks of proteins such as laminins and collagens) and stromal cell sheath (i.e., pericytes).[11] Similar to endothelial cells, the underlying extracellular matrix (ECM) are subject to remodeling processes during inflammation. In particular, there are regions of low protein deposition within the ECM and gaps between adjacent pericytes where leukocytes can easily pass through. This process is further facilitated by activated integrins on leukocytes such as $\alpha_6\beta_1$-integrin (which is a main receptor for laminin) and proteases (i.e., matrix metalloproteinases [MMPs]) that degrade ECM components.

8.3   STEM CELL HOMING

8.3.1   INTRODUCTION TO STEM CELL HOMING

As leukocyte homing is critical for leukocytes to exert their biological functions (i.e., immune defense), systemically infused stem cells should also find their desirable sites before a therapeutic effect can be applied. However, unlike the well-established leukocyte homing cascade, little is known about the stem cell homing process. It is not clear whether there is anything unique about stem cell homing, although it seems that leukocytes and stem cells might share similar trafficking molecules and cues.[12,24,25] Among various stem cells, HSCs are by far the most investigated because of their successful history in the clinic. In contrast, the study of homing for other stem cells, including MSCs and endothelial progenitor cells (EPCs), are still in their infancy. Similar to the leukocyte homing cascade, the homing of stem cells, particularly HSCs, may also include cell capture and rolling (mainly mediated by selectins and some integrins), cytokine activation and integrin-regulated cell arrest, crawling along endothelium, transendothelial migration, and invasion of underlying extracellular matrix. We will first discuss the techniques to study stem cell homing in Section 8.3.2 and subsequently summarize the adhesion molecules, cytokines, and proteolytic enzymes that are involved in homing of each specific stem cell in detail in Sections 8.3.3 through 8.3.6.

8.3.2   TECHNIQUES TO STUDY STEM CELL HOMING

A variety of *in vitro* and *in vivo* methods have been applied to investigate leukocyte and stem cell homing, however, the accurate and sensitive monitoring of stem cell trafficking and homing, particularly in real-time *in vivo* models, is still an unmet challenge. Commonly used *in vitro* techniques to study stem cell trafficking include flow chamber cell rolling/adhesion assay,[26] chemotaxis/migration assay,[27] trans-matrigel chemoinvasion assay,[27] together with other standard bioassays including DNA microarray, polymerase chain reaction (PCR), flow cytometry, enzyme-linked immunosorbent assay (ELISA), antibody blocking, and western blotting.[27]

In vivo animal models are ultimately used to verify *in vitro* results and investigate how stem cells traffic and home under physiological conditions. In such studies, stem cells are typically labeled with fluorescent and luminescent dyes, magnetic particles, or radioisotopes.[2,28] These probes allow cells to be analyzed histologically in different organs after animals are sacrificed or to be monitored in live animals via *in vivo* imaging techniques. Magnetic resonance imaging (MRI) and x-ray computed tomography (CT) have been used in animal models and the clinic to monitor cell homing but lack sufficient resolution to detect individual cells.[29] Intravital confocal fluorescent microscopy is a powerful tool to monitor stem cell trafficking and homing in real time at a single cell resolution in animal models,[29] but cannot be applied in humans at the present time.

8.3.3   HEMATOPOIETIC STEM/PROGENITOR CELL HOMING

Hematopoietic stem and progenitor cells (collectively referred to as HSCs) are a rare population of precursor cells in the bone marrow that maintain blood cell homeostasis and facilitate innate immune responses. HSC therapy, comprising of bone marrow and peripheral blood stem cell transplantation, has been successfully applied in the clinic for treatment of leukemia and other blood diseases since the 1970s, a major success for stem cell therapy.[4] HSC homing to the bone marrow is a prerequisite for both endogenous HSC functions under healthy physiological conditions and repopulation of transplanted HSCs in the blood system of the patient. However, it wasn't until the 1990s when investigators began to understand the molecular basis for HSC homing to the bone marrow.[30] Research of HSC homing and engraftment continues to this day as HSC transplantation has become an effective treatment and is now routinely used to treat patients with disorders of the blood and immune system.[4,30]

8.3.3.1 HSC Homing: The Rolling, Adhesion Molecules, and Proteolytic Enzymes

Previous studies indicated that HSC homing closely resembles the classical leukocyte homing cascade.[25,31] In the first cell capture/rolling step, cell rolling ligands including PSGL-1 and HCELL, a specialized sialofucosylated glycoform of CD44, expressed on HSCs engage with P-selectin and E-selectin on endothelium to initiate cell rolling (Figure 8.2, Table 8.5). In particular, E- and P-selectin expressions on marrow microvascular endothelium are upregulated under inflammatory conditions. *In vitro* flow chamber assays demonstrated that primary HSCs roll on P- and E-selectins.[32,33] Intravital microscopy experiments further confirmed the important role of P- and E-selectins in HSC rolling. Wild-type mice treated with E-/P-selectin antibodies or in E- and P-selectin knockout mice, HSC rolling was significantly reduced.[33] Interestingly, HSCs also express L-selectin and roll on L-selectin coated substrate. This suggests an important role of L-selectin in the secondary capture of HSCs by endothelium adherent HSCs.[12]

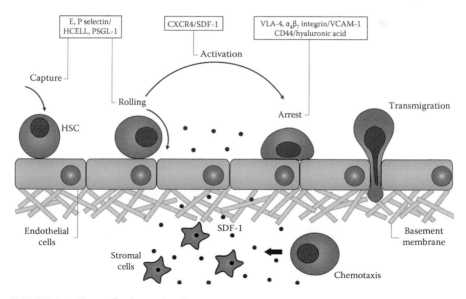

FIGURE 8.2 (See color insert.) Schematic illustration of HSC homing.

TABLE 8.5

Phenotypic Characterization of HSCs Adhesion Molecules That are Known to Be Important in HSC Homing

Adhesion Molecules Expressed on HSCs	Ligands
ICAM-1, CD54	LFA-1
PECAM-1, CD31	PECAM-1, sulfated glycosaminoglycans
LFA-3, CD58	LFA-2
$\alpha_4\beta_1$ (VLA-4, CD49d/CD29)	VCAM-1, fibronectin
$\alpha_5\beta_1$ (VLA-5, CD49e/CD29)	VCAM-1, fibronectin
$\alpha_L\beta_2$ (LFA-1, CD11a/CD18)	ICAM-1,2,3
L-selectin	SLeX, CD34
HCAM, CD44	Hyaluronan, E-selectin, collagen
CD43, leucosialin	ICAM-1
CD34	Bone marrow stromal
SLeX, PSGL-1	E- and P- selectins
HCELL	E-selectin
CD45 and Thy-1	Heparin sulfate, cell membrane-associated proteoglycans; on fibronectin

Source: Data from Sweetenham, J., *Brit. J Haematol.*, 94, 592, 1996.

In addition to the selectin family, it was found that some integrins can also mediate HSC rolling on endothelium. Strikingly, it was found that in irradiated mice marrow, the rolling of murine HSCs was supported only by VCAM-1/VLA-4 without contribution from selectins, as shown by injection of antibodies to P- and E- selectins, which failed to impair rolling.[32] In contrast to murine cells, human CD34+ cells transplanted in P- and E-selectin knockout mice showed dramatically decreased (more than 95%) interactions with marrow endothelium. This limited interaction could not be compensated by the presence of VCAM-1.[34] Overall, it is suggested that human cells use selectins to initiate rolling interactions with bone marrow endothelium, whereas murine cells utilize both integrins and selectins.[35]

In addition to cell rolling, the subsequent stages of HSC homing are also mediated by key adhesion molecules as summarized in Table 8.5. Once HSCs slow down on the endothelium during the initial rolling steps, integrins firmly arrest cells and participate in the crawling, transmigration, and lodging of cells within the bone marrow niche (the activation of integrins will be discussed in the next section). The integrins on HSCs and their counterparts on endothelium and underlying stroma that are known to be crucial for homing include VLA-4/VCAM-1, LFA-1/ICAM-1, VLA-4 / fibronectin, and VLA-5/fibronectin. For instance, it has been shown that HSC homing was inhibited by more than 90% when antibodies were used to block VLA-4 or VCAM-1.[35] Upon firm arrest, a series of complex interactions between HSCs and endothelium lead to transendothelium migration through the action of JAM, cadherins, and PECAM-1. Subsequently, HSCs invade ECMs, migrate on ECM components (i.e., hyaluronic acid, laminin, collagen and fibronectin), and interact with stromal cells through the use of MMPs, integrins, CD44, and chemokines. In particular, MMPs, including MMP-2 and MMP-9, are expressed by HSCs and play important roles in the degradation of ECMs, which facilitates migration through the ECM.[36] After HSCs home to the underlying "niche" microenvironment, they may exert their functions through a complex interplay with neighboring cells, ECM components, cytokines, and growth factors. The stem cell niche, which is not the focus of this chapter, is the topic of several recent excellent reviews.[37,38]

8.3.3.2 HSC Homing: SDF-1/CXCR4 AXIS

Chemokines produced in the bone marrow microenvironment, particularly under "stressed" conditions including injury, inflammation, hypoxia, and bone marrow irritation, attract HSCs to the bone marrow through a transendothelial gradient and support migration across the endothelium and within underlying tissue. In particular, chemokine signaling via their GPCR[11,23] activate important cell adhesion molecules, particularly integrins that are required for HSC homing.

Among all the chemokines and GPCRs, the SDF-1 and CXCR4 axis is known as a key pathway regulating the trafficking and homing of a variety of stem cells including HSCs.[39] SDF-1 is produced, and often upregulated during injury and inflammation, by a variety of cells including endothelial cells and bone marrow stromal cells. SDF-1/CXCR4 signaling plays a predominant role in HSC trafficking and homing, evidenced by incubation of CXCR4 antibodies with CD34+ cells virtually nullifying bone marrow homing in NOD/SCID mice.[40,41]

The complex SDF-1/CXCR4 signaling pathway, which has been the subject of several recent reviews,[25,39] is known to regulate cell adhesion, motility, chemotactic responses, and secretion of MMPs and other cytokines/growth factors. In particular, SDF-1/CXCR4 signaling activates a number of integrins including LFA-1 and VLA-4 that regulate HSC adhesion and transmigration by interacting with ICAM-1 and VCAM-1, respectively. Moreover, SDF-1/CXCR4 signaling affects cell motility and chemotaxis by inducing changes and rearrangement of cytoskeletal proteins. Furthermore, SDF-1/CXCR4 signaling leads to secretion of various growth factors, cytokines, and particularly MMPs (i.e., MMP-2 and MMP-9), which enable arrested HSCs to cross the endothelial and ECM barriers.[25,39]

Interestingly, SDF-1/CXCR4 signaling can be upregulated or downregulated by a variety of positive and negative factors, respectively.[39] Factors that are produced or upregulated during tissue damage and injury including C3a anaphylatoxin, platelet-derived membrane microvesicles, hyaluronic acid, thrombin, fibrinogen, fibronectin, soluble VCAM-1 and ICAM-1, SCF + IL-6, HGF, and hypoxia-induced factors are found to significantly increase SDF-1/CXCR4 axis.[39] This supports the observation that stem cells specifically home to inflamed and injured tissues.[39] In contrast, proteases downregulate SDF-1/CXCR4 signaling through cleavage of functional domains on SDF-1 and/or CXCR4.[39]

In addition to SDF-1/CXCR4, other important chemotactic factors involved in HSC trafficking and homing include SCF/c-kit, granulocyte colony-stimulating factor (G-CSF), GROβ/CXCR2, PTH receptor, granulocyte/macrophage colony-stimulating factor (GM-CSF), IL-8, VEGF, and cysteinyl leukotriene receptor (cysLT1), S1P lyase and receptor, and proteases (i.e., neutrophil elastase; cathepsin G).[42]

8.3.4 Mesenchymal Stem Cell Homing

MSCs, also referred to as mesenchymal multipotent stromal cells, or connective tissue progenitors, have pro-angiogenic and immunomodulatory effects and are capable of differentiating into multiple cell types that can form connective tissues including bone, cartilage, fat, and muscle.[2] MSC therapy, which is currently the focus of clinical trials, have potential utility for treating a variety of diseases and disorders including GVHD, organ transplantation, cardiovascular disease, brain and spinal cord injury, lung, liver and kidney diseases, and skeletal injuries.[2]

Similar to HSC therapy, MSC therapy often depends on systemic administration, which requires homing and transmigration of the cells in order for the cells to persist in the patient. Compared to leukocyte and HSC homing, the knowledge in MSC homing is extremely limited.[2] It has been shown that a small population of systemically infused MSCs home specifically to inflamed/injured/hypoxic tissues (i.e., myocardial infarction, stroke), which suggests that MSC trafficking is regulated by chemokines and adhesion molecules expressed in these regions, although the process is poorly understood.[2,43] MSCs are highly heterogeneous, and their surface marker expression and homing capacity are dependent on the origin of MSCs, culture conditions, confluence, and passage number, which contributes to the large inconsistency regarding MSC properties in the literature.[2] For example, *in vitro* culture-expanded MSCs gradually lose several key homing receptors including SH3, ICAM-1, integrin $\beta1$, which results in loss of MSC's ability to home after repeated passaging.[44]

8.3.4.1 MSC Rolling, Adhesion on and Transmigration through Endothelial Cells

To date, very few reports have investigated MSC rolling and adhesion on endothelial cells *in vitro*. Ruster and coworkers found that MSCs roll on P-selectin-coated glass and on P-selectin-expressing human umbilical vein endothelial cells (HUVECs).[45] It is interesting to note that such rolling is only effective up to 1 dyne/cm^2 with a relatively high rolling velocity (100–500 µm/s), which is in sharp contrast to leukocyte rolling. By using intravital microscopy, the authors observed MSCs rolling on postcapillary venules in wild-type mice while rolling events were significantly reduced on P-selectin$^{-/-}$ mice. Unfortunately, the P-selectin ligands on MSCs that mediated MSC rolling remain to be determined.[45] By contrast, Brooke *et al.*[46] reported that MSCs derived from human bone marrow and placenta do not functionally bind to P- or E-selectins. Another report by Sackstein and coworkers enhanced MSC rolling by enzymatically converting a native CD44 glycoform (a non-E-selectin ligand) on MSCs to HCELL, a potent ligand to E-selectin.[47] They showed that HCELL$^+$ MSCs roll on activated E-selectin-expressing HUVECs *in vitro* and, when intravenously infused *in vivo*, infiltrated marrow within hours of infusion. Interestingly, MSCs express L-selectin, as assessed by RT-PCR; however, the expression of L-selectin on MSC membrane was very low and often absent.[45] Consequently, whether L-selectin plays a role in MSC rolling and homing remains unknown.

MSCs are known to express various integrin molecules (Table 8.6).[48] Pittenger *et al.*[49] reported that MSCs express integrin α_1, α_2, α_3, α_4, α_v, β_1, β_3, and β_4 together with other adhesion molecules ICAM-1, ICAM-3, VCAM-1, ALCAM, CD44, and

TABLE 8.6

Expression of Adhesion Molecules, Growth Factor/Cytokine Receptors, and Digestive Enzymes on Human MSCs That May Be Involved in MSC Homing

Adhesion molecules: ALCAM (activated leukocyte cell adhesion molecule), ICAM-1, ICAM-2, ICAM-3, LFA-3, HCAM, VCAM-1, α smooth muscle actin, STRO-1, BST-1, MUC18, TLR1, Endoglin, Thy-1, tetraspan, vitronectin R β chain, β_4 integrin, VLA-β chain, VLA-$\alpha_{1,2,3,4,5,6}$

Growth factor/cytokine receptors: IL-1R (α and β) (interleukin-1R), IL-3R, IL-4R, IL-6R, IL-7R, Inteferon γ R, TNF-α-1R (tumor necrosis factor-α-1R), TNF-α-2R, FGFR (fibroblast growth factor receptor), PDGFRα (platelet-derived growth factor receptor α), PDGFRβ (platelet-derived growth factor receptor β), transferrin receptor, Flk-1 (fetal liver kinase-1), EGFR (epidermal growth factor receptor), HGFR (hepatocyte growth factor receptor—c-met), IGF1R (insulin-like growth factor-1 receptor), VEGFR1 (vascular endothelial growth factor receptor 1), VEGFR2, Fibroblast growth factor receptor 2 (FGFR2), Tie-1 (tyrosine kinase with immunoglobulin-like and EGF-like domains), CCR1 (chemokine (C-C motif) receptor-1), CCR2, CCR3, CCR4, CCR5, CCR6, CCR7, CCR8, CCR9, CCR10, CXCR1 (cytokine (C-X-C Motif) receptor-1), CXCR2, CXCR3-A/B, CXCR4, CXCR5, CXCR6, CX3CR1, CX3CR6, XCR1, TLR2 (toll like receptor2), TLR3, TLR4, TLR5, TLR6, NGFR (nerve growth factor receptor), TNFRSF6 (tumor necrosis factor receptor super family, member 6)

Digestive enzymes: Aminopeptidase N, 5' terminal nucleotidase, MMP-2, and membrane type 1 MMP (MT1-MMP)

endoglin/CD105. The presence of β_1 integrins is in agreement with Ruster *et al.*[45] who observed expression of VLA-4 on MSCs, which mediate the binding of MSCs to VCAM-1 on HUVECs. Antibody blocking of either VLA-4 or VCAM-1 significantly reduced the adhesion of MSCs on HUVECs. The crucial role of VLA-4/VCAM-1 in MSC adhesion on endothelium has been confirmed by a number of other groups.[50] For instance, Segers *et al.*[51] showed that rat MSCs adhere to cultured cardiac microvascular endothelium (CMECs) in static and dynamic conditions, which is regulated almost completely by VLA-4/VCAM-1 and not by ICAM-1. Moreover, they found that the treatment of MSCs and/or CMECs with inflammatory factors such as TNF-α and IL-1β, which upregulates VCAM-1 and VLA-1 expression, significantly increases *in vivo* cardiac homing of MSCs.[51] In addition to their roles in regulating MSC rolling and adhesion on endothelium, integrins, together with other adhesion molecules particularly CD44, also participate in MSC transendothelial migration (see the following text) and the interactions between MSCs and ECM components.

MSCs are also capable of transmigrating across the endothelium. In a coculture system of MSC and endothelial cell monolayer, Schmidt *et al.*[52,53] revealed that within the first 30 min, ~30% of MSCs transmigrated through the endothelial barrier. The percentage of MSCs transmigrated increased to about 50% at 60 min, after which the percentage of transmigrated MSCs remain unchanged. When MSCs were perfused into an isolated heart and investigated using electron microscopy, the authors observed that the tight junctions between endothelial cells became abolished and MSCs interacted with the endothelial cell layer with tight cell–cell contacts. More recently, the same group revealed that such transmigration process required the interaction between VCAM-1 and VLA-4 as verified by antibody blocking assay.[52]

Proteases serve a critical role in MSC's tranendothelial migration and the invasion into ECM matrix due to their ability to degrade ECM components (e.g., collagen IV).[52,54] MSCs constitutively express MMP-2 and membrane type 1 MMP (MT1-MMP).[54,55] It has been shown that synthetic MMP inhibitors effectively blocked MSCs from traversing reconstituted human basement membranes.[54,55] In addition, knockdown of MMP-2 and MT1-MMP genes substantially impaired MSC invasion. In particular, hypoxia conditions and inflammatory cytokines (TGF-β1, IL-1β, and TNF-α) upregulate MMP-2, MT1-MMP, and/or MMP-9 production in MSCs, resulting in a strong stimulation of chemotactic migration through ECM.[56] Therefore, production of specific MMPs in MSCs under inflammatory conditions promotes cell migration across reconstituted basement membrane *in vitro*, providing a potential mechanism for MSC homing and extravasation into injured tissues *in vivo*.[55]

8.3.4.2   Cytokines

Several chemokines and growth factors have been demonstrated to be important chemotactic factors mediating MSC trafficking and homing (Table 8.6).[57,58] These include SDF-1/CXCR4,[54] monocyte-chemoattractant protein 1 (MCP-1, CCL2)/CCR2, RANTES (CCL5)/CCR3 and CCR5, macrophage-derived cytokine (MDC, CCL22)/CCR4, MCP-3/CCR1 and CCR2, MIP-3α/CCR6, MIP-1α, 1b (CCR5), fractalkine (CX3CL1)/CX3CR1, Tie-1/Ang-1, c-met/HGF, IGF-1, PDGF-bb, bFGF, and VEGF.[54,59] In particular, the preconditioning of MSCs with proinflammatory cytokines (i.e., TNFα) is known to upregulate many chemokine receptors and therefore enhance their migration and homing efficiency.[27]

It has been demonstrated that at least a subset of MSCs expresses CXCR4 and use the SDF-1/CXCR4 axis for trafficking and homing. Abbott *et al.* demonstrated that IV-administered bone marrow–derived MSCs migrate to damaged myocardium by SDF-1/CXCR4 signaling as confirmed with the finding that administration of a specific CXCR4 receptor antagonist, AMD3100, significantly inhibited MSC migration.[60] Furthermore, when the myocardium was transduced with an adenoviral vector containing SDF-1, the number of MSCs detected in the heart was significantly increased.[60] On the contrary, Ip and coworkers have recently shown that blocking the CXCR4 receptor had no impact on MSC migration whereas integrin β1 is primarily responsible for MSC's migration and engraftment to myocardium.[61]

MCP-1 is another chemokine that plays an important role in MSC trafficking. It was demonstrated that the brain tissue extract, obtained from an ischemic brain injury model, contained MCP-1 and was chemotactic for MSCs *in vitro*, which can be blocked by MCP-1 antibody.[62] MCP-1 may exert its effect by engaging its receptor CCR2, which is expressed on MSCs.[62] Braun and coworkers found that MCP-1/CCR2 is indeed important for MSC homing to the heart.[63] Interestingly, they found that this process is critically dependent on the intracellular adaptor molecule FROUNT, which interacts with CCR2. FROUNT was required for clustering CCR2 and cytoskeletal rearrangement. Furthermore, CCL2 is known to be secreted by a variety of cancer cells, which may in fact regulate the migration of MSCs to tumor sites.[63]

Moreover, monocyte chemotactic protein-3 (MCP-3, ligand of CCR1 and CCR2) is known as a myocardial MSC homing factor. Schenk *et al.* found that the local overexpression of MCP-3 recruits MSCs to sites of injured tissue and improves cardiac regeneration.[64] Other chemokines that are involved in MSC trafficking, particularly after TNFα stimulation, include RANTES and MDC.[65] It has been shown that TNFα-primed MSCs express upregulated CCR3 and CCR4 and showed increased migration in response to RANTES and MDC.

8.3.4.3 Growth Factors

MSCs express several growth factor receptors including PDGF-R, FGF-R2, EGF-R, TIE-1, HGF-R (c-met), and IGF-R and demonstrate chemotaxis toward growth factors.[65] Domenech *et al.* systematically studied the *in vitro* migration capacity of human bone marrow–derived MSCs, preincubated with the inflammatory cytokines (IL-1 and TNFα), in response to 16 growth factors.[65] They showed that platelet-derived growth factor-AB (PDGF-AB) and insulin-like growth factor 1 (IGF-1) are the most potent chemoattractants of MSCs.

In support of these data, Li and coworkers demonstrated that IGF-1-induced upregulation of CXCR4 expression on MSCs increased the migratory capacity of the cells toward an *in vitro* SDF-1 gradient.[66] IGF-1 is also capable of stimulating the expression of other chemokine receptors such as CCR5—the RANTES (CCL5) receptor.[65] Other growth factors including basic fibroblast growth factor and VEGF may also participate in MSC homing.[65] For instance, VEGF, secreted by tumor cells and other tumor stromal cells, is known to promote migration of MSCs toward tumor sites.[67]

It has also been reported that MSCs express c-met (receptors for HGF) and are strongly attracted by HGF gradients.[54,56] In particular, under hypoxic conditions, the expression of c-Met is upregulated on MSCs. *In vivo* experiments have showed that local expression of HGF in hind limb ischemia attracts hypoxia preconditioned MSCs, which enhances revascularization. Interestingly, the chemokines/cytokines/growth factors produced by MSCs may facilitate MSC migration and homing through an autocrine mechanism by engaging their respective receptors on MSCs.[65] For example, MSCs cultured in the presence of irradiated tumor cell conditioned media increased expression of VEGFa, VEGFc, PDGFb, and SDF-1.[65] The expression of these growth factors by MSCs may interact with MSC receptors (i.e., SDF-1/CXCR4, VEGF/VEGFR) and regulate their migratory capacity.[65]

8.3.5 Homing of Endothelial Progenitor Cells

EPCs are another progenitor cells found in the bone marrow and peripheral circulation, and they have great potential therapeutic applications particularly in angiogenic cell therapy (i.e., formation of new blood vessels in injured tissues). EPCs are typically isolated using antibody-attached beads and are typically positive for CD34, VEGFR-2, CD133, CD31, CD144 (cadherin), and von Willebrand factor (vWF).[68,69] Vajkoczy *et al.*[70] reported that the homing of ex vivo–expanded embryonic EPCs to tumor sites is mediated by PSGL-1 and P- and E-selectins, suggesting EPCs share a selectin-mediated homing mechanism similar to HSCs. Integrins also play important roles in EPC adhesion and extravasation. It has been shown that β_2 integrins and their interactions with ICAM-1 mediate the adhesion of EPCs to endothelial cell monolayer.[71] In addition, β_2 integrins seem to serve a role in chemokine-induced transendothelial migration of EPCs.[71] The important role of β_2 integrins is further evidenced by *in vivo* studies where the administration of β_2 integrin antibody nearly eliminated EPC engraftment following acute myocardial infarction.[71] Furthermore, in a tumor model, the inhibition of VLA-4 significantly reduced the homing of EPCs to sites of active tumor neovascularization, indicating an important role of VLA-4 in EPC homing.[72] Finally, proteases are known to be involved in EPC invasion of underlined ECM. In particular, it has been shown that cathepsin L is required for EPC invasion and proteolytic matrix-degrading activity.[69] Specifically, the improvement of neovascularization after hind limb ischemia was significantly impaired in cathepsin L$^{-/-}$ mice, and infused cathepsin L$^{-/-}$ progenitor cells failed to home to sites of ischemia and to augment neovascularization.[69]

In addition to adhesion molecules and proteases, chemokines and growth factors also participate in EPC homing. CXCR4/SDF-1 is recognized as an important factor for the trafficking of EPCs to ischemic tissues.[69] It has been shown that SDF-1 secreted from carcinoma-associated fibroblasts promotes angiogenesis by recruiting EPCs to tumor tissue.[69] The inhibition of CXCR4 using antibodies significantly reduced SDF-1-induced adhesion of EPCs on confluent endothelial cell monolayers, the migration of EPCs *in vitro*,[73] and the *in vivo* homing of EPCs to ischemic sites.[74] Moreover, VEGF acts as a chemoattractant to EPCs.[75] Interestingly, VEGF can induce recruitment of EPCs from bone marrow to ischemic sites via induction of SDF-1 expression by perivascular myofibroblasts, indicating that different cytokines may cooperate during EPC homing.[76] Other chemokines and growth factors that are known to mediate EPC trafficking include IL-8, MCP-1, high-mobility group box-1, and IGF2.[69]

8.3.6   Homing of Circulating Cancer (Stem) Cells

Cancer cells acquire genetic mutations that enable them to spread to secondary locations via blood and lymphatic systems. Cancer cells are capable of shedding from the primary tumor sites, intravasating into the blood circulation, and can home to a new niche to form secondary tumors, a process called cancer metastasis.[77] Among cancer cells, there might be a subset of so-called cancer stem cells that share common characteristics of normal stem cells and, in particular, are more resistant to chemotherapeutic drugs and are capable of supporting regrowth of new tumors after unsuccessful treatment.[24] Therefore, the understanding of the homing process of circulating cancer (stem) cells (CSCs) may be of significant importance in treating cancer metastasis.[24] Unfortunately, our knowledge of CSC homing is very limited. Regardless, the available data revealed that CSC homing may involve similar mechanisms with that of HSCs.[24] In particular, the SDF-1/CXCR4 axis appears to play a pivotal role in CSC homing. Ratajczak and coworkers[39] suggested that most, if not all, malignancies originate in the stem/progenitor cell compartment, CSCs also express CXCR4 on their surface, as a result, the SDF-1/CXCR4 axis is also responsible for CSC homing to organs that highly express SDF-1 (e.g., lymph nodes, bones, liver, lungs, etc.). It is important to note that, as mentioned in Section 8.3.3, in addition to its chemotactic and mobilization effect, SDF-1/CXCR4 signaling has a complex impact on regulating cell trafficking that includes activation/upregulation of adhesion molecules (i.e., integrins) and secretion of MMPs and other cytokines, which may collectively regulate CSC homing.[24,39]

8.3.7 HOMING OF OTHER STEM/PROGENITOR CELLS

Other stem/progenitor cells that can find direct therapeutic applications include neuron progenitor cells, cardiac stem cell, and skeletal muscle satellite cell.[78] After systemic administration, these cells may employ similar homing cascades and molecular cues that have been described for leukocytes, HSCs, and MSCs.[78] It is likely that each cell type adopts distinct trafficking patterns that dependent on both markers expressed on the cell surface and the target niche microenvironment. Interested readers are referred to some excellent reviews on the homing of each individual stem/progenitor cell.[1,78]

8.4 ENGINEERED STEM CELL HOMING

While stem cells appear to possess some native homing capability, the delivery of systemically infused stem cells to target sites is typically not efficient, which represents a challenge in stem cell–based therapy.[2] Therefore, a number of strategies have been proposed to engineer stem cells to achieve optimal homing efficiency.

First, stem cells can be conditioned by certain cytokines in the culture medium to upregulate key adhesion molecules or receptors required for efficient homing. For instance, treatment of human CD34+ HSCs with SDF-1 increases surface expression of integrins, VLA-4, VLA-5, and LFA-1 and increases adhesion of cells to bone marrow endothelium.[79] In addition, as stated previously, preconditioning of stem cells with proinflammatory factors such as TNFα often leads to upregulation of homing molecules, which enhances their homing capability.

Second, homing receptors can be introduced on stem cells through genetic engineering and enzymatic modification. For example, retrovirus vectors encoding homing receptors such as CXCR4,[80] or the α4 subunit of the VLA-4-integrin[50] have recently been used. Alternatively, Sackstein et al. have shown enzymatic engineering of surface glycans of MSCs enable them to home to bone more efficiently than unmodified MSCs.[47]

Finally, homing receptors can be directly conjugated onto stem cells using chemical approaches.[26,81] For instance, Sarkar et al.[26] have recently developed a simple platform technology to chemically attach cell adhesion molecules to MSC surface to improve homing efficiency to specific tissues. Specifically, as shown in Figures 8.3 and 8.4, the chemical approach involves a stepwise process including (1) treatment of cells with sulfonated biotinyl-N-hydroxy-succinimide to introduce biotin groups on the cell surface, (2) addition of streptavidin that binds to the biotin on the cell surface and presents unoccupied binding sites, and (3) attachment of biotinylated targeting ligands. In this model system, a biotinylated cell rolling ligand, SLeX, is conjugated on the MSC surface. The SLeX-engineered MSCs exhibit a rolling response on P-selectin-coated substrates under shear stress conditions, indicating their potential utility in targeting P-selectin-expressing endothelium in the bone marrow or at sites

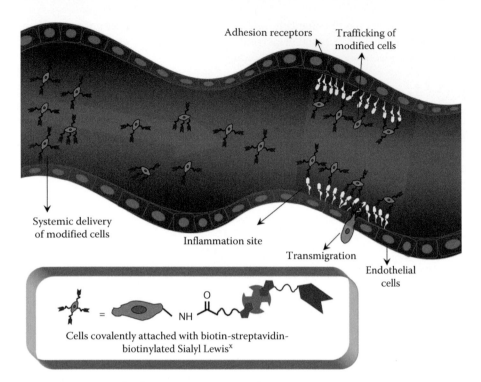

FIGURE 8.3 (See color insert.) Schematic of the homing process of systemically infused SLeX-engineered MSCs to P-selectin-expressing endothelium under inflammatory conditions. (Reproduced with permission from Sarkar, D., Vemula, P.K., Teo, G.S.L., Spelke, D., Karnik, R., Wee, L.Y., Karp, J.M., Chemical engineering of mesenchymal stem cells to induce a cell rolling response, *Bioconj. Chem.*, 19(11), 2105–2109. Copyright 2008 American Chemical Society.)

of inflammation.[26] In another report, Dennis *et al.* coated MSC surface with ICAM-1 antibody using palmitate protein G as a bridge.[81] They showed that the antibody-modified MSCs can target TNFα-activated HUVECs that express ICAM-1. Note that these approaches to modify the cell surface and immobilize required ligands are not limited to MSCs or the SLeX ligand/ICAM-1 antibodies. Rather it could have broad implications on cell therapies that utilize systemic administration and require targeting of cells to specific tissues.[26]

8.5 CONCLUSIONS AND PERSPECTIVES

Stem cell homing is an important phenomenon in stem cell biology and stem cell–based therapy. In particular, a better understanding of stem cell homing and stem cell engineering has enormous potential in regenerative medicine. Regarding the homing mechanisms, the traditional well-established leukocyte homing cascade is often used as a reference frame to gauge stem cell homing. However, whether stem cell homing has anything unique (for example, associated to their "stemness"), compared to

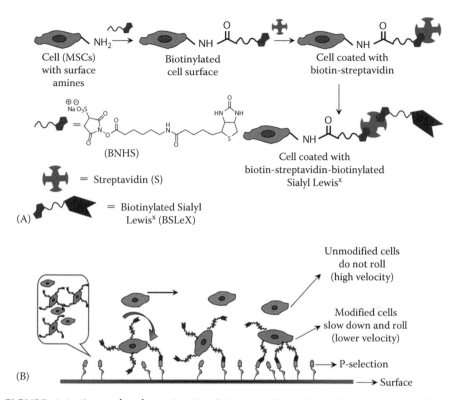

FIGURE 8.4 (See color insert.) (A) Schematic illustration of preparation of SLeX-engineered MSCs and (B) SLeX modification induces a MSC rolling response on P-selectin coated surface. (Reproduced with permission from Sarkar, D., Vemula, P.K., Teo, G.S.L., Spelke, D., Karnik, R., Wee, L.Y., Karp, J.M., Chemical engineering of mesenchymal stem cells to induce a cell rolling response, *Bioconj Chem.*, 19(11), 2105–2109. Copyright 2008 American Chemical Society.)

leukocyte homing, is unclear. Regardless, it seems that, based on the available data, stem cell homing employs similar molecular cues to home (i.e., adhesion molecules, cytokines and proteases), with perhaps distinct combinations and timing patterns. It is important to note that in most, if not all, cases, stem cell homing is not regulated by a single factor, but by a complex interplay of adhesion molecules, cytokines, chemokines, growth factors, proteases, ECM components, and niche cells. The study of stem cell homing in the future will need to take all these factors into account.

Of particular interest, stem cells often exhibit enhanced homing efficiency to the injured and inflamed tissues and tumor sites. This is apparently due to the specific "cues" in damaged sites, i.e., activation and secretion of adhesion molecules and cytokines, which recruit stem cells. This not only gets stem cells to the damaged tissues for repair or regeneration, but also provides a unique means of using stem cells as a gene/drug delivery vehicle. For instance, Marini and coworkers have applied MSCs to deliver anticancer agents (IFN-β expression vector) to tumor sites taking advantage of the tumor homing tropism of MSCs.[82]

In contrast to stem cell therapy where an increased homing efficiency is preferred, for cancer metastasis, the goal would be to interfere with the homing process to prevent cancer (stem) cells from seeding secondary tumor sites. Therefore, the further understanding of cell homing has implications for the development of new therapies that have the potential to treat a multitude of diseases.

REFERENCES

1. Laird, D. J.; von Andrian, U. H.; Wagers, A. J., Stem cell trafficking in tissue development, growth, and disease. *Cell* 2008, *132*(4), 612–630.
2. Karp, J. M.; Teo, G., Mesenchymal stem cell homing: The devil is in the details. *Cell Stem Cell* 2009, *4*(3), 206–216.
3. Lee, R. H.; Pulin, A. A.; Seo, M. J.; Kota, D. J.; Ylostalo, J.; Larson, B. L.; Semprun-Prieto, L.; Delafontaine, P.; Prockop, D. J., Intravenous hMSCs improve myocardial infarction in mice because cells embolized in lung are activated to secrete the anti-inflammatory protein TSG-6. *Cell Stem Cell* 2009, *5*(1), 54–63.
4. Ho, A.; Haas, R.; Champlin, R., *Hamatopoietic Stem Cell Transplantation*. Marcel Dekker Inc., New York, 2005.
5. Thomson, J. A.; Itskovitz-Eldor, J.; Shapiro, S. S.; Waknitz, M. A.; Swiergiel, J. J.; Marshall, V. S.; Jones, J. M., Embryonic stem cell lines derived from human blastocysts. *Science* 1998, *282*(5391), 1145–1147.
6. Takahashi, K.; Yamanaka, S., Induction of pluripotent stem cells from mouse embryonic and adult fibroblast cultures by defined factors. *Cell* 2006, *126*, 663–677.
7. Discher, D. E.; Mooney, D. J.; Zandstra, P. W., Growth factors, matrices, and forces combine and control stem cells. *Science* 2009, *324*(5935), 1673–1677.
8. Muschler, G.; Nakamoto, C.; Griffith, L., Engineering principles of clinical cell-based tissue engineering. *J. Bone Joint Surg. Am.* 2004, *86*, 1541–1558.
9. Yamanaka, S., Strategies and new developments in the generation of patient-specific pluripotent stem cells. *Cell Stem Cell* 2007, *1*, 39–49.
10. Lee, H.; Park, J.; Forget, B.; Gaines, P., Induced pluripotent stem cells in regenerative medicine: An argument for continued research on human embryonic stem cells. *Regen. Med.* 2009, *4*, 759–769.
11. Ley, K.; Laudanna, C.; Cybulsky, M. I.; Nourshargh, S., Getting to the site of inflammation: The leukocyte adhesion cascade updated. *Nat. Rev. Immunol.* 2007, *7*(9), 678–689.
12. Sackstein, R., The bone marrow is akin to skin: Hcell and the biology of hematopoietic stem cell homing. *J. Investig. Dermatol.* 2004, *122*(5), 1061–1069.
13. Kansas, G., Selectins and their ligands: Current concepts and controversies. *Blood* 1996, *88*(9), 3259–3287.
14. Springer, T., Traffic signals on endothelium for lymphocyte recirculation and leukocyte emigration. *Annu. Rev. Physiol.* 1995, *57*, 827–872.
15. Bargatze, R. F; Kurk, S.; Butcher, E. C.; Jutila, M. A., Neutrophils roll on adherent neutrophils bound to cytokine-induced endothelial cells via L-selectin on the rolling cells. *J. Exp. Med.* 1994, *180*, 1785–1792.
16. Hidalgo, A.; Peired, A. J.; Wild, M. K.; Vestweber, D.; Frenette, P. S., Complete identification of E-selectin ligands on neutrophils reveals distinct functions of PSGL-1, ESL-1, and CD44. *Immunity* 2007, *26*(4), 477–489.
17. Springer, T. A., Traffic signals for lymphocyte recirculation and leukocyte emigration the multistep paradigm. *Cell* 1994, *76*, 301–314.
18. Luster, A. D.; Alon, R.; von Andrian, U. H., Immune cell migration in inflammation: Present and future therapeutic targets. *Nat. Immunol.* 2005, *6*, 1182–1190.

19. Chesnutt, B. C.; Smith, D. F.; Raffler, N. A.; Smith, M. L.; White, E. J.; LEY, K., Induction of LFA-1-dependent neutrophil rolling on ICAM-1 by engagement of E-selectin. *Microcirculation* 2006, *13*(2), 99–109.

20. Kadono, T.; Venturi, G. M.; Steeber, D. A.; Tedder, T. F., Leukocyte rolling velocities and migration are optimized by cooperative L-selectin and intercellular adhesion molecule-1 functions. *J. Immunol.* 2002, *169*(8), 4542–4550.

21. Kunkel, E. J.; Ley, K., Distinct phenotype of E-selectin–deficient mice: E-selectin is required for slow leukocyte rolling in vivo. *Circ. Res.* 1996, *79*(6), 1196–1204.

22. Zlotnik, A.; Yoshie, O., Chemokines: A new classification system and their role in immunity. *Immunity* 2000, *12*(2), 121–127.

23. Imhof, B. A.; Aurrand-Lions, M., Adhesion mechanisms regulating the migration of monocytes. *Nat. Rev. Immunol.* 2004, *4*(6), 432–444.

24. Gazitt, Y., Homing and mobilization of hematopoietic stem cells and hematopoietic cancer cells are mirror image processes, utilizing similar signaling pathways and occurring concurrently: Circulating cancer cells constitute an ideal target for concurrent treatment with chemotherapy and antilineage-specific antibodies. *Leukemia* 2003, *18*(1), 1–10.

25. Lapidot, T.; Dar, A.; Kollet, O., How do stem cells find their way home? *Blood* 2005, *106*(6), 1901–1910.

26. Sarkar, D.; Vemula, P. K.; Teo, G. S. L.; Spelke, D.; Karnik, R.; Wee, L. Y.; Karp, J. M., Chemical engineering of mesenchymal stem cells to induce a cell rolling response. *Bioconj. Chem.* 2008, *19*(11), 2105–2109.

27. Honczarenko, M.; Le, Y.; Swierkowski, M.; Ghiran, I.; Glodek, A.; Silberstein, L., Human bone marrow stromal cells express a distinct set of biologically functional chemokine receptors. *Stem Cells* 2006, *24*, 1030–1041.

28. Fox, J. M.; Chamberlain, G.; Ashton, B. A.; Middleton, J., Recent advances into the understanding of mesenchymal stem cell trafficking. *Brit. J. Haematol.* 2007, *137*(6), 491–502.

29. Halin, C.; Mora, J.; Sumen, C.; von Andrian, U., In vivo imaging of lymphocyte trafficking. *Annu. Rev. Cell Dev. Biol.* 2005, *21*, 581–603.

30. Sweetenham, J., Haempoietic progenitor homing and mobilization. *Brit. J. Haematol.* 1996, *94*, 592–596.

31. Tsvee, L.; Isabelle, P., Current understanding of stem cell mobilization: The roles of chemokines, proteolytic enzymes, adhesion molecules, cytokines, and stromal cells. *Exp. Hematol.* 2002, *30*(9), 973–981.

32. Mazo, I. B.; Gutierrez-Ramos, J.-C.; Frenette, P. S.; Hynes, R. O.; Wagner, D. D.; von Andrian, U. H., Hematopoietic progenitor cell rolling in bone marrow microvessels: Parallel contributions by endothelial selectins and vascular cell adhesion molecule 1. *J. Exp. Med.* 1998, *188*(3), 465–474.

33. Mazo, I. B.; Quackenbush, E. J.; Lowe, J. B.; von Andrian, U. H., Total body irradiation causes profound changes in endothelial traffic molecules for hematopoietic progenitor cell recruitment to bone marrow. *Blood* 2002, *99*(11), 4182–4191.

34. Hidalgo, A.; Weiss, L. A.; Frenette, P. S., Functional selectin ligands mediating human CD34+ cell interactions with bone marrow endothelium are enhanced postnatally. *J. Clin. Invest.* 2002, *110*(4), 559–569.

35. Papayannopoulou, T., Bone marrow homing: The players, the playfield, and their evolving roles. *Curr. Opin. Hematol.* 2003, *10*(3), 214–219.

36. Heissig, B.; Hattori, K.; Dias, S.; Friedrich, M.; Ferris, B.; Hackett, N. R.; Crystal, R. G.; Besmer, P.; Lyden, D.; Moore, M. A. S.; Werb, Z.; Rafii, S., Recruitment of stem and progenitor cells from the bone marrow niche requires MMP-9 mediated release of kit-ligand. 2002, *109*(5), 625–637.

37. Moore, K. A.; Lemischka, I. R., Stem cells and their niches. *Science* 2006, *311*(5769), 1880–1885.

38. Li, L.; Xie, T., Stem cell niche: Structure and function. *Annu. Rev. Cell Dev. Biol.* 2005, *21*, 605–631.

39. Kucia, M.; Reca, R.; Miekus, K.; Wanzeck, J.; Wojakowski, W.; Janowska-Wieczorek, A.; Ratajczak, J.; Ratajczak, M. Z., Trafficking of normal stem cells and metastasis of cancer stem cells involve similar mechanisms: Pivotal role of the SDF-1-CXCR4 Axis. *Stem Cells* 2005, *23*(7), 879–894.

40. Sipkins, D.; Wei, X.; Wu, J.; Runnels, J.; Cote, D.; Means, T.; Luster, A.; Scadden, D.; Lin, C., In vivo imaging of specialized bone marrow endothelial microdomains for tumour engraftment. *Nature* 2005, *435*, 969–973.

41. Kollet, O.; Spiegel, A.; Peled, A.; Petit, I.; Byk, T.; Hershkoviz, R.; Guetta, E.; Barkai, G.; Nagler, A.; Lapidot, T., Rapid and efficient homing of human CD34+CD38{-}/low-CXCR4+ stem and progenitor cells to the bone marrow and spleen of NOD/SCID and NOD/SCID/B2mnull mice. *Blood* 2001, *97*(10), 3283–3291.

42. Schulz, C.; von Andrian, U.; Massberg, S., Hematopoietic stem and progenitor cells: Their mobilization and homing to bone marrow and peripheral tissue. *Immunol. Res.* 2009, *44*(1), 160–168.

43. Spaeth, E.; Klopp, A.; Dembinski, J.; Andreeff, M.; Marini, F., Inflammation and tumor microenvironments: Defining the migratory itinerary of mesenchymal stem cells. *Gene. Ther.* 2008, *15*(10), 730–738.

44. Rombouts, W. J. C.; Ploemacher, R. E., Primary murine MSC show highly efficient homing to the bone marrow but lose homing ability following culture. *Leukemia* 2003, *17*(1), 160–170.

45. Ruster, B.; Gottig, S.; Ludwig, R. J.; Bistrian, R.; Muller, S.; Seifried, E.; Gille, J.; Henschler, R., Mesenchymal stem cells display coordinated rolling and adhesion behavior on endothelial cells. *Blood* 2006, *108*(12), 3938–3944.

46. Stock, J. K.; Giadrossi, S.; Casanova, M.; Brookes, E.; Vidal, M.; Koseki, H.; Brockdorff, N.; Fisher, A. G.; Pombo, A., Ring1-mediated ubiquitination of H2A restrains poised RNA polymerase II at bivalent genes in mouse ES cells. *Nat. Cell Biol.* 2007, *9*(12), 1428–1435.

47. Sackstein, R.; Merzaban, J. S.; Cain, D. W.; Dagia, N. M.; Spencer, J. A.; Lin, C. P.; Wohlgemuth, R., Ex vivo glycan engineering of CD44 programs human multipotent mesenchymal stromal cell trafficking to bone. *Nat. Med.* 2008, *14*(2), 181–187.

48. Meirelles, L. d. S.; Caplan, A. I.; Nardi, N. B., In search of the in vivo identity of mesenchymal stem cells. *Stem Cells* 2008, *26*(9), 2287–2299.

49. Pittenger, M.; Mackay, A.; Beck, S.; Jaiswal, R.; Douglas, R.; Mosca, J.; Moorman, M.; Simonetti, D.; Craig, S.; Marshak, D., Multilineage potential of adult human mesenchymal stem cells. *Science* 1999, *284*, 143–147.

50. Kumar, S.; Ponnazhagan, S., Bone homing of mesenchymal stem cells by ectopic alpha 4 integrin expression. *FASEB J.* 2007, *21*(14), 3917–3927.

51. Segers, V. F. M.; Van Riet, I.; Andries, L. J.; Lemmens, K.; Demolder, M. J.; De Becker, A. J. M. L.; Kockx, M. M.; De Keulenaer, G. W., Mesenchymal stem cell adhesion to cardiac microvascular endothelium: Activators and mechanisms. *Am. J. Physiol. Heart Circ. Physiol.* 2006, *290*(4), H1370–H1377.

52. Steingen, C.; Brenig, F.; Baumgartner, L.; Schmidt, J.; Schmidt, A.; Bloch, W., Characterization of key mechanisms in transmigration and invasion of mesenchymal stem cells. *J. Mole. Cell Cardiol.* 2008, *44*(6), 1072–1084.

53. Schmidt, A.; Ladage, D.; Steingen, C.; Brixius, K.; Schinköthe, T.; Klinz, F.-J.; Schwinger, R. H. G.; Mehlhorn, U.; Bloch, W., Mesenchymal stem cells transmigrate over the endothelial barrier. *Eur. J. Cell Biol.* 2006, *85*(11), 1179–1188.

54. Son, B.-R.; Marquez-Curtis, L. A.; Kucia, M.; Wysoczynski, M.; Turner, A. R.; Ratajczak, J.; Ratajczak, M. Z.; Janowska-Wieczorek, A., Migration of bone marrow and cord blood mesenchymal stem cells in vitro is regulated by stromal-derived

factor-1-CXCR4 and hepatocyte growth Factor-c-met axes and involves matrix metal-loproteinases. *Stem Cells* 2006, *24*(5), 1254–1264.

55. Ries, C.; Egea, V.; Karow, M.; Kolb, H.; Jochum, M.; Neth, P., MMP-2, MT1-MMP, and TIMP-2 are essential for the invasive capacity of human mesenchymal stem cells: Differential regulation by inflammatory cytokines. *Blood* 2007, *109*(9), 4055–4063.

56. Rosová, I.; Dao, M.; Capoccia, B.; Link, D.; Nolta, J. A., Hypoxic preconditioning results in increased motility and improved therapeutic potential of human mesenchymal stem cells. *Stem Cells* 2008, *26*(8), 2173–2182.

57. Chamberlain, G.; Fox, J. M.; Ashton, B. A.; Middleton, J., Concise review: Mesenchymal stem cells: Their phenotype, differentiation capacity, immunological features, and potential for homing. *Stem Cells* 2007, *25*(11), 2739–2749.

58. Sordi, V.; Malosio, M. L.; Marchesi, F.; Mercalli, A.; Melzi, R.; Giordano, T.; Belmonte, N.; Ferrari, G.; Leone, B. E.; Bertuzzi, F.; Zerbini, G.; Allavena, P.; Bonifacio, E.; Piemonti, L., Bone marrow mesenchymal stem cells express a restricted set of functionally active chemokine receptors capable of promoting migration to pancreatic islets. *Blood* 2005, *106*(2), 419–427.

59. Ji, J. F.; He, B. P.; Dheen, S. T.; Tay, S. S. W., Interactions of chemokines and chemokine receptors mediate the migration of mesenchymal stem cells to the impaired site in the brain after hypoglossal nerve injury. *Stem Cells* 2004, *22*(3), 415–427.

60. Abbott, J.; Huang, Y.; Liu, D.; Hickey, R.; Krause, D.; Giordano, F., Stromal cell-derived factor-1alpha plays a critical role in stem cell recruitment to the heart after myocardial infarction but is not sufficient to induce homing in the absence of injury. *Circulation* 2004, *110*, 3300–3305.

61. Ip, J. E.; Wu, Y.; Huang, J.; Zhang, L.; Pratt, R. E.; Dzau, V. J., Mesenchymal stem cells use integrin beta1 not CXC chemokine receptor 4 for myocardial migration and engraftment. *Mol. Biol. Cell* 2007, *18*(8), 2873–2882.

62. Wang, L.; Li, Y.; Chen, J.; Gautam, S.; Zhang, Z.; Lu, M., Ischemic cerebral tissue and MCP-1 enhance rat bone marrow stromal cell migration in interface culture. *Exp. Hematol.* 2002, 30(7), 831–836.

63. Belema-Bedada, F.; Uchida, S.; Martire, A.; Kostin, S.; Braun, T., Efficient homing of multipotent adult mesenchymal stem cells depends on FROUNT-mediated clustering of CCR2. *Cell Stem Cell* 2008, *2*(6), 566–575.

64. Schenk, S.; Mal, N.; Finan, A.; Zhang, M.; Kiedrowski, M.; Popovic, Z.; McCarthy, P. M.; Penn, M. S., Monocyte chemotactic protein-3 is a myocardial mesenchymal stem cell homing factor. *Stem Cells* 2007, *25*(1), 245–251.

65. Ponte, A. L.; Marais, E.; Gallay, N.; Langonné, A.; Delorme, B.; Hérault, O.; Charbord, P.; Domenech, J., The *in vitro* migration capacity of human bone marrow mesenchymal stem cells: Comparison of chemokine and growth factor chemotactic activities. *Stem Cells* 2007, *25*(7), 1737–1745.

66. Li, Y.; Yu, X.; Lin, S.; Li, X.; Zhang, S.; Song, Y. H., Insulin-like growth factor 1 enhances the migratory capacity of mesenchymal stem cells. *Biochem. Biophys. Res. Commun.* 2007, *356*(3), 780–784.

67. Brower, V., Search and destroy: Recent research exploits adult stem cells' attraction to cancer. *J. Natl. Cancer Inst.* 2005, *97*, 414–416.

68. Hristov, M.; Erl, W.; Weber, P. C., Endothelial progenitor cells: Mobilization, differentiation, and homing. *Arterioscler Thromb. Vasc. Biol.* 2003, *23*(7), 1185–1189.

69. Urbich, C.; Chavakis, E.; Dimmeler, S., Homing and differentiation of endothelial progenitor cells. In *Tumor Angiogenesis,* Springer, Berlin, Heidelberg, Germany, 2008.

70. Vajkoczy, P.; Blum, S.; Lamparter, M.; Mailhammer, R.; Erber, R.; Engelhardt, B.; Vestweber, D.; Hatzopoulos, A. K., Multistep nature of microvascular recruitment of ex vivo-expanded embryonic endothelial progenitor cells during tumor angiogenesis. *J. Exp. Med.* 2003, *197*(12), 1755–1765.

71. Chavakis, E.; Aicher, A.; Heeschen, C.; Sasaki, K.; Kaiser, R.; Makhfi, N.; Urbich, C.; Peters, T.; Scharffetter-Kochanek, K.; Zeiher, A.; Chavakis, T.; Dimmeler, S., Role of ß2-integrins for homing and neovascularization capacity of endothelial progenitor cells. *J. Exp. Med.* 2005, *201*, 63–72.

72. Li, C.; Zhou, J.; Shi, G.; Ma, Y.; Yang, Y.; Gu, J.; Yu, H.; Jin, S.; Wei, Z.; Chen, F.; Jin, Y., Pluripotency can be rapidly and efficiently induced in human amniotic fluid-derived cells. *Hum. Mol. Genet.* 2009, *18*(22), 4340–4349.

73. Ceradini, D.; Kulkarni, A.; Callaghan, M.; Tepper, O.; Bastidas, N.; Kleinman, M.; Capla, J.; Galiano, R.; Levine, J.; Gurtner, G., Progenitor cell trafficking is regulated by hypoxic gradients through HIF-1 induction of SDF-1. *Nat. Med.* 2004, *10*, 858–864.

74. Armstrong, L.; Hughes, O.; Yung, S.; Hyslop, L.; Stewart, R.; Wappler, I.; Peters, H.; Walter, T.; Stojkovic, P.; Evans, J.; Stojkovic, M.; Lako, M., The role of PI3K/AKT, MAPK/ERK and NFkappabeta signalling in the maintenance of human embryonic stem cell pluripotency and viability highlighted by transcriptional profiling and functional analysis. *Hum. Mol. Genet.* 2006, *15*(11), 1894–1913.

75. Kalka, C.; Masuda, H.; Takahashi, T.; Gordon, R.; Tepper, O.; Gravereaux, E.; Pieczek, A.; Iwaguro, H.; Hayashi, S.; Isner, J.; Asahara, T., Vascular endothelial growth Factor165 gene transfer augments circulating endothelial progenitor cells in human subjects. *Circ. Res.* 2000, *86*, 1198–1202.

76. Grunewald, M.; Avraham, I.; Dor, Y.; Bachar-Lustig, E.; Itin, A.; Jung, S.; Chimenti, S.; Landsman, L.; Abramovitch, R.; Keshet, E., VEGF-induced adult neovascularization: Recruitment, retention, and role of accessory cells. *Cell* 2006, *124*, 175–189.

77. Alix-Panabières, C.; Riethdorf, S.; Pantel, K., Circulating tumor cells and bone marrow micrometastasis. *Clin. Cancer Res.* 2008, *14*(16), 5013–5021.

78. Smart, N.; Riley, P., The stem cell movement. *Circ. Res.* 2008, *102*, 1155–1168.

79. Peled, A.; Kollet, O.; Ponomaryov, T.; Petit, I.; Franitza, S.; Grabovsky, V.; Slav, M.; Nagler, A.; Lider, O.; Alon, R.; Zipori, D.; Lapidot, T., The chemokine SDF-1 activates the integrins LFA-1, VLA-4, and VLA-5 on immature human CD34+ cells: Role in transendothelial/stromal migration and engraftment of NOD/SCID mice. *Blood* 2000, *95*, 3289–3296.

80. Mali, P.; Ye, Z.; Hommond, H. H.; Yu, X.; Lin, J.; Chen, G.; Zou, J.; Cheng, L., Improved efficiency and pace of generating induced pluripotent stem cells from human adult and fetal fibroblasts. *Stem Cells* 2008, *26*(8), 1998–2005.

81. Ko, I.; Kean, T.; Dennis, J., Targeting mesenchymal stem cells to activated endothelial cells. *Biomaterials* 2009, *30*, 3702–3710.

82. Studeny, M.; Marini, F. C.; Champlin, R. E.; Zompetta, C.; Fidler, I. J.; Andreeff, M., Bone Marrow-derived mesenchymal stem cells as vehicles for interferon-beta delivery into tumors. *Cancer Res.* 2002, *62*(13), 3603–3608.

9 *In Vitro* Vascular Tissue Engineering

Jeffrey J.D. Henry and Song Li

CONTENTS

9.1 INTRODUCTION

Atherosclerosis, which causes narrowing of the artery wall, is one of the most severe forms of cardiovascular disease. Atherosclerosis accounts for more than 50% of all mortalities in the United States, Europe, and Japan.[1] Bypass surgery is the most common surgical intervention for advanced atherosclerosis and each year more than 500,000 vascular graft surgeries are performed. Autologous vessels, namely the saphenous vein or internal mammary artery, are commonly used for bypass surgeries. However, at least 30% of patients lack a suitable vessel for bypass due to size mismatch or removal during a previous surgery.[2] Synthetic graft materials such as Dacron and expanded polytetrafluoroethylene (ePTFE) have been successfully used in vascular bypass graft applications, but only when vessel diameters are *greater than 6 mm*. When the graft diameter is *smaller than 6 mm,* as is the case for coronary artery bypass and peripheral artery anastomosis, these grafts are prone to clogging due to thrombosis and stenosis. Thus, there remains a pressing need to engineer functional small diameter arterial replacements.

Advancements made in the field of tissue engineering have made it possible to create living vascular grafts that mimic native vessels in structure and function. In the past two decades, several approaches have been taken to construct vascular grafts by using cells alone, matrix alone, or a combination of matrix and cells. Despite the progress in many aspects, the development of a functional small diameter vascular graft has continued to be one of the most sought after goals within the field. This chapter serves to discuss the requirements for tissue-engineered arteries, the individual cell and matrix components used for construction as well as some of the enabling technologies.

9.2   DESIRED PROPERTIES OF TISSUE-ENGINEERED VASCULAR GRAFTS

9.2.1   REPLICATION OF BLOOD VESSEL STRUCTURE

A straight forward approach to construct tissue-engineered vascular grafts (TEVGs) is to engineer the structure similar to that of native blood vessels. Arteries and veins are composed of three distinct layers known as the *tunica intima, tunica media, and tunica adventitia.* All three layers can be characterized by their distinctive embodiment of extracellular matrix (ECM) and cells.

The *tunica adventitia,* the outermost layer, is mostly connective tissue and serves to protect the vessel and strengthen its walls. Adventitial layer components are primarily fibroblast cells and collagen; however, undifferentiated pericytes are also found in this layer. The middle layer, known as the *tunica media*, is a network of circumferentially aligned smooth muscle cells (SMCs) and ECM, namely collagen and elastin. The majority of the mechanical strength of a blood vessel is from this layer. SMCs in the medial layer provide muscle tone to initiate changes in vessel diameter, a key event in regulating circulatory flow resistance. The innermost layer, known as the *tunica intima,* is comprised of a confluent layer of longitudinally aligned endothelial cells (ECs) that are in direct contact with circulating blood. The vascular endothelium plays a central regulatory role in a variety of blood vessel functions, including regulation of vessel wall permeability, prevention of thrombus formation, and paracrine control over smooth muscle cell function.

9.2.2   BLOOD COMPATIBILITY

In order to achieve blood compatibility, it is widely accepted that tissue-engineered arteries must possess a functional and confluent endothelium. Obtaining complete luminal coverage by ECs is important since exposure of subendothelial tissue to blood readily leads to platelet aggregation. A normal, functional endothelium inhibits thrombosis through the action of several inherent mechanisms (Figure 9.1). ECs express heparan sulfate proteoglycans (HSPGs), which bind and activate the potent thrombin inhibitor, antithrombin III. Thrombomodulin is a functional thrombin receptor expressed on the EC surface. Thrombomodulin activation by thrombin triggers events that lead to the inactivation of factors Va and VIIa, thereby decreasing thrombin production. Platelet adhesion and activation is prevented by the expression of nitric oxide, prostacyclin, and the negatively charged cell surface due to HSPGs. If clotting occurs despite these upstream antithrombotic mechanisms, ECs secrete

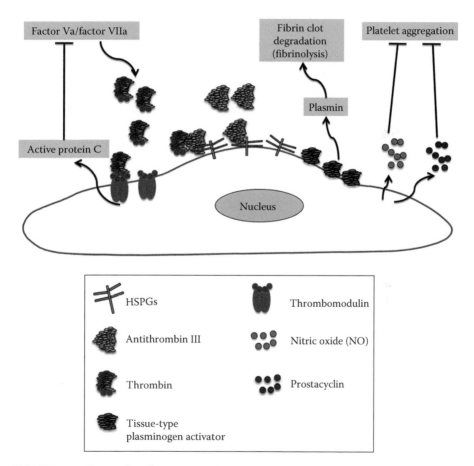

FIGURE 9.1 (See color insert.) Antithrombogenic mechanisms of ECs. Membrane-associated thrombomodulin is activated upon thrombin binding. Activation of thrombomodulin leads to activation of protein C. Active protein C inactivates factors Va and VIIa that are responsible for upstream thrombin production. Heparan sulfate proteoglycans (HSPG) on the EC surface bind to antithrombin III (ATIII), which inactivates thrombin. ECs also secrete nitric oxide (NO) and prostacyclin that inhibit platelet aggregation. EC expression of tissue-type-plasminogen activator (t-PA) causes plasminogen conversion to produce plasmin to degrade fibrin in the clot.

tissue-type-plasminogen activator (t-PA), which leads to plasmin-mediated degradation of the fibrin clot.[3,4]

9.2.3 Mechanical Properties

Obtaining the appropriate mechanical properties is critical when engineering a vascular graft. Engineered vessels should have adequate mechanical strength to withstand the deformations due to pulsatile blood flow and pressure. Burst pressure is defined as the internal pressure at which rupture occurs. In addition to achieving the required burst pressure, engineered vessels must exhibit adequate compliance as well as vasoactivity to be fully functional. Vascular compliance defines the change in vessel diameter in response to changes in internal pressure. Matching vascular compliance allows the graft wall to undergo cyclic deformations similar to the surrounding native vessels. Vasoactivity enables the vessel to contract and relax in order to alter its diameter to meet the physiological requirements for blood flow. Vasoactive function is provided directly from medial smooth muscle cells, which respond to biochemical cues originating from the neighboring endothelium.

9.3 MATRIX MATERIALS FOR TEVGs

9.3.1 Natural Matrix Materials

9.3.1.1 Collagen

Collagen is one of the primary matrix proteins found in the vessel wall. Cross-linked collagen fibrils provide mechanical stiffness to the vascular wall and act to maintain the overall integrity against physiological pressure. For this reason, collagen has long been a matrix material candidate for tissue-engineered blood vessels.

An early model involved the incorporation of vascular cells into a reconstituted type-I collagen gel, which was molded into a tubular construct.[5] One advantage of using a reconstituted collagen gel is that cells can be directly incorporated into the solution prior to gel casting. However, reconstituted collagen gels are not strong enough and have suboptimal burst pressures when used as matrix materials for TEVGs. This was evidenced by the requirement of a reinforcing Dacron sleeve to withstand physiological pressures.[5]

9.3.1.2 Elastin

Elastin is the second most abundant matrix protein found in the vascular wall. Elastin fibers provide compliance to the vascular wall and allow the vessel to stretch and recoil under physiological pressure. The ability to recoil when subjected to pulsatile flow is important since the presence of permanent creep can be associated with aneurysm formation. Mismatches in vessel compliance have been a key issue in the failure of engineered vascular grafts, and for this reason, mature elastin fibers could serve as an ideal matrix material. It has also been discovered that elastin plays an important regulatory function during arterial development, serving to regulate SMC overproliferation and possibly prevent obstructive arterial diseases.[6]

9.3.1.3 Fibrin

Fibrin is a natural matrix protein physiologically derived from its blood circulating precursor fibrinogen. Polymeric fibrin has been previously used as a tissue engineering matrix and was also shown to be a natural in-growth matrix for ECs. Additionally, cross-linked fibrin is relatively nonthrombogenic, making it a potential matrix candidate for TEVGs.[7] Since cells can also be incorporated prior to gel polymerization, fabrication of cellularized fibrin constructs is relatively simple. Reconstituted fibrin scaffolds have been reported to show higher levels of ECM synthesis by embedded cells than the reconstituted collagen scaffolds. One study reported up to a fivefold increase in collagen synthesis by SMCs seeded in fibrin gels when compared to SMCs seeded in collagen gels. The increase in ECM synthesis has led to more desirable burst strengths for fibrin-based TEVGs.[8]

9.3.1.4 Decellularized Tissues

Decellularized tissues represent another means of obtaining natural matrices for blood vessel tissue engineering. Recognition of foreign cellular antigens by the immune system prevents the efficacious use of xenogenic and allogeneic tissues for use as vascular grafts. ECM components, however, are relatively conserved across species and could possibly be well accepted during transplantation. Xenogeneic and allogeneic tissues can be processed in order to remove all cellular components and debris while preserving the native ECM structure and function. The most common sources of decellularized tissue for blood vessel tissue engineering include xenogeneic blood vessels and tissue from the small intestinal submucosa.

There are a variety of existing methods for obtaining acellular tissue and they can be separated into three categories: *physical methods, chemical methods, and enzymatic methods.*[9] These methods are summarized in Table 9.1. In order to achieve efficient decellularization without damaging ECM components, tissue engineers have generally utilized a combination of methods. For example, one possible method uses mechanical force to strip the mucosal and membranous layers of the small intestine to yield a submucosal layer. Subsequently, the submucosal layer is processed using a variety of chemical methods such as treatment with ethylenediaminetetraacetic acid (EDTA), detergents, and various salt solutions.[2]

TABLE 9.1
Summary of Decellularization Methods

Physical Methods	Chemical Methods	Enzymatic Methods
Freezing	Alkaline; acid	Trypsin
Mechanical force	Ionic detergents (e.g., sodium dodecyl sulfate SDS)	Endonucleases
Mechanical agitation (e.g., sonication)	Nonionic detergents (e.g., Triton-X100)	Exonucleases
	Zwitterionic detergents	
	Hypertonic and hypotonic solutions	
	Chelating agents (e.g., EDTA)	

9.3.2   SYNTHETIC MATRIX MATERIALS

Bioadsorbable synthetic materials such as poly*(glycolic acid)* (PGA) and poly*(lactic acid)* (PLA) can be used as a temporary artificial matrix for TEVGs. As an artificial matrix, bioadsorbable materials can provide temporary mechanical support while allowing cells to grow, reorganize, and synthesize natural ECM proteins.

PGA and PLA are linear aliphatic polyesters. PGA, PLA, and their copolymers have been widely investigated and have been approved for use in a variety of medical applications. For example, the first bioadsorbable sutures were developed using PGA and have been commercially available since the 1970s. In addition to an extensive safety profile, PGA and PLA polymers exhibit high mechanical strength, which has made them the most common bioadsorbable materials for vascular graft tissue engineering.[10–14]

A key concern when using bioadsorbable materials is matching the rate of degradation to the rate of cell/tissue growth. Degradation periods that are too short can result in premature vessel structures that lack sufficient strength to resist bursting. PGA has a degradation period ranging from 2 to 4 weeks, and over this period, the polymer experiences a rapid decrease in its mechanical strength. Therefore, PGA is an appropriate tentative matrix if the rate of matrix synthesis is high, either in a bioreactor or *in vivo*. PLA has a degradation rate of more than 1 year, which allows gradual matrix remodeling upon implantation. Copolymers such as poly(lactic-co-glycolic acid) and poly(lactic acid-caprolactone) can also be used to engineer the degradation rate and the mechanical property of the synthetic matrix.

9.3.3   PROCESSING TECHNIQUES FOR MATRIX MATERIALS

It is important for a matrix material to be capable of supporting cellular activities such as adhesion, migration, proliferation, and matrix synthesis. Matrix materials must have the proper physical and chemical features to support such cellular behavior. For example, in order to facilitate cell infiltration and matrix remodeling, the material must have adequate porosity. This section will highlight some of the processing techniques used for the fabrication of matrix materials for TEVGs.

9.3.3.1   Physical Processing Techniques

To make a three-dimensional (3D) porous matrix for vascular grafts, synthetic polymers have been processed using techniques such as solvent casting, particulate leaching, and bonding.[15,16] The porosity and mechanical property of the matrix can be manipulated by varying parameters such as salt weight fraction, polymer content, and the extent of bonding. Recently, several research groups have used electrospinning techniques to fabricate fibrous scaffolds that simulate the native ECM's micro/nanostructure (Figure 9.2). Electrospinning has been applied to create scaffolds from many synthetic matrix materials as well as natural matrix materials such as collagen and elastin.[14,17] During electrospinning, an electric field is created between a grounded target and a positively charged capillary or syringe pump filled with polymer solution. When the electrostatic charge is greater than the surface tension between the polymer and the capillary tip, a jet

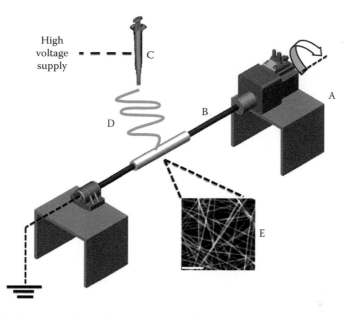

FIGURE 9.2 (See color insert.) Making nanofibrous scaffolds by using an electrospinning technique. An electric field is created between charged capillary tip carrying a polymer solution and a grounded mandrel. (A) Motor; (B) rotating mandrel (grounded); (C) polymer solution delivered from a syringe pump to a charged capillary; the capillary tip is connected to a high voltage source (~15kV); (D) nanofibers exiting charged capillary to be collected on rotating mandrel; (E) nanofibers collected on the rotating mandrel.

stream of polymer is driven out and collected as an interconnected web of fibers. The web of fibers can be collected on a flat surface to form a fibrous membrane, or it can be collected on a rotating mandrel to form a fibrous tubular structure (Figure 9.2). Additionally, orientation and alignment of electrospun fibers can be controlled during the process.

Thermally induced phase separation (TIPS) can be used to yield highly porous polymer matrix structures.[18,19] During the TIPS process, a polymer is dissolved in a solvent and placed into an appropriate mold. The mold is then rapidly cooled until the solvent becomes frozen. Finally, the solvent is removed using freeze-drying, leaving behind a porous polymer structure. The scaffold for TEVGs can be prepared using a tubular mold.

9.3.3.2 Chemical Modifications for Acellular (Matrix Only) Grafts

When used in smaller diameter applications such as bypass procedure for coronary arteries and peripheral arteries, Dacron and ePTFE grafts experience rapid failure due to thrombotic events. In native blood vessels, ECs play the key role in regulating thrombotic events. For this reason, the majority of *in vitro* TEVGs have utilized a combination of matrix and ECs. However, some approaches have aimed to create acellular grafts by improving the thrombotic properties of materials through chemical modification. Most chemical modifications have hinged on trying to mimic the specific antithrombogenic features inherent to ECs (Figure 9.1). Although chemical modification approaches have exhibited some degree of success, they have not approached the antithrombogenic effectiveness of native ECs.

Surfaces containing immobilized heparin molecules have been shown to effectively reduce the extent of thrombotic events.[20] As a result, immobilized heparin has been widely used to modify the surfaces of various blood contacting devices, including vascular grafts. Methods for modifying vascular graft materials with heparin include covalent grafting, electrostatic binding to negatively charged heparin sulfate groups, and direct loading into a reconstituted material (i.e., collagen gels). A recent study investigated the performance of an acellular collagen graft derived from decellularized intestinal layers. When the acellular graft was treated with heparin, researchers observed improved blood compatibility and patency.[2] However, there are some inherent limitations to the heparin surface modification approach. For example, heparin can bind to multiple growth factors and matrix proteins in the circulating blood, resulting in nonspecific activities.

Heparin can neutralize thrombin, but is incapable of curtailing upstream thrombin production. Thrombomodulin, which can limit upstream thrombin generation (Figure 9.1), can be immobilized to enhance the antithrombogenic properties of materials.[21] When tested using a specialized *in vitro* flow model, thrombomodulin-modified surfaces exhibited superior capability to lower thrombin production and thus overall circulating thrombin levels.[22] However, the *in vivo* effectiveness of immobilized thrombomodulin remains to be determined.

Another limitation of the heparin modification approach is that heparin is not a direct thrombin inhibitor. Heparin requires the cofactor antithrombin III to exert its effects. Additionally, heparin is not capable of inhibiting fibrin bound thrombin found in the clot. Hirudin, a small polypeptide derived from the medicinal leech, is capable of directly inhibiting both soluble and bound thrombin.[23] Although hirudin is not naturally expressed by vascular ECs, its immobilization could be a more efficient and direct method to prevent graft thrombosis.[24,25]

9.4   CELL SOURCES FOR TEVGs

9.4.1   Vascular Cells

To meet biocompatibility requirements, an engineered blood vessel must be antithrombogenic, nonimmunogenic, and resistant to infection. To achieve these criteria, engineered vessels must possess a confluent and nonactivated endothelium derived from autologous cells. The most obvious source for obtaining autologous ECs is to harvest them from arterial biopsies.[13,26,27] This method however may not be optimal as it can result in morbidity and is limited due to a potential shortage of healthy autologous vessels. ECs found within subcutaneous fat, known as microvascular ECs, represent an alternative to arterial biopsies. However, it is unclear whether these cells have the same functionality as arterial ECs. Currently, there is no way to circumvent the high immunogenicity of using non-autologous ECs. Autologous human ECs have decreased proliferative potential, making the expansion of isolated cells a long and difficult process. Issues related to the sourcing and culture of human autologous ECs has created difficulties in transferring TEVG technologies into clinical use.

SMCs can also be obtained from blood vessel biopsies. However, it is unclear whether these cells must be autologous in order to maintain biocompatibility. It is suggested that since SMCs are not in direct contact with circulating blood, the requirement for autologous SMCs in not as stringent as it is for ECs.[28]

The vast majority of TEVG construction relies on the combination of both matrix materials and cells. A new approach, which is referred to as "sheet-based" tissue engineering, relies solely on cultured autologous vascular cells. Sheet-based tissue engineering takes advantage of the abundant and endogenous synthesis of ECM by fibroblasts and SMCs when cultured in the presence of ascorbic acid. Sheets of fibroblasts or SMCs are cultured for about a month, after which they can be manually peeled from the culture flask. To fabricate the engineered artery, researchers create an acellular inner membrane by dehydrating a fibroblast sheet and subsequently wrapping it around a central mandrel. A medial layer is constructed by wrapping the inner membrane with a sheet of SMCs. This is followed by 1 week of culture in a bioreactor with luminal flow and mechanical support. The adventitial layer is constructed by subsequently wrapping a sheet of fibroblasts around the construct, followed by an 8 week maturation period in the bioreactor. Cultured autologous ECs are then seeded into the lumen of the vessel and allowed to mature for 1 week. After artery construction, the seeded endothelium remained confluent and active. A later approach using sheet-based tissue engineering eliminated the need to use SMCs to develop arteries. Arteries that were fabricated without an SMC sheet layer were tested using an *in vivo* model and were shown to develop a regenerated neo-medial SMC layer.[26,27] Studies suggest that an autologous cell-synthesized matrix exhibits a minimal host inflammatory response. Additionally, an endogenous cell-synthesized ECM allows for the production of "living" blood vessels that are capable of undergoing normal remodeling in response to environmental cues. Currently, however, sheet-based tissue-engineered arteries require several months, which may prohibit their use when emergency bypass procedures are necessary.

9.4.2 STEM CELLS

Stem cells and progenitor cells are promising cell sources for engineering functional cardiovascular tissues. A stem cell is a cell that can self-renew as well as give rise to daughter cells with more specialized function, whereas progenitor cells only give rise to daughter cells with more specialized function.

Embryonic stem cells (ESCs) are pluripotent and have unlimited self-renewal capacity.[29] They can give rise to cell types of all somatic cell lineages, including vascular cell types.[30] Having the ability to propagate in mass and differentiate into mature vascular cell types, ESCs may potentially be an unlimited ex vivo cell source for TEVGs.

Adult and fetal stem cells are another class of stem cells that have shown therapeutic potential. Compared to ESCs, these stem cells are more limited in their differentiation capacity, but raise less ethical concerns for research use. Adult stem cells and progenitors can be found in many types of tissues such as the bone marrow, peripheral blood, adipose tissue, and umbilical cord. Among the various types of adult stem cells, the two categories most studied for tissue-engineered blood vessel applications are mesenchymal stem cells (MSCs) and endothelial progenitor cells (EPCs).

MSCs are multipotent stem cells that reside in the stroma of adult bone marrow. Although they represent only 0.01% of the total nucleated cell population in the bone marrow, they can be expanded *in vitro* to more than a million-fold while maintaining multipotency. The advantages of MSCs include their ease of isolation by a bone marrow aspiration in the iliac crest, high expansion potential, and reproducible characteristics among isolations.[31] Unlike other cell types, MSCs do not express the major histocompatibility complex II (MHC II) antigens that are responsible for immune rejection, making MSCs a candidate cell source for allogeneic cell transplantation. Recent reports indicate that MSCs do not acquire MHC II cell surface antigens upon differentiation along adipogenic, chondrogenic, and osteogenic lineages, and it is possible that MHC II antigens would not be expressed upon vascular differentiation.[32]

Our recent work has investigated the use of human MSCs (hMSCs) in conjunction with an electrospun biodegradable matrix for the construction of TEVGs. Vascular grafts were prepared by seeding hMSCs onto electrospun PLA sheets that were subsequently rolled around a small diameter mandrel to form a tubular structure. When implanted in rats as a carotid artery bypass, the hMSC seeded grafts showed high patency rates and low intimal thickening. Interestingly, it was shown that MSCs resisted platelet adhesion and aggregation by expressing heparan sulfate proteoglycans (HSPGs), which is similar to the platelet resistive mechanism exhibited by ECs.[14] This suggests that MSCs may be able to serve an antithrombogenic role when used in a TEVG.

Recently, a small diameter arterial wall was engineered by seeding hMSCs onto a tubular PGA scaffold, followed by culture using a biomimetic perfusion system. Arterial walls were cultured over an 8 week period, which was divided into a 4 week proliferation phase followed by a 4 week differentiation phase. Culture conditions for the proliferation and differentiation phases were optimized to promote hMSC proliferation and differentiation respectively. The arterial walls showed strong histological similarities to native arteries, which confirmed that MSCs are a potential source of TEVG fabrication.[12]

EPCs can be found within bone marrow as well as within blood during normal circulation. EPCs have also been proven to enhance angiogenesis and differentiate into cardiovascular cell types.[33] There is evidence that EPCs could replace ECs in forming patent small-diameter vascular grafts.[34] Autologous EPCs from sheep were seeded on decellularized arteries in a bioreactor, and the grafts were implanted into carotid arteries. Long-term *in vivo* study showed that the EPC-seeded grafts remained patent for over 130 days, whereas the acellular control grafts were occluded within 15 days. The EPC-seeded grafts developed vessel-like characteristics, including contractility and nitric oxide–mediated vascular relaxation.[34]

Induced pluripotent stem cells (iPSC) are a novel stem cell type derived from adult somatic cells by reprogramming with a defined group of transcription factors.[35–37] iPSCs exhibit the critical characteristics of embryonic stem cells and thus are capable of differentiating into all vascular cell types. Unlike human ES cells, human iPSCs can be derived from autologous cells, which bypass the immune rejection issue and make iPSCs an unlimited cell source for patient specific therapies.

9.5   ENABLING TECHNOLOGIES FOR TEVGs

9.5.1   CELL SEEDING TECHNOLOGY

The most common and simple method of cell seeding is the "drip" method, whereby cell suspensions are manually pipetted into the lumen of a tubular matrix. Although simple, the "drip" method is rather inefficient and lends itself toward low, nonuniform cellularization. More efficient cell seeding methodologies include vacuum cell seeding and rotational cell seeding.[38]

Recently, a novel cell seeding device has been developed that utilizes the benefits of vacuum, rotation, and flow to achieve fast, reliable, and uniform cell seeding.[19] The cell seeding process takes place inside an evacuated, airtight chamber. Inside the chamber, a cell suspension is delivered through both ends of the tubular scaffold from a syringe pump. The combination of vacuum and luminal flow induces rapid transmural movement of the suspension throughout the tube wall, allowing cells to become embedded in the porous matrix. Inside the chamber, the scaffold is also mounted onto a pair of concentric tees that are coupled with an external DC motor to provide rotation. The use of rotation helps to increase uniformity of seeding circumferentially. This multifaceted seeding device can achieve efficient and reproducible bulk seeding, and the seeded grafts show a similar cellular distribution and density as native arteries.

9.5.2   BIOREACTORS AND MECHANICAL CONDITIONING

The appropriate environmental cues can trigger cells to remodel their own microenvironment, leading to the construction and organization of new tissue. The *in vivo* vascular environment exposes blood vessels to both biochemical and biophysical factors. Hemodynamic forces include shear stress, the tangential frictional force that acts primarily on ECs, and cyclic radial tension caused by pulsatile blood pressure. These physiological forces have been found to regulate cellular behavior and tissue remodeling. For example, cyclic strain has been found to affect the phenotype, orientation and ECM deposition of SMCs.[39] In addition to providing mechanical stimulation, bioreactors also serve to facilitate and enhance the delivery of nutrients to tissue constructs.[40] Investigators have developed bioreactors to mimic the *in vivo* vascular environment and promote the remodeling of TEVGs. In one study, SMCs were manually seeded onto tubular, biodegradable PGA scaffolds that were mounted onto silicon tubing inside a bioreactor.[13] The bioreactor was filled with culture medium containing supplements that aided ECM production. Pulsatile pumps delivered flow of medium through the silicon tubing, thereby imparting pulsatile radial strains to the developing construct. Using this approach, engineered vessels cultured under pulsatile conditions have superior mechanical strength and integrity when compared to vessels developed under static conditions.

9.6 FUTURE DIRECTIONS

Although significant progress has been made in the past two decades, there are still several hurdles to overcome in order to translate TEVG technologies into clinical applications. Currently, only acellular grafts are being used clinically. Engineering more effective nonthrombogenic surfaces by chemical modification can further improve the patency of vascular grafts. However, endothelialization is required for the long-term patency. Although endothelialization can be achieved in animal models, it has been difficult to realize complete endothelialization of the grafts in human. Bioactive surfaces that recruit ECs and EPCs, enhance EC growth, and promote endothelialization may make it possible to develop acellular vascular grafts with long-term patency and off-the-shelf availability. Alternatively, in situ tissue engineering is a promising approach to generate autologous functional vascular grafts. This approach involves the implantation of an engineered scaffold that delivers the proper biophysical and biochemical cues in order to trigger the organized growth of new arterial tissue. For cellular grafts, one of the most critical issues is the source of immune acceptable ECs. Advances in stem cell technology potentially offer a solution to this problem. For patient-specific therapies, ECs can be obtained from EPCs or iPSCs. How to increase the efficiency of deriving ECs from pluripotent iPSCs still needs further investigation. Furthermore, by using the technologies in stem cell engineering and genetic engineering, one could explore the potential of creating universally immune acceptable ECs for TEVG construction in the near future.

REFERENCES

1. Ross, R., The pathogenesis of atherosclerosis: A perspective for the 1990s. *Nature* 1993, *362* (6423), 801–809.
2. Huynh, T.; Abraham, G.; Murray, J.; Brockbank, K.; Hagen, P.; Sullivan, S., Remodeling of acellular collagen graft into a physiologically responsive neovessel. *Nature Biotechnology* 1999, *17*, 1083–1086.
3. Mitchell, S.; Niklason, L., Requirements for growing tissue-engineered vascular grafts. *Cardiovascular Pathology* 2003, *12* (2), 59–64.
4. van Hinsbergh, V., The endothelium: Vascular control of haemostasis. *European Journal of Obstetrics and Gynecology* 2001, *95* (2), 198–201.
5. Weinberg, C.; Bell, E., A blood vessel model constructed from collagen and cultured vascular cells. *Science* 1986, *231* (4736), 397–400.
6. Li, D. Y.; Brooke, B.; Davis, E. C.; Mecham, R. P.; Sorensen, L. K.; Boak, B. B.; Eichwald, E.; Keating, M. T., Elastin is an essential determinant of arterial morphogenesis. *Nature* 1998, *393* (6682), 276–280.
7. Cummings, C. L.; Gawlitta, D.; Nerem, R. M.; Stegemann, J. P., Properties of engineered vascular constructs made from collagen, fibrin, and collagen-fibrin mixtures. *Biomaterials* 2004, *25* (17), 3699–3706.
8. Grassl, E. D.; Oegema, T. R.; Tranquillo, R. T., A fibrin-based arterial media equivalent. *Journal of Biomedical Materials Research Part A* 2003, *66A* (3), 550–561.
9. Gilbert, T.; Sellaro, T.; Badylak, S., Decellularization of tissues and organs. *Biomaterials* 2006, *27* (19), 3675–3683.
10. Mooney, D. J.; Breuer, C.; McNamara, K.; Vacanti, J. P.; Langer, R., Fabricating tubular devices from polymers of lactic and glycolic acid for tissue engineering. *Tissue Engineering* 1995, *1* (2), 107–118.

11. Gong, Z.; Niklason, L., Blood vessels engineered from human cells. *Trends in Cardiovascular Medicine* 2006, *16* (5), 153–156.

12. Gong, Z.; Niklason, L. E., Small-diameter human vessel wall engineered from bone marrow-derived mesenchymal stem cells (hMSCs). *The FASEB Journal* 2008, *22* (6), 1635–1648.

13. Niklason, L. E.; Gao, J.; Abbott, W. M.; Hirschi, K. K.; Houser, S.; Marini, R.; Langer, R., Functional arteries grown in vitro. *Science* 1999, *284* (5413), 489–493.

14. Hashi, C.; Zhu, Y.; Yang, G.; Young, W.; Hsiao, B.; Wang, K.; Chu, B.; Li, S., Antithrombogenic property of bone marrow mesenchymal stem cells in nanofibrous vascular grafts. *Proceedings of the National Academy of Sciences* 2007, *104* (29), 11915.

15. Mooney, D. J.; Organ, G.; Vacanti, J. P.; Langer, R., Design and fabrication of bio-degradable polymer devices to engineer tubular tissues. *Cell Transplant* 1994, *3* (2), 203–210.

16. Wake, M. C.; Gupta, P. K.; Mikos, A. G., Fabrication of pliable biodegradable polymer foams to engineer soft tissues. *Cell Transplant* 1996, *5* (4), 465–473.

17. Sell, S. A.; McClure, M. J.; Garg, K.; Wolfe, P. S.; Bowlin, G. L., Electrospinning of collagen/biopolymers for regenerative medicine and cardiovascular tissue engineering. *Advanced Drug Delivery Reviews* 2009, *61* (12), 1007–1019.

18. Aubert, J. H.; Clough, R. L., Low-density, microcellular polystyrene foams. *Polymer* 1985, *26* (13), 2047–2054.

19. Soletti, L.; Nieponice, A.; Guan, J.; Stankus, J.; Wagner, W.; Vorp, D., A seeding device for tissue engineered tubular structures. *Biomaterials* 2006, *27* (28), 4863–4870.

20. Gott, V. L.; Whiffen, J. D.; Dutton, R. C., Heparin bonding on colloidal graphite surfaces. *Science* 1963, *142* (3597), 1297–1298.

21. Kishida, A.; Ueno, Y.; Fukudome, N.; Yashima, E.; Maruyama, I.; Akashi, M., Immobilization of human thrombomodulin onto poly(ether urethane urea) for developing antithrombogenic blood-contacting materials. *Biomaterials* 1994, *15* (10), 848–852.

22. Tseng, P.-Y.; Rele, S. S.; Sun, X.-L.; Chaikof, E. L., Membrane-mimetic films containing thrombomodulin and heparin inhibit tissue factor-induced thrombin generation in a flow model. *Biomaterials* 2006, *27* (12), 2637–2650.

23. Di Nisio, M.; Middeldorp, S.; Buller, H. R., Direct thrombin inhibitors. *New England Journal of Medicine* 2005, *353* (10), 1028–1040.

24. Phaneuf, M. D.; Berceli, S. A.; Bide, M. J.; Quist, W. G.; LoGerfo, F. W., Covalent linkage of recombinant hirudin to poly(ethylene terephthalate) (Dacron): Creation of a novel antithrombin surface. *Biomaterials* 1997, *18* (10), 755–765.

25. Hashi, C. K.; Derugin, N.; Janairo, R. R.; Lee, R.; Schultz, D.; Lotz, J.; Li, S., Antithrombogenic modification of small-diameter microfibrous vascular grafts. *Arteriosclerosis, Thrombosis, and Vascular Biology* 2010, *30* (8), 1621–1627.

26. L'Heureux, N.; Dusserre, N.; Konig, G.; Victor, B.; Keire, P.; Wight, T. N.; Chronos, N. A. F.; Kyles, A. E.; Gregory, C. R.; Hoyt, G.; Robbins, R. C.; McAllister, T. N., Human tissue-engineered blood vessels for adult arterial revascularization. *Natural Medicine* 2006, *12* (3), 361–365.

27. L'heureux, N.; Paquet, S.; Labbe, R.; Germain, L., A completely biological tissue-engineered human blood vessel. *The FASEB Journal* 1998, *12* (1), 47–56.

28. Nerem, R. M.; Seliktar, D., Vascular tissue engineering. *Annual Review of Biomedical Engineering* 2001, *3* (1), 225–243.

29. Thomson, J. A.; Itskovitz-Eldor, J.; Shapiro, S. S.; Waknitz, M. A.; Swiergiel, J. J.; Marshall, V. S.; Jones, J. M., Embryonic stem cell lines derived from human blastocysts. *Science* 1998, *282* (5391), 1145–1147.

30. Levenberg, S.; Golub, J. S.; Amit, M.; Itskovitz-Eldor, J.; Langer, R., Endothelial cells derived from human embryonic stem cells. *Proceedings of the National Academy of Sciences of the United States of America* 2002, *99* (7), 4391–4396.

31. Pittenger, M. F.; Mackay, A. M.; Beck, S. C.; Jaiswal, R. K.; Douglas, R.; Mosca, J. D.; Moorman, M. A.; Simonetti, D. W.; Craig, S.; Marshak, D. R., Multilineage potential of adult human mesenchymal stem cells. *Science* 1999, *284* (5411), 143–147.

32. Le Blanc, K.; Tammik, C.; Rosendahl, K.; Zetterberg, E.; RingdÈn, O., HLA expression and immunologic properties of differentiated and undifferentiated mesenchymal stem cells. *Experimental Hematology* 2003, *31* (10), 890–896.

33. Asahara, T.; Murohara, T.; Sullivan, A.; Silver, M.; van der Zee, R.; Li, T.; Witzenbichler, B.; Schatteman, G.; Isner, J. M., Isolation of putative progenitor endothelial cells for angiogenesis. *Science* 1997, *275* (5302), 964–966.

34. Kaushal, S.; Amiel, G. E.; Guleserian, K. J.; Shapira, O. M.; Perry, T.; Sutherland, F. W.; Rabkin, E.; Moran, A. M.; Schoen, F. J.; Atala, A.; Soker, S.; Bischoff, J.; Mayer, J. E., Jr., Functional small-diameter neovessels created using endothelial progenitor cells expanded ex vivo. *Natural Medicine* 2001, *7* (9), 1035–1040.

35. Takahashi, K.; Yamanaka, S., Induction of pluripotent stem cells from mouse embryonic and adult fibroblast cultures by defined factors. *Cell* 2006, *126* (4), 663–676.

36. Park, I. H.; Zhao, R.; West, J. A.; Yabuuchi, A.; Huo, H.; Ince, T. A.; Lerou, P. H.; Lensch, M. W.; Daley, G. Q., Reprogramming of human somatic cells to pluripotency with defined factors. *Nature* 2008, *451* (7175), 141–146.

37. Yu, J.; Vodyanik, M. A.; Smuga-Otto, K.; Antosiewicz-Bourget, J.; Frane, J. L.; Tian, S.; Nie, J.; Jonsdottir, G. A.; Ruotti, V.; Stewart, R.; Slukvin, I. I.; Thomson, J. A., Induced pluripotent stem cell lines derived from human somatic cells. *Science* 2007, *318* (5858), 1917–1920.

38. van Wachem, P. B.; Stronck, J. W. S.; Koers-Zuideveld, R.; Dijk, F.; Wildevuur, C. R. H., Vacuum cell seeding: A new method for the fast application of an evenly distributed cell layer on porous vascular grafts. *Biomaterials* 1990, *11* (8), 602–606.

39. Kim, B.-S.; Nikolovski, J.; Bonadio, J.; Mooney, D. J., Cyclic mechanical strain regulates the development of engineered smooth muscle tissue. *Natural Biotechnology* 1999, *17* (10), 979–983.

40. Pei, M.; Solchaga, L. A.; Seidel, J.; Zeng, L.; Vunjak-Novakovic, G.; Caplan, A. I.; Freed, L. E., Bioreactors mediate the effectiveness of tissue engineering scaffolds. *FASEB Journal* 2002, *16* (12), 1691–1694.

Part IV

Stem Cell Transplantation Strategies

10 Scaffold-Based Approaches to Maintain the Potential of Transplanted Stem Cells

Dmitry Shvartsman and David J. Mooney

CONTENTS

10.1 INTRODUCTION

Stem cells are now believed to exist in virtually all parts of the human body and to contribute to tissue and organ development during all stages of life.[1–8] The specific and unique responses required from stem cells are controlled by a vast number of growth factors, surrounding cells, and interactions with extracellular matrices.[9,10] This microenvironment, called the "stem cell niche," is crucial to stem cell fate determination and specifies their responses to the environmental signals.[11–20] Furthermore, stem cells are often considered to be truly pluri- or multipotent and

259

self-renewing only within the context of their niches. The adult stem cell (ASC) niche can be separated into four major components: ASCs, supporting cells, extracellular matrix (ECM), and the vascular network.[10,20] Each of these components has an important role in stem cell survival and proliferation, and current biomedical research has to consider all of them in order to create successful strategies for stem cell engraftment and tissue repair.[21] Here, we discuss bioengineering approaches for restoration and creation of a stem cell niche based on polymeric scaffolding methods, specific microenvironmental designs, and growth factor incorporation.[22] Stem cell–based tissue engineering utilizes the unique properties of self-renewal and directional differentiation of stem cells, combined with development of synthetic scaffolds required for cell incorporation and growth. This chapter is divided into several topics: first, it defines the term "stem cell" and its subsets; second, it addresses specific ASC properties, using as examples bone marrow, muscle, and nerve stem cells; third, it stipulates the design environment for systems mimicking stem cell niches; finally, a description of the specific types of polymeric systems used to form artificial niches are reviewed. This chapter addresses two coexisting fields of current biomedical research: stem cell therapy and tissue restoration with engineered scaffolds, as they are substantially merging in current bioengineering applications and medical procedures.

10.2   STEM CELLS

A description of stem cell biology and applications requires usage of specific terms and definitions. This has been done extensively in the previous chapters, but we would like to remind the reader of the types of stem cells that are currently of interest in the field. These include embryonic stem cells (ESCs),[23,24] ASCs,[13,25] and induced pluripotent stem cells (iPSCs).[26,27] Stem cell classification and discovery first emerged as a concept in the early 1960s of the twentieth century and has evolved into one of the most intriguing, controversial, and exciting fields of biomedical research.[1–8]

   This chapter focuses on the use of ASCs for tissue repair, regeneration, and tissue engineering. ESCs are attractive for tissue engineering and repair due to their proliferative potential and ability to differentiate into virtually several types of tissue or organ of the human body. However, while ESCs and the more recently developed iPSCs have a higher development potential than ASCs, they also pose challenges of uncontrolled transformation and growth, potentially leading to tumor formation.[28,29] Furthermore, there are significant concerns raised by those who argue against their uses due to moral or religious concerns. In response to these issues, the therapeutic use of ASCs, typically obtained from the patient, has been the emphasis of most translational research. In contrast to ESCs or iPSCs, the more limited potential of ASCs are believed to allow more specific control and characterization of their development and proliferation. They have the ability to proliferate, duplicate, and restore damaged tissues, typically via proceeding down predefined paths of development.[30] However, these cells vary significantly in their distribution and numbers between various tissues and are often much more difficult to rigorously characterize than ESCs or iPSCs, which often makes it very hard to identify and study the ASC populations.[28,31]

Noticeable progress in clinical use of ASCs has been made over the past four decades, although this concept is still limited by the complexities of the tissue composition, architecture, and specificity of interactions within the host. The most successful clinical usage of stem cells is bone marrow transplantation (BMT) for treatment of hematopoietic system deficiencies and diseases. ASCs within the bone marrow, termed hematopoietic stem cells (HSCs), have been well defined and extensively studied over the last 50 years, allowing their transplantation to be highly specific.[3,4,32–38] The overall efficiency of HSCs engraftment is represented by the finding that only two transplanted HSCs are sufficient to completely restore the hematopoietic system of an irradiated mouse.[39] More broadly, many routine therapies are based on the transplantation of donor or host tissue or organs, and many of these tissues contain self-expanding stem cell populations. Examples include skin grafts and liver transplants. Skin epithelial cells are shed in vast numbers each day, which requires large number of stem cells in the epidermis to self-renew, expand, and proliferate. The long-term success of liver transplants likely also is dependent on a resident stem cell population that maintains the organ, although the definitive identification of liver stem cells remains to be shown. In contrast to skin or liver engraftment, therapeutic use of other ASCs, such as the intestinal epithelium or hair follicle stem cells, remains to be developed, although the knowledge of their biology and their identification is increasing rapidly.[40,41]

Some ASCs are likely rare, making definitive conclusions of their roles in tissue and organs complex, and often there is scientific debate regarding the specific subpopulations of ASC involved in tissue regeneration. The next subsections details examples of a few of the aforementioned ASCs and their properties.

10.2.1 HEMATOPOIETIC STEM CELLS

The only widely used stem cell therapy today is the use of BMT for the regeneration of the hematopoietic system. The success of this method likely relates to the well-developed body of knowledge for HSCs, and the fairly easy cell isolation and implantation procedures required for patient treatment. The overall success rates for BMT vary between 70% and 90%, which is a very high success rate for tissue and organ transplantation, but there still remain significant health complications due to the immunosuppressant treatments and chance of relapse.[42] It has been known for more than five decades that to be successful, the bone marrow needs to be extracted from the donor's bone cavities and placed into the bone cavities of the recipient, upon radioactive eradication of the recipient bone marrow cells. This approach serves as a first example of the usage of specific cell populations (HSCs) and a specific microenvironment (bone marrow) needed for their proliferation. In the course of these early transplantation studies, it was found that only a small fraction (~1%) of transplanted cells contributed to bone marrow restoration.[43] Further characterization revealed that the key cells in the normal tissue remain in an inactive (G_0) state and become active upon external signaling (e.g., injury, blood loss, inflammation, or other events requiring replenishing or increasing the cell pools of hematopoietic origin). The original pool of HSCs was estimated at around 100,000 cells, and these generate multipotent progenitors, which in turn produce a billion blood cells per day in humans.[44]

Recent models, based on lineage-tracing experiments and purification studies of HSCs, are leading to the conclusion that the small numbers of freely circulating or adherent HSCs require a specific and highly integrated environment in order to respond swiftly to body needs. There are several specific cell protein markers for HSC, which allow identification of these cells *in vitro*, but their specific location *in vivo* remains highly debated due to the low number of HSC, and complexity of their environment.[44–46]

10.2.2   Skeletal Muscle Stem Cells

Muscle injury and repair have been a focus of biomedical research since the dawn of medical thought. In contrast to the established BMT, muscle stem cell transplants are being actively developed, and still in early clinical trials, and have yet to become a standard and successful medical procedure. It is believed that muscle regeneration is based on specific stem cells, called muscle satellite cells.[47] These cells are considered to be in a quiescent state in the healthy adult muscle and proliferate during muscle development, or following injury or muscle degradation.[47] Adult muscles are highly specialized tissues, with specific structural and functional components. They are composed of muscle fibers, formed by fusing myogenic cells, bundled together by the basal membrane. The muscle satellite cells are mononucleated cells, as opposed to the multinuclear muscle fibers, and reside between the fibers and surrounding basal lamina. They were first identified morphologically and were demonstrated to have the regenerative capacity to restore muscle structure.[48,49] Further tracing studies *in vivo* and tissue culture experiments with isolated single muscle fibers and cells have confirmed their role and shown that the number of stem cells remains relatively stable and that these cells are capable of self-renewal.[49–51] The specific and morphologically distinct location of these cells provides a good example and study model for stem cell microenvironments. Moreover, muscle satellite cells have specific protein expression markers at different stages of differentiation, allowing their sorting and characterization.[51] Satellite cells are considered to be ASCs, although there is a significant amount of controversy and scientific debate, regarding the specificity of these cells and the role of their subpopulations in muscle regeneration.[52,53] Satellite cell numbers and composition varies significantly between different muscles, and they can have high heterogeneity in the expression profiles for their markers, making it difficult to specify exact roles for each subtype of the satellite cells in muscle growth, differentiation, and repair.[53–55]

10.2.3   Neural Stem Cells

Brain damage and degeneration are some of most widespread human diseases and injuries. There are no tissue repair methods currently known to successfully restore the function and structure of the central (CNS) or peripheral (PNS) nerve systems. Current medical treatments are focused either on prevention of further injuries and degeneration caused by trauma, or as in case of neurodegenerative diseases (e.g., Parkinson, Alzheimer, Gauche) on improving the quality of life of the patient or slowing the progress of the disease. The use of stem cells for the restoration of CNS or PNS is a highly attractive alternative approach. Although some progress has been made using direct implantations of ESCs or ASCs in host brain or spinal cord, these cells poses significant risks of malignant cancers and uncontrolled growth.

Certain proposed therapies are based on the unique subset of ASCs called the neural stem cells (NSCs) of the CNS. The existence of these cells and their potential for nerve repair and brain damage restoration were overlooked for years, due to the early concepts of brain development and growth of the nervous system, which postulated that these processes were limited to the early stages of development and could not be modulated in the adult. However, proliferation assays of the adult brain demonstrated significant numbers of actively proliferating and dividing cells, followed by further identification of these cells as newly generated neurons.[56] By isolating these cells and identifying specific growth factors required for their survival and proliferation, it has been possible to culture these cells and show their ability for self-replication, and with it the potential of progenitors to become both neuron and glia cells.[57] These NSCs are not restricted to a specific brain location, and can be identified and isolated from many regions. NSCs can proliferate in tissue culture with significant plasticity and may give rise to lineages not expected for brain-derived cells. Research has focused on the specific architectural and morphological context in which the NSC is residing, based on the hypothesis that unique features of specific brain regions determine the fate of NSCs. Normal brain function requires a complex and highly controlled three-dimensional environment comprised of neurons, glia cells, oligodendrocytes, nerve cell axons, and blood vessels, and each of the regions in the brain provides specificity due to both its cellular composition and spatial organization.[13,58]

There are several other examples of ASCs with specific tissue localization, structural requirements, and morphology, such as the germ cells in male reproductive organs, hair follicle cells, and epithelial cells of intestines. The common factor for all of these ASCs is their requirement for a specific niche within the tissue where they can proliferate, self-renew, and give rise to predetermined cellular progeny.

10.3   SYNTHETIC STEM CELL NICHE

The initial view in stem cell research was that upon transplantation, stem cells would find their own ways into the appropriate tissue, repopulate the tissue and restore organ functions. Although this approach has been clinically successful in the case of bone marrow transplants, even there it is highly inefficient. Further, the results of many studies using various types of ASCs has revealed that this approach, whether involving injection directly into the tissue or into the blood stream, requires huge numbers of cells, and only 1%–5% of the transplanted cells typically survive and engraft in the target tissue or organ.[30,59] This greatly reduces the potency of the approach, and clinical trial results of ASC transplantation are frequently disappointing.

The focus of ASC research has shifted in the past decade toward understanding not only the biology of single stem cells, but also the role of the stem cell surroundings.[10] Many researchers view the stem cell niche as the dividing factor between the "regular" adult cell and ASCs. It is not just a place of residence for stem cells; it has unique structural and protein compositions, which are responsible for delivering and modulating outside signals and challenges to stem cells, leading the niche to play as crucial role in tissue regeneration as the stem cell itself. The niche contains multiple components, including growth factors, supporting cells and blood vessels, all in a specific architecture that is likely required to ensure stem cell survival (Figure 10.1). The niche is also responsible for stem cell activation, division, and eventual release of the daughter cells to the surrounding tissue or to the blood stream. A significant amount of research has resulted in characterization of several ASC niches (Table 10.1), including the hematopoietic (HSC) niche, skeletal muscle niche, hair follicle niche, intestinal epithelium niche, and spermatogonial niche.[20]

It has been proposed that artificial scaffolds may promote the survival and engraftment of transplanted ASCs, and control their proliferation and daughter cell fate. Such materials would in many ways provide an artificial mimic of the native stem cell niche.[60–63] While direct ASC injections lead to large-scale cell death at least in part due to poor attachment and anoikis (a form of programmed cell death) of the implanted cells, the artificial scaffolds can be designed to retain the ASC and provide them with adhesive sites. A number of functionalities may be built into these scaffolds, and major design considerations include the chemistry and architecture of the scaffolding material, its physical properties, release of morphogens and other signaling molecules from the scaffold, and the ability of the scaffolds to disperse daughter cells throughout the surrounding tissue.[10] Large-scale clinical implementation of this

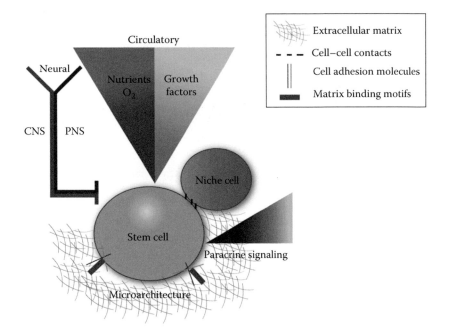

FIGURE 10.1 (See color insert) Microenvironmental inputs regulating stem cell niche. Local environmental cues regulating the stem cell life cycle are depicted. These include mechanical cues, physical interactions with microarchitectural features of ECM, feedback signaling loops from supporting cells, metabolic regulation and control by the circulatory system, neutral input from the surrounding tissues, and the central nervous system.

TABLE 10.1

Representative Mammalian Stem Cell Niches

Stem Cell	Niche Location	Supporting Cells	Reviewed in
Hematopoietic stem cells	Bone marrow (endosteal, perivascular)	Osteoblasts, osteoclasts, mesenchymal cells, reticular cells, vascular cells	[122]
Satellite muscle cells	Muscle fiber (adjacent to basal lamina)	Myofiber	[55]
Neural stem cells	Brain (lateral ventricle subventricular zone)	Oligodendrocytes, astrocytes, ependymal cells	[13]
Intestinal epithelial stem cells	Intestines (base of crypt)	Fibroblasts, hematopoietic cells	[123]
Hair follicle stem cells	Epidermis (hair bulge)	Vascular	[124]
Epithelial stem cells	Dermis (basal layer)	Dermis	[125]
Spermatogonial stem cells	Male testis (basal layer, seminefrous tubules)	Leydig, sertoli, vascular	[126]

approach has been successful in the context of engineered skin tissues, and significant progress has been made in other applications. These procedures utilized one or two types of tissue-forming cells (e.g., fibroblasts to form a dermal component of skin, combined with keratinocytes to engineer an epidermal compartment), and scaffolds with an isotropic architecture.[64] Another example is studies utilizing bone-forming cells with a scaffold of uniform microarchitecture. Upon the seeding of the scaffold with ASCs, it was implanted and monitored for the mineralization, with significant rates of success.[65] As these methods were successfully implemented, large blood vessel replacement was done using arterial vessels from human donors or animals, stripped of any cellular content and reseeded with host smooth epithelial cells that formed a functional blood vessel wall.[66,67] Successful clinical trials of small diameter blood vessels engineered with this approach have also been reported, as have trials of urologic tissues, demonstrating the potentially broad relevance of this concept.[68,69]

In spite of significant early success, many challenges remain for the synthetic niches. These include providing a well-controlled mechanism for cell adhesion, and resultant control over gene expression, allowing scaffold degradation to occur in concert with cellular deposition of their matrix, and providing topographical cues to the cells from the scaffolding. There has been tremendous interest and research on scaffold fabrication using a wide variety of synthetic polymers, and self-assembling peptides and recombinant proteins, in order to address these various challenges.

In this section, we present an overview of types of natural and synthetic materials, used in developing synthetic niches, methods of creation of stem cell scaffolds, the

roles of their mechanical properties, control over biodegradation, and functionalization with signaling molecules. Appropriate combinations of all these variables are
likely required for the creation of functional synthetic stem cell niches.

10.3.1 MATERIALS FOR SYNTHETIC STEM CELL NICHES

Both the chemical composition and physical form are major variables in the design
of synthetic niches. For the initial design of cell-supporting scaffolds, researchers
turned to natural polymers present in the ECM surrounding tissues and individual
cells, since they are biocompatible.[60,70] This approach resulted in the development
of successful commercial products derived from animal by-products, decellularized
intact ECM, or polymeric proteins purified from the ECM.[71–73] Due to the animal
origin of these products, there may be noticeable risks of immunogenicity, disease
transfer (e.g., prionic infections), and compatibility with the individual patient. To
partially address these issues, highly purified ECM molecules, including collagen,
hyaluronic acid, and fibrin, have been used alone and in various combinations to create scaffolds for cell adhesion and tissue repair.[74]

Scaffolds are also often fabricated from nonmammalian ECM, including alginate
and chitosan-based hydrogels and silk-based scaffolds.[75] Alginate and chitosan are
derived from brown algae and shellfish, respectively, and are easy to process and load
with cells and small molecules. Alginate's main structural component is alginic acid,
a viscous gum that is abundant in the cell walls of kelp (Laminaria) in the sea. The
physical properties of alginate gels are dependent on the abundance in the polymer of
two sugar residues, β-D-mannuronate (M) and α-L-guluronate (G). Gels are formed
by ionic binding with divalent cations (e.g., Ca^{2+}) by blocks of consecutive residues
of G sugars, but not the M sugars. These gels are widely used in both humans and
animals, as the gentle phase transition mediated by ionic cross-linking allows drugs
and cells to be encapsulated without damage.[76] Alginate does not degrade in our
bodies, due to lack of appropriate enzymes, but its dissolution can be manipulated
through control over molecular weight distribution and chemical modification.[76] Silk-
based materials provide an alternative to soft hydrogels, as they can exhibit high
elastic modulus, combined with the ability to be modified with growth and adhesion
factors.[75]

Concurrently to the development of natural scaffolds, significant research has been
focused on the application of a variety of biocompatible synthetic polymers, such
as poly(L-lactic acid, PLLA), poly(glycolic acid, PGA), poly(ethylene glycol, PEG)/
poly(caprolactones), poly(orthoesters), poly(anhydrides) and polycarbonates, as scaffolding materials.[75] The ease of production and quality control for synthetic polymers
combined with history of use of several of these polymers in FDA approved medical
devices motivated use of these materials.[63] Poly (lactide-co-glycolide) (PLGA) is an
FDA approved biocompatible polymer, successfully implemented as a suture material
and other devices. PLGA is synthesized by means of random ring-opening copoly-
merization of two different monomers, the cyclic dimers (1,4-dioxane-2,5-diones) of
glycolic acid and lactic acid. Common catalysts used in the preparation of this polymer include tin(II) 2-ethylhexanoate, tin(II) alkoxides, or aluminum isopropoxide.

During polymerization, successive monomeric units (of glycolic or lactic acid) are linked together in PLGA by ester linkages, thus yielding a linear aliphatic polyester as a product. The common theme for many biodegradable synthetic polymers is ester linkages in the main chain, allowing for hydrolysis in the body and clearance. Simple changes in polymer composition can lead to easily varied degradation rates; for example, variation of the ratio of lactide to glycolide used to form PLGA allows one to easily generate polymers with widely varying degradation rates and mechanical properties that may be suitable for a variety of applications.[77] Certain of these polymers allow simple mixing of dissolved polymers with ASCs or progenitor cells prior to solidification, and subsequent introduction into the tissue. Alternatively, scaffolds are fabricated and then seeded with cells, frequently with a subsequent development period *in vitro* using bioreactor technology. In these approaches, cells typically proliferated and tissue developed supporting this ability of the synthetic niche to guide tissue formation.[63,70]

10.3.2 Mechanical Properties of Synthetic Niche and Transmission of Stresses to Cells

The mechanical properties of synthetic scaffolds is a key design criteria, as they will often define the stability of the new tissue *in vitro* and *in vivo*, control the transmission of external forces to the adherent cells, and may directly determine the developmental fate of the embedded stem cells.[61,63] They also affect the disappearance rate of the synthetic niche and stem cell dispersal. Cells in tissue are exposed to a variety of loads, including tension, compression, bending moments, and shear forces from circulating fluids.[10] The cellular microenvironment, particularly the ECM, transfers and modulates these signals, and cells actively respond to these cues.[78] Synthetic scaffolds are often designed to mimic specific mechanical properties of the native ECM and target tissue, and these properties determine the success of regeneration and scaffold incorporation into the surrounding tissues and organs.[63] Modulation of the mechanical interactions between ASCs and the scaffold can serve as an important tool to control cell fate and phenotype, and the success of tissue restoration and function.[10] The mechanical interactions between cells and growth substrates are modulated via adhesive interactions between ECM structural proteins, or peptide mimics and adhesion receptors on cells (e.g., integrin family proteins).[79] The response of the cell is related to both the number of binding interactions and the scaffold stiffness. It also has been shown that the stiffness alone is sufficient to control multiple aspects of cell phenotype. For example, it has been shown previously that mesenchymal stem cells grown on synthetic substrates, which mimicked either rigid bone-like or softer muscle-like surfaces, proliferate and show elevated expression of protein markers specific to the growth substrate phenotype.[80] The specific cell signaling pathways and biophysical dynamics of these interactions remain to be fully described and are a focus of current studies. Attention to date has focused on material elasticity, but other mechanical properties are also likely important. In addition to the mechanics of cell–scaffold binding interactions, external mechanical stresses and strains conveyed to cells via the scaffold also regulate cell function.

Cyclic strain has been shown to alter the proliferation and gene expression of cells in tissue-engineered blood vessels, and the function of ASCs have also been shown to vary with the type and magnitude of applied mechanical loading.[81–83]

10.3.3 STEM CELL NICHE ARCHITECTURE AND TOPOGRAPHY

The capabilities developed in the semiconductor industry to create materials with defined micro- and nanoscale features are being utilized to create new synthetic niches with the features mimicking the natural microarchitectures of tissues and stem cell niches.[84] This is partially motivated by the findings that cells attached to specific topography respond differently and have the ability to probe and sense the scale, type, and density of the cues. Micropatterning and nanofabrication techniques allow patterning of micro- and nanoscaled highly reproducible structures that enable specific spatial distribution and densities of growth factors, chemicals, and immobilized cells, as shown in Figure 10.2.[85] The physical dimensions of the attachment sites control stem cell proliferation, and proliferating cells can "sense" the area of the attachment surface, and follow different developmental pathways as a function of the of the surface area.[86] The growth and proliferation of smooth muscle cells has also been shown to respond to topography cues in micropatterned PLG scaffolds.[87] The calcium dynamics in cultured cardiomyocytes has been similarly demonstrated to be influenced by nanotopograhy of their adhesion substrate.[88] Motivation for the development of patterning of scaffold mechanical properties arises from injury models that have shown fibrotic scars have a higher modulus than the surrounding tissues.[89] The mechanical property gradient appears to attract mesenchymal stem cells by "durotaxis," where the stem cells migrate toward the site of injury, gradually modify the scar, and restore healthy tissue.[90] Micron-scale chambers can also be encapsulated in the overall scaffold structure, with specific factor release profiles that respond to microenvironmental cues, including changes in osmotic pressure, electrostatic fields, and binding of signaling molecules. Overall, cell fate decisions can be modulated by changing the 3D landscape in which cells are immersed.

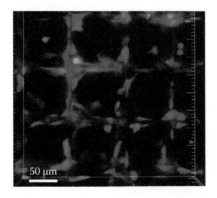

FIGURE 10.2 (See color insert)   Fluorescence image showing isometric 3D-rendered view of HT1080 cells inside a 110 mm pore-sized scaffold 24 h after seeding cells. (Modified from Tayalia, P., et al., *Adv. Mater.*, 20 (23), 4494, 2008.)

10.3.4 DEGRADATION AND INFLAMMATORY RESPONSES TO SYNTHETIC SCAFFOLDS

Successful tissue restoration and organ regeneration is likely to be controlled by
the rate of degradation and clearing of implanted artificial scaffolds, as infiltration
by host cells generally will be impacted by scaffold remodeling and disappearance. The degradation rate is critical to prevent scarring, to promote penetration
by blood vessels and nerves, and allow the spread of daughter cells toward the surrounding tissues.[65] Scaffolds with controlled rates of degradation can be fabricated,
via modulation of main chain susceptibility to hydrolytic or proteolytic cleavages
(e.g., synthesizing copolymers with different monomeric units or blocks with variable degradation rates). There are several examples of such scaffolds. Biomineralized
PLGA scaffolds for bone restoration have been fabricated.[91] PEG hydrogels with
controlled rates of enzymatic degradation by matrix metaloproteases (MMP) have
also been synthesized, and both encapsulated cells and cells penetrating from the
surrounding tissues can degrade these polymers via secreted MMPs.[92] Varied numbers of MMP cleavage sites have been introduced in these PEG hydrogels, giving the
one the ability to shorten or to prolong hydrogel degradation.[92] Bone-like biomineralized PLGA scaffolds degrade within the months, while other materials mimicking the mineral phase of bone tissue (i.e., sintered porous hydroxyapatite scaffolds,
coralline matrices, hydroxyapatite bone cements) typically degrade incompletely,
even over very long time scales.[91] The importance of the scaffold degradation rate in
tissue regeneration *in vivo* was demonstrated with alginate hydrogels with variable
degradation rates in a bone regeneration model.[65] The more rapidly degrading gels
led to a dramatic increase in the extent and quality of bone formation.

Controlling the immunogenic properties of polymers and their breakdown products can provide a handle to induce different responses from the host. The foreign
body response initiated by scaffold implantation and degradation, and possible
immune responses, are typically considered to be negative factors in the success of
engineered or regenerated tissues. Fibrotic scar tissue formation is clearly negative for
new tissue formation. However, inflammatory responses also can lead to angiogenesis and host cell recruitment, which are often desirable.[70,93] Immune responses are
directed by recruitment and activation of macrophages and dendritic cells, and influence the severity of inflammation resulting from scaffold introduction. The material
chosen to fabricate scaffolds may have a significant impact on the inflammatory/
immune response. Hyaluronic acid, known for its anti-inflammatory properties, has
been suggested to result in scarless wound healing.[94,95] Alginate gels are considered
noninflammatory as long as they are formed from highly purified polymers.[76] In
contrast, there are reports of significant dendritic cell activation upon contact with
PLGA.[96,97] However, these responses could be modulated by incorporation of immu-
nomodulating agents in the scaffolds, e.g., anti-inflammatory molecules.[98]

10.3.5 FACTOR RELEASE FROM SYNTHETIC NICHES

The soluble molecules in the cell stem niche play key roles in promoting ASC proliferation, regulating daughter cell differentiation and dispersion, and coordinating the
response of surrounding cells (e.g., angiogenesis). One approach to control in space

and time the presentation of these signals is to immobilize these factors within the materials comprising the synthetic niche, for their local and regulated presentation, typically utilizing the technologies of controlled drug delivery. Polymeric systems have been developed that have the ability to control presentation of small molecules, proteins, and even large molecules (e.g., plasmid DNA or antibodies) for these purposes.[99] Small to large molecules have been encapsulated in polymers such as PLGA for the local release, and hydrogels are also often used to deliver these factors.[63] In both cases, specific release profiles, controlled by the degradation characteristics of the scaffold, can be achieved by controlled factor encapsulation, diffusion, and degradation of the polymer. For example, alginate gels have the protein-binding property of ECM components such as proteoglycans (e.g., heparan sulfate), allowing the release of encapsulated factors (e.g., VEGF, IGF, or PDGF) with predefined and sustained profiles over time periods of days or weeks.[100,101] These materials allow one to maintain high levels of signaling ligands in specific sites, and establish signaling gradients, mimicking the developmental cues in growing tissues.

Scaffolds with encapsulated growth factors have been observed to promote the growth of both transplanted cells and host cells. Controlled release of insulin-like growth factor (IGF-1) was shown to improve the function of cardiomyocytes in the heart,[102,103] whereas bone formation was supported by the action of transforming growth factor (TGFβ) on bone-forming mesenchymal cells.[103] Scaffolds releasing angiogenic molecules, such as vascular endothelial growth factor (VEGF), can lead to the creation of networks of small blood vessels, and increase local blood flow and oxygen supply to hypoxic tissues.[104] In this approach, VEGF gradients maintain the directionality of blood vessel growth within and outside of the scaffold.[105] The ability of transplanted endothelial cell progenitor populations and mature endothelial cells to form new capillaries can also be modulated by local angiogenic factor delivery.[106] It has been successfully shown that biomineralized PLG scaffolds loaded with angiogenic ligand VEGF promote bone formation and that sustained local VEGF delivery enhances the blood vessel growth and penetration into newly formed bone tissue.[91] Importantly, the degradation rate of these PLGA scaffolds has a critical role in blood vessel penetration and lowering the spatial constraints on the expansion of the new bone/tissue.[91]

Similar strategies can be employed to promote innervation of synthetic niches. Neural progenitor cells have been combined with neurotrophic factors such as nerve growth factor (NGF), brain-derived neurotrophic factor (BDNF), neurotrophin (NT), and the netrin families of proteins. These ligands activate growth of PNS axons and their gradients control directional migration.[107–112]

Scaffolds can also be loaded with small circular DNA plasmids encoding for the growth factor of interest (e.g., VEGF), and the plasmid can be taken up by surrounding cells following release from the polymer, leading to elevated local expression of these factors.[113] One of the first methods for gene delivery with synthetic scaffolds was demonstrated and successfully implemented for bone regeneration *in vivo*.[114] Additional proof of principle can be found in the early work where plasmid DNA, encoding the expression of pleiotrophin (PTN), was embedded in a fibrin-based scaffold and injected into ischemic myocardium.[115] It was shown to increase vascularization, as compared to the bolus delivery of plasmid DNA. Furthermore, the power of this approach was demonstrated by comparing the delivery of β-galactosidase reporter plasmid by subcutaneously implanted scaffolds in rats. Reporter gene expression was one order of magnitude higher than bolus delivery at the 2 week time point, and nearly two orders of magnitude higher at the 8 and 15 week time points. This study demonstrated that the sustained exterior delivery of plasmid DNA from PLGA scaffolds led to a long-term and high level of gene expression in cells infiltrating the implants *in vivo*, and this system may be applicable for vascularizing engineered bone tissue.[116]

Multiple families of additional signaling molecules have been identified within the context of ASC niches, and the controlled release of these factors from synthetic niches is likely to be important in the future. These signaling molecules include WNT and Notch ligands, which regulate proliferation in the hair follicle stem cell niche, intestinal epithelium, satellite muscle cells, skin epidermis, and CNS.[20] Combinations of these growth factors with cytokines to create gradients within the niche and surrounding tissue will likely be implemented in future cell stem cell therapies.[20]

10.3.6 CELL DISPERSION FROM SYNTHETIC NICHES

While most work to date in the field has focused on designing scaffolds to hold transplanted ASCs, in many situations, one instead desires to disperse the cells within a large tissue volume to promote regeneration. The two approaches to ASC transplantation that are most widely pursued are injection of cell suspensions, with the hope that the transplanted cells will appropriately disperse in the target tissue, or transplantation on a synthetic niche. As noted earlier, transplantation of cell solutions leads to poor tissue engraftment, while classic scaffold approaches hold the cells in the material and do not allow significant cell mobility. However, scaffolds that mimic the ability of stem cell niches to mobilize daughter cells at an appropriate stage of differentiation and efficiently disperse them in surrounding tissue have recently been pursued.

A key aspect of effective ASC release into surrounding tissue will likely involve the controlled activation of resident stem cells to increase stem cell survival, proliferation, and formation of daughter cells at a stage of differentiation appropriate for dispersion. The incorporation of specific adhesive cues in scaffolds, such as small protein sequences resembling binding targets of cellular attachment proteins may enable these functions. An extensively used adhesion motif is the RGD-binding sequence, composed of Arginine (R)-Glycine (G)-Aspartic Acid (D), which serves as a major binding site for cellular integrin receptors. The RGD-binding sequence can be attached covalently to the polymers in synthetic niches to bind cells with sufficient force to retain the cells in place during the implantation of the scaffold,[117,118] and to provide sufficient adhesion for subsequent daughter cell outward migration in response to soluble factors promoting cell migration (Figure 10.3).[119–121]

The potential of this concept has been shown in both skeletal muscle regeneration and vascularization of ischemic limbs. For muscle regeneration, satellite cells have been embedded in alginate hydrogels functionalized with RGD adhesion ligands and growth factors (HGF and FGF2). The myoblasts in RGD-modified hydrogels

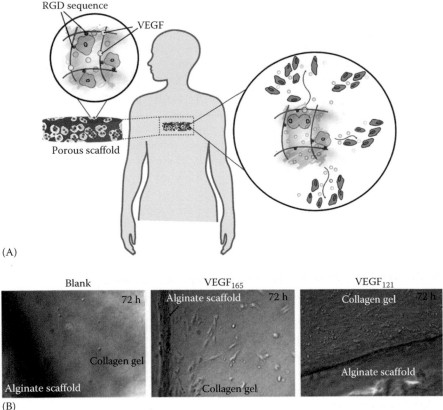

FIGURE 10.3 (See color insert)   Proposed cell delivery approach, characterization of cell migration from macroporous alginate scaffolds. (A) Diagram of approach to present cell adhesion ligands (RGD-containing peptides) and local morphogens (VEGF) in the material to maintain cell viability and to activate and induce cell migration out of scaffold. (B) Phase-contrast micrographs of OECs that have migrated out from scaffolds that contain no VEGF (blank), VEGF121, or VEGF165 and populated the surrounding tissue mimic (collagen gel) after 72 h.

demonstrated a significant increase in survival, outward migration, and the ability to regenerate damaged muscle to a much greater extent than cells simply injected at the site of injury.[119,120] Similarly, synthetic scaffolds can serve as carriers of endothelial progenitor cells, which upon activation by angiogenic factors proliferate and migrate outward from the implant, creating networks of small capillary tubes interconnected with the vascular system of the host.[106]

Synthetic niches may also be useful in activating already existing ASCs within the body, by the presentation of the attracting signals and directing the migration of these cells to the target sites. Beyond the release of growth factors and cytokines from the niches, presenting specific microarchitectural features may be important in the resulting cell trafficking. This technique is very attractive for clinical use, mainly due to the lack of ex vivo cell manipulation and lower cost, and decreased complexity as compared to approaches requiring cell transplantation.[63]

10.4 SUMMARY AND FUTURE OF SYNTHETIC STEM CELL NICHE

Significant progress has been made over last decades in developing new materials to serve as synthetic niches in tissue engineering research. Starting from the first bulk scaffolds designed to provide a structural role with minimal biological activity, researchers subsequently developed versatile, biodegradable, and multifunctional materials, which can be modified with growth factors and cells for the specific type of tissue desired. These scaffolds have the potential to interact with cells in the surrounding tissues and lead to minimal levels of inflammation. Several principles related to the structural and chemical design of synthetic niches utilized for stem cell delivery and proliferation have been developed, and these principles provide a basis for the future work in developing much more sophisticated scaffolds and tissue regeneration methods. Importantly, our constantly increasing knowledge of the structure and function of the stem cells niches, both in early and adult stages of tissue development will dramatically impact the design of these systems.

The current view of this field postulates that the future of stem cell delivery and tissue regeneration will involve multifunctional scaffolds, capable of regulating and responding to complex arrays of intra and extracellular signals, and facilitating and altering new and existing paracrine and autocrine signaling loops. These scaffolds will provide mechanical support to the growing tissue and respond to the mechanical and environmental stimuli from the host tissues, present adhesion cues for the progeny cells within the scaffold, and facilitate cell dispersal from it.[10] In general, synthetic scaffolds for stem cell delivery have certain essential structural and functional properties. These properties include biocompatibility, induction of a functional capillary network and innervation to transfer nerve signals from the CNS or PNS, and appropriate immunological response from the host. Moreover, the spatiotemporal control over the scaffold properties and interactions with the surrounding cells and tissues will likely be key factors for success. These could be achieved by external activation, and or preprogrammed changes over time. For example, scaffolds may have a multilayered structure, leading to changes in microarchitecture or chemistry as the layers degrade. The controlled release of multiple encapsulated growth factors may drive a programmed timeline for stem cell activation, divisions, and fate determination. Advances in chemical synthesis, and exploration of new types of scaffolds, such as peptides or oligonucleotides with self-assembly properties, gives tissue engineers the ability of design scaffolds with properties controlled at varied length scales. It is also important to note that proliferating stem cells and their progeny may actively modify the materials in specific temporal and localized manner.[10,63,70]

ACKNOWLEDGMENTS

The authors acknowledge funding from the National Institutes of Health (National Institute of Dental and Craniofacial Research). Dmitry Shvartsman is supported by a European Molecular Biology Organization (EMBO) Long-Term Fellowship.

REFERENCES

1. Bessis, M.; Thiery, J. P., Electron microscopy of human white blood cells and their stem cells. *Int Rev Cytol* 1961, *12*, 199–241.
2. Goodman, J. W.; Hodgson, G. S., Evidence for stem cells in the peripheral blood of mice. *Blood* 1962, *19*, 702–714.
3. Lewis, J. P.; Trobaugh, F. E., Jr., Haematopoietic stem cells. *Nature* 1964, *204*, 589–590.
4. Bryant, B. J.; Cole, L. J., Evidence for pluripotentiality of marrow stem cells: Modification of tissue distribution of in vivo I-125-UdR labeled transplanted marrow. USNRDL-TR-1028. *Res Dev Tech Rep* 1966, 1–26.
5. Niewisch, H.; Vogel, H.; Matioli, G., Concentration, quantitation, and identification of hemopoietic stem cells. *Proc Natl Acad Sci USA* 1967, *58* (6), 2261–2267.
6. Bartlett, P. F., Pluripotential hemopoietic stem cells in adult mouse brain. *Proc Natl Acad Sci USA* 1982, *79* (8), 2722–2725.
7. Alvarez-Buylla, A.; Lois, C., Neuronal stem cells in the brain of adult vertebrates. *Stem Cells* 1995, *13* (3), 263–272.
8. Tropepe, V.; Coles, B. L.; Chiasson, B. J.; Horsford, D. J.; Elia, A. J.; McInnes, R. R.; van der Kooy, D., Retinal stem cells in the adult mammalian eye. *Science* 2000, *287* (5460), 2032–2036.
9. Brooker, G. J.; Kalloniatis, M.; Russo, V. C.; Murphy, M.; Werther, G. A.; Bartlett, P. F., Endogenous IGF-1 regulates the neuronal differentiation of adult stem cells. *J Neurosci Res* 2000, *59* (3), 332–341.
10. Discher, D. E.; Mooney, D. J.; Zandstra, P. W., Growth factors, matrices, and forces combine and control stem cells. *Science* 2009, *324* (5935), 1673–1677.
11. Raaijmakers, M. H.; Scadden, D. T., Evolving concepts on the microenvironmental niche for hematopoietic stem cells. *Curr Opin Hematol* 2008, *15* (4), 301–306.
12. Scadden, D. T., The stem-cell niche as an entity of action. *Nature* 2006, *441* (7097), 1075–1079.
13. Ma, D. K.; Bonaguidi, M. A.; Ming, G. L.; Song, H., Adult neural stem cells in the mammalian central nervous system. *Cell Res* 2009, *19* (6), 672–682.
14. Takao, T.; Tsujimura, A., Prostate stem cells: The niche and cell markers. *Int J Urol* 2008, *15* (4), 289–294.
15. Li, W.; Hayashida, Y.; Chen, Y. T.; Tseng, S. C., Niche regulation of corneal epithelial stem cells at the limbus. *Cell Res* 2007, *17* (1), 26–36.
16. Arai, F.; Suda, T., Maintenance of quiescent hematopoietic stem cells in the osteoblastic niche. *Ann N Y Acad Sci* 2007, *1106*, 41–53.
17. Naveiras, O.; Daley, G. Q., Stem cells and their niche: A matter of fate. *Cell Mol Life Sci* 2006, *63* (7–8), 760–766.
18. Rizvi, A. Z.; Wong, M. H., Epithelial stem cells and their niche: There's no place like home. *Stem Cells* 2005, *23* (2), 150–165.
19. Schofield, R., The relationship between the spleen colony-forming cell and the haemopoietic stem cell. *Blood Cells* 1978, *4* (1–2), 7–25.
20. Morrison, S. J.; Spradling, A. C., Stem cells and niches: Mechanisms that promote stem cell maintenance throughout life. *Cell* 2008, *132*, 598–611.
21. Guilak, F., Control of stem cell fate by physical interactions with the extracellular matrix. *Cell Stem Cell* 2009, *5*, 17–26.
22. Burdick, J. A.; Vunjak-Novakovic, G., Engineered microenvironments for controlled stem cell differentiation. *Tissue Eng A* 2009, *15*, 205–219.
23. *Essentials of stem cell biology*. Edited by Lanza, R. P (Elsevier/Academic Press, 2006). San Diego. CA USA.
24. *Encyclopedia of stem cell research*. Edited by Svendsen, C. & Ebert, A. D (SAGE Publications, 2008). Thousand Oaks. CA USA.

25. Basak, O.; Taylor, V., Stem cells of the adult mammalian brain and their niche. *Cell Mol Life Sci* 2009, *66* (6), 1057–1072.
26. Nakagawa, M.; Koyanagi, M.; Tanabe, K.; Takahashi, K.; Ichisaka, T.; Aoi, T.; Okita, K.; Mochiduki, Y.; Takizawa, N.; Yamanaka, S., Generation of induced pluripotent stem cells without Myc from mouse and human fibroblasts. *Nat Biotechnol* 2008, *26* (1), 101–106.
27. Aoi, T.; Yae, K.; Nakagawa, M.; Ichisaka, T.; Okita, K.; Takahashi, K.; Chiba, T.; Yamanaka, S., Generation of pluripotent stem cells from adult mouse liver and stomach cells. *Science* 2008, *321* (5889), 699–702.
28. Daley, G. Q.; Scadden, D. T., Prospects for stem cell-based therapy. *Cell* 2008, *132*, 544–548.
29. Amariglio, N.; Hirshberg, A.; Scheithauer, B. W.; Cohen, Y.; Loewenthal, R.; Trakhtenbrot, L.; Paz, N.; Koren-Michowitz, M.; Waldman, D.; Leider-Trejo, L.; Toren, A.; Constantini, S.; Rechavi, G., Donor-derived brain tumor following neural stem cell transplantation in an ataxia telangiectasia patient. *PLoS Med* 2009, *6* (2), e1000029.
30. Pittenger, M. F.; Mackay, A. M.; Beck, S. C.; Jaiswal, R. K.; Douglas, R.; Mosca, J. D.; Moorman, M. A.; Simonetti, D. W.; Craig, S.; Marshak, D. R., Multilineage potential of adult human mesenchymal stem cells. *Science* 1999, *284* (5411), 143–147.
31. Vogel, G., Stem cell policy. Can adult stem cells suffice? *Science* 2001, *292* (5523), 1820–1822.
32. Goodman, J. W., Stem cells of the hematopoietic and lymphatic tissues. *Natl Cancer Inst Monogr* 1964, *14*, 151–168.
33. Nowell, P. C.; Hirsch, B. E.; Fox, D. H.; Wilson, D. B., Evidence for the existence of multipotential lympho-hematopoietic stem cells in adult rat. *J Cell Physiol* 1970, *75* (2), 151–158.
34. de Vries, P.; Brasel, K. A.; McKenna, H. J.; Williams, D. E.; Watson, J. D., Thymus reconstitution by c-kit-expressing hematopoietic stem cells purified from adult mouse bone marrow. *J Exp Med* 1992, *176* (6), 1503–1509.
35. Crosby, H. A.; Strain, A. J., Adult liver stem cells: Bone marrow, blood, or liver derived? *Gut* 2001, *48* (2), 153–154.
36. Hofmeister, C. C.; Zhang, J.; Knight, K. L.; Le, P.; Stiff, P. J., Ex vivo expansion of umbilical cord blood stem cells for transplantation: Growing knowledge from the hematopoietic niche. *Bone Marrow Transpl* 2007, *39* (1), 11–23.
37. Lutolf, M. P.; Doyonnas, R.; Havenstrite, K.; Koleckar, K.; Blau, H. M., Perturbation of single hematopoietic stem cell fates in artificial niches. *Integr Biol* 2009, *1*, 59–69.
38. Marion, R. M.; Strati, K.; Li, H.; Murga, M.; Blanco, R.; Ortega, S.; Fernandez-Capetillo, O.; Serrano, M.; Blasco, M. A., A p53-mediated DNA damage response limits reprogramming to ensure iPS cell genomic integrity. *Nature* 2009, *460* (7259), 1149–1153.
39. Till, J. E.; McCulloch, E. A.; Siminovitch, L., A Stochastic Model of Stem Cell Proliferation, Based on the Growth of Spleen Colony-Forming Cells. *Proc Natl Acad Sci U S A* 1964, *51*, 29–36.
40. Till, J. E.; Mc C. E., A direct measurement of the radiation sensitivity of normal mouse bone marrow cells. *Radiat Res* 1961, *14*, 213–222.
41. Levy, V.; Lindon, C.; Harfe, B. D.; Morgan, B. A., Distinct stem cell populations regenerate the follicle and interfollicular epidermis. *Dev Cell* 2005, *9* (6), 855–861.
42. Burt, R. K.; Loh, Y.; Pearce, W.; Beohar, N.; Barr, W. G.; Craig, R.; Wen, Y.; Rapp, J. A.; Kessler, J., Clinical applications of blood-derived and marrow-derived stem cells for nonmalignant diseases. *JAMA* 2008, *299* (8), 925–936.
43. Orkin, S. H.; Zon L. I., Hematopoiesis: an evolving paradigm for stem cell biology. *Cell* 2008, *132* (4), 631–644.
44. Purton, L. E.; Scadden, D. T., Limiting factors in murine hematopoietic stem cell assays. *Cell Stem Cell* 2007, *1* (3), 263–270.

45. Adams, G. B., Therapeutic targeting of a stem cell niche. *Nat Biotechnol* 2007, *25*, 238–243.
46. Adams, G. B.; Scadden, D. T., The hematopoietic stem cell in its place. *Nat Immunol* 2006, *7* (4), 333–337.
47. Mauro, A., Satellite cell of skeletal muscle fibers. *J Biophys Biochem Cytol* 1961, *9*, 493–495.
48. Bischoff, R., Regeneration of single skeletal muscle fibers in vitro. *Anat Rec* 1975, *182* (2), 215–235.
49. Konigsberg, U. R.; Lipton, B. H.; Konigsberg, I. R., The regenerative response of single mature muscle fibers isolated in vitro. *Dev Biol* 1975, *45* (2), 260–275.
50. Lipton, B. H.; Schultz, E., Developmental fate of skeletal muscle satellite cells. *Science* 1979, *205* (4412), 1292–1294.
51. Kuang, S.; Kuroda, K.; Le Grand, F.; Rudnicki, M. A., Asymmetric self-renewal and commitment of satellite stem cells in muscle. *Cell* 2007, *129* (5), 999–1010.
52. Sherwood, R. I.; Christensen, J. L.; Conboy, I. M.; Conboy, M. J.; Rando, T. A.; Weissman, I. L.; Wagers, A. J., Isolation of adult mouse myogenic progenitors: Functional heterogeneity of cells within and engrafting skeletal muscle. *Cell* 2004, *119* (4), 543–554.
53. Lepper, C.; Conway, S. J.; Fan, C. M., Adult satellite cells and embryonic muscle progenitors have distinct genetic requirements. *Nature* 2009, *460*, 627–631.
54. Dhawan, J.; Rando, T. A., Stem cells in postnatal myogenesis: Molecular mechanisms of satellite cell quiescence, activation and replenishment. *Trends Cell Biol* 2005, *15* (12), 666–673.
55. Kuang, S.; Gillespie, M. A.; Rudnicki, M. A., Niche regulation of muscle satellite cell self-renewal and differentiation. *Cell Stem Cell* 2008, *2* (1), 22–31.
56. Kuhn, H. G.; Dickinson-Anson, H.; Gage, F. H., Neurogenesis in the dentate gyrus of the adult rat: Age-related decrease of neuronal progenitor proliferation. *J Neurosci* 1996, *16* (6), 2027–2033.
57. Reynolds, B. A.; Weiss, S., Generation of neurons and astrocytes from isolated cells of the adult mammalian central nervous system. *Science* 1992, *255* (5052), 1707–1710.
58. Doetsch, F., A niche for adult neural stem cells. *Curr Opin Genet Dev* 2003, *13* (5), 543–550.
59. Pittenger, M. F.; Martin, B. J., Mesenchymal stem cells and their potential as cardiac therapeutics. *Circ Res* 2004, *95* (1), 9–20.
60. Lutolf, M. P.; Hubbell, J. A., Synthetic biomaterials as instructive extracellular microenvironments for morphogenesis in tissue engineering. *Nat Biotechnol* 2005, *23*, 47–55.
61. Lutolf, M. P.; Gilbert, P. M.; Blau, H. M., Designing materials to direct stem-cell fate. *Nature* 2009, *462* (7272), 433–441.
62. Ratner, B. D.; Bryant, S. J., Biomaterials: Where we have been and where we are going. *Annu Rev Biomed Eng* 2004, *6*, 41–75.
63. Huebsch, N.; Mooney, D. J., Inspiration and application in the evolution of biomaterials. *Nature* 2009, *462* (7272), 426–432.
64. Metcalfe, A. D.; Ferguson, M. W. J., Tissue engineering of replacement skin: The crossroads of biomaterials, wound healing, embryonic development, stem cells and regeneration. *J Roy Soc Interface* 2007, *4* (14), 413–437.
65. Alsberg, E.; Kong, H. J.; Hirano, Y.; Smith, M. K.; Albeiruti, A.; Mooney, D. J., Regulating bone formation via controlled scaffold degradation. *J Dent Res* 2003, *82* (11), 903–908.
66. Campbell, J. H.; Efendy, J. L.; Campbell, G. R., Novel vascular graft grown within recipient's own peritoneal cavity. *Circ Res* 1999, *85* (12), 1173–1178.
67. Weinberg, C. B.; Bell, E., A blood vessel model constructed from collagen and cultured vascular cells. *Science* 1986, *231* (4736), 397–400.

68. Viswanathan, S. R.; Daley, G. Q.; Gregory, R. I., Selective blockade of microRNA processing by Lin28. *Science* 2008, *320* (5872), 97–100.
69. Isenberg, B. C.; Williams, C.; Tranquillo, R. T., Small-diameter artificial arteries engineered in vitro. *Circ Res* 2006, *98* (1), 25–35.
70. Chan, G.; Mooney, D. J., New materials for tissue engineering: Towards greater control over the biological response. *Trends Biotechnol* 2008, *26* (7), 382–392.
71. Badylak, S. F.; Kropp, B.; McPherson, T.; Liang, H.; Snyder, P. W., Small intestinal submucosa: A rapidly resorbed bioscaffold for augmentation cystoplasty in a dog model. *Tissue Eng* 1998, *4* (4), 379–387.
72. Matthews, J. A.; Wnek, G. E.; Simpson, D. G.; Bowlin, G. L., Electrospinning of collagen nanofibers. *Biomacromolecules* 2002, *3* (2), 232–238.
73. Hubbell, J. A., Materials as morphogenetic guides in tissue engineering. *Curr Opin Biotechnol* 2003, *14* (5), 551–558.
74. Lutolf, M. P.; Weber, F. E.; Schmoekel, H. G.; Schense, J. C.; Kohler, T.; Muller, R.; Hubbell, J. A., Repair of bone defects using synthetic mimetics of collagenous extracellular matrices. *Nat Biotechnol* 2003, *21* (5), 513–518.
75. Chang, T. C.; Yu, D.; Lee, Y. S.; Wentzel, E. A.; Arking, D. E.; West, K. M.; Dang, C. V.; Thomas-Tikhonenko, A.; Mendell, J. T., Widespread microRNA repression by Myc contributes to tumorigenesis. *Nat Genet* 2008, *40* (1), 43–50.
76. Augst, A. D.; Kong, H. J.; Mooney, D. J., Alginate hydrogels as biomaterials. *Macromol Biosci* 2006, *6* (8), 623–633.
77. Chesne, P.; Adenot, P. G.; Viglietta, C.; Baratte, M.; Boulanger, L.; Renard, J. P., Cloned rabbits produced by nuclear transfer from adult somatic cells. *Nat Biotechnol* 2002, *20* (4), 366–369.
78. Hynes, R. O., The extracellular matrix: Not just pretty fibrils. *Science* 2009, *326* (5957), 1216–1219.
79. Alsberg, E.; Anderson, K. W.; Albeiruti, A.; Franceschi, R. T.; Mooney, D. J., Cell-interactive alginate hydrogels for bone tissue engineering. *J Dent Res* 2001, *80* (11), 2025–2029.
80. Engler, A. J.; Sen, S.; Sweeney, H. L.; Discher, D. E., Matrix elasticity directs stem cell lineage specification. *Cell* 2006, *126*, 677–689.
81. Kurpinski, K.; Chu, J.; Hashi, C.; Li, S., Anisotropic mechanosensing by mesenchymal stem cells. *Proc Natl Acad Sci USA* 2006, *103* (44), 16095–16100.
82. Kim, B. S.; Nikolovski, J.; Bonadio, J.; Mooney, D. J., Cyclic mechanical strain regulates the development of engineered smooth muscle tissue. *Nat Biotechnol* 1999, *17* (10), 979–983.
83. Niklason, L. E.; Gao, J.; Abbott, W. M.; Hirschi, K. K.; Houser, S.; Marini, R.; Langer, R., Functional arteries grown in vitro. *Science* 1999, *284* (5413), 489–493.
84. Lutolf, M. P., Artificial ECM: Expanding the cell biology toolbox in 3D. *Integr Biol* 2009, *1*, 235–241.
85. Prakriti, T.; Cleber, R. M.; Tommaso, B.; David, J. M.; Eric, M., 3D cell-migration studies using two-photon engineered polymer scaffolds. *Adv Mater* 2008, *20* (23), 4494–4498.
86. Underhill, G. H.; Bhatia, S. N., High-throughput analysis of signals regulating stem cell fate and function. *Curr Opin Chem Biol* 2007, *11*, 357–366.
87. Khademhosseini, A.; Langer, R.; Borenstein, J.; Vacanti, J. P., Microscale technologies for tissue engineering and biology. *Proc Natl Acad Sci USA* 2006, *103* (8), 2480–2487.
88. Yin, L.; Bien, H.; Entcheva, E., Scaffold topography alters intracellular calcium dynamics in cultured cardiomyocyte networks. *Am J Physiol Heart Circ Physiol* 2004, *287* (3), H1276–H1285.

89. Berry, M. F.; Engler, A. J.; Woo, Y. J.; Pirolli, T. J.; Bish, L. T.; Jayasankar, V.; Morine, K. J.; Gardner, T. J.; Discher, D. E.; Sweeney, H. L., Mesenchymal stem cell injection after myocardial infarction improves myocardial compliance. *Am J Physiol Heart Circ Physiol* 2006, *290* (6), H2196–H2203.

90. Lo, C. M.; Wang, H. B.; Dembo, M.; Wang, Y. L., Cell movement is guided by the rigidity of the substrate. *Biophys J* 2000, *79* (1), 144–152.

91. Murphy, W. L.; Simmons, C. A.; Kaigler, D.; Mooney, D. J., Bone regeneration via a mineral substrate and induced angiogenesis. *J Dent Res* 2004, *83* (3), 204–210.

92. April, M. K.; Mark, W. T.; Andrea, M. K.; Jonathan, A. F.; Kristi, S. A., Tunable hydrogels for external manipulation of cellular microenvironments through controlled photodegradation. *Adv Mater* 2009, *22* (1), 61–66.

93. Kyriakides, T. R.; Leach, K. J.; Hoffman, A. S.; Ratner, B. D.; Bornstein, P., Mice that lack the angiogenesis inhibitor, thrombospondin 2, mount an altered foreign body reaction characterized by increased vascularity. *Proc Natl Acad Sci USA* 1999, *96* (8), 4449–4454.

94. Gerecht, S.; Burdick, J. A.; Ferreira, L. S.; Townsend, S. A.; Langer, R.; Vunjak-Novakovic, G., Hyaluronic acid hydrogel for controlled self-renewal and differentiation of human embryonic stem cells. *Proc Natl Acad Sci USA* 2007, *104* (27), 11298–11303.

95. Gerecht, S.; Townsend, S. A.; Pressler, H.; Zhu, H.; Nijst, C. L.; Bruggeman, J. P.; Nichol, J. W.; Langer, R., A porous photocurable elastomer for cell encapsulation and culture. *Biomaterials* 2007, *28* (32), 4826–4835.

96. Ali, O. A.; Huebsch, N.; Cao, L.; Dranoff, G.; Mooney, D. J., Infection-mimicking materials to program dendritic cells in situ. *Nat Mater* 2009, *8*, 151–158.

97. Yoshida, M.; Mata, J.; Babensee, J. E., Effect of poly(lactic-co-glycolic acid) contact on maturation of murine bone marrow-derived dendritic cells. *J Biomed Mater Res A* 2007, *80* (1), 7–12.

98. Irene, C.; Robert, M.; Mirren, C.; Keith, A. B.; Calogero, F.; Anthony, J. R.; Sheila, M., Development of an Ibuprofen-releasing biodegradable PLA/PGA electrospun scaffold for tissue regeneration. *Biotechnol Bioeng 105* (2), 396–408.

99. Sands, R. W.; Mooney, D. J., Polymers to direct cell fate by controlling the microenvironment. *Curr Opin Biotechnol* 2007, *18* (5), 448–453.

100. Chen, R. R.; Silva, E. A.; Yuen, W. W.; Mooney, D. J., Spatio-temporal VEGF and PDGF delivery patterns blood vessel formation and maturation. *Pharm Res* 2007, *24* (2), 258–264.

101. Silva, E. A.; Mooney, D. J., Spatiotemporal control of vascular endothelial growth factor delivery from injectable hydrogels enhances angiogenesis. *J Thromb Haemost* 2007, *5* (3), 590–598.

102. Davis, M. E.; Motion, J. P.; Narmoneva, D. A.; Takahashi, T.; Hakuno, D.; Kamm, R. D.; Zhang, S.; Lee, R. T., Injectable self-assembling peptide nanofibers create intramyocardial microenvironments for endothelial cells. *Circulation* 2005, *111* (4), 442–450.

103. Park, H.; Temenoff, J. S.; Tabata, Y.; Caplan, A. I.; Mikos, A. G., Injectable biodegradable hydrogel composites for rabbit marrow mesenchymal stem cell and growth factor delivery for cartilage tissue engineering. *Biomaterials* 2007, *28* (21), 3217–3227.

104. Trentin, D.; Hall, H.; Wechsler, S.; Hubbell, J. A., Peptide-matrix-mediated gene transfer of an oxygen-insensitive hypoxia-inducible factor-1alpha variant for local induction of angiogenesis. *Proc Natl Acad Sci USA* 2006, *103* (8), 2506–2511.

105. Chen, R. R.; Silva, E. A.; Yuen, W. W.; Brock, A. A.; Fischbach, C.; Lin, A. S.; Guldberg, R. E.; Mooney, D. J., Integrated approach to designing growth factor delivery systems. *FASEB J* 2007, *21* (14), 3896–3903.

106. Silva, J.; Barrandon, O.; Nichols, J.; Kawaguchi, J.; Theunissen, T. W.; Smith, A., Promotion of reprogramming to ground state pluripotency by signal inhibition. *PLoS Biol* 2008, *6* (10), e253.

107. Li, X.; Yang, Z.; Zhang, A., The effect of neurotrophin-3/chitosan carriers on the proliferation and differentiation of neural stem cells. *Biomaterials* 2009, *30* (28), 4978–4985.

108. Zhou, X. M.; Yuan, H. P.; Wu, D. L.; Zhou, X. R.; Sun, D. W.; Li, H. Y.; Shao, Z. B., Study of brain-derived neurotrophic factor gene transgenic neural stem cells in the rat retina. *Chin Med J (Engl)* 2009, *122* (14), 1642–1649.

109. Ii, M.; Nishimura, H.; Sekiguchi, H.; Kamei, N.; Yokoyama, A.; Horii, M.; Asahara, T., Concurrent vasculogenesis and neurogenesis from adult neural stem cells. *Circ Res* 2009, *105* (9), 860–868.

110. Denham, M.; Dottori, M., Signals involved in neural differentiation of human embryonic stem cells. *Neurosignals* 2009, *17* (4), 234–241.

111. Blurton-Jones, M.; Kitazawa, M.; Martinez-Coria, H.; Castello, N. A.; Muller, F. J.; Loring, J. F.; Yamasaki, T. R.; Poon, W. W.; Green, K. N.; LaFerla, F. M., Neural stem cells improve cognition via BDNF in a transgenic model of Alzheimer disease. *Proc Natl Acad Sci USA* 2009, *106* (32), 13594–13599.

112. Horne, M. K.; Nisbet, D. R.; Forsythe, J. S.; Parish, C., Three dimensional nanofibrous scaffolds incorporating immobilized BDNF promote proliferation and differentiation of cortical neural stem cells. *Stem Cells Dev* 2009, *19* (6), 843–852.

113. Shea, L. D.; Smiley, E.; Bonadio, J.; Mooney, D. J., DNA delivery from polymer matrices for tissue engineering. *Nat Biotechnol* 1999, *17* (6), 551–554.

114. Levy, R. J.; Goldstein, S. A.; Bonadio, J., Gene therapy for tissue repair and regeneration. *Adv Drug Deliv Rev* 1998, *33* (1–2), 53–69.

115. Sridharan, R.; Tchieu, J.; Mason, M. J.; Yachechko, R.; Kuoy, E.; Horvath, S.; Zhou, Q.; Plath, K., Role of the murine reprogramming factors in the induction of pluripotency. *Cell* 2009, *136* (2), 364–377.

116. Huang, Y. C.; Riddle, K.; Rice, K. G.; Mooney, D. J., Long-term in vivo gene expression via delivery of PEI-DNA condensates from porous polymer scaffolds. *Hum Gene Ther* 2005, *16* (5), 609–617.

117. Alsberg, E.; Anderson, K. W.; Albeiruti, A.; Rowley, J. A.; Mooney, D. J., Engineering growing tissues. *Proc Natl Acad Sci USA* 2002, *99* (19), 12025–12030.

118. Comisar, W. A.; Kazmers, N. H.; Mooney, D. J.; Linderman, J. J., Engineering RGD nanopatterned hydrogels to control preosteoblast behavior: A combined computational and experimental approach. *Biomaterials* 2007, *28* (30), 4409–4417.

119. Hill, E.; Boontheekul, T.; Mooney, D. J., Regulating activation of transplanted cells controls tissue regeneration. *Proc Natl Acad Sci USA* 2006, *103* (8), 2494–2499.

120. Hill, E.; Boontheekul, T.; Mooney, D. J., Designing scaffolds to enhance transplanted myoblast survival and migration. *Tissue Eng* 2006, *12* (5), 1295–1304.

121. Silva, E. A.; Kim, E. S.; Kong, H. J.; Mooney, D. J., Material-based deployment enhances efficacy of endothelial progenitor cells. *Proc Natl Acad Sci USA* 2008, *105* (38), 14347–14352.

122. Howe, S. J.; Mansour, M. R.; Schwarzwaelder, K.; Bartholomae, C.; Hubank, M.; Kempski, H.; Brugman, M. H.; Pike-Overzet, K.; Chatters, S. J.; de Ridder, D.; Gilmour, K. C.; Adams, S.; Thornhill, S. I.; Parsley, K. L.; Staal, F. J.; Gale, R. E.; Linch, D. C.; Bayford, J.; Brown, L.; Quaye, M.; Kinnon, C.; Ancliff, P.; Webb, D. K.; Schmidt, M.; von Kalle, C.; Gaspar, H. B.; Thrasher, A. J., Insertional mutagenesis combined with acquired somatic mutations causes leukemogenesis following gene therapy of SCID-X1 patients. *J Clin Invest* 2008, *118* (9), 3143–3150.

123. Scoville, D. H.; Sato, T.; He, X. C.; Li, L., Current view: Intestinal stem cells and signaling. *Gastroenterology* 2008, *134* (3), 849–864.

124. Tiede, S.; Kloepper, J. E.; Bodo, E.; Tiwari, S.; Kruse, C.; Paus, R., Hair follicle stem cells: Walking the maze. *Eur J Cell Biol* 2007, *86* (7), 355–376.

125. Verstappen, J.; Katsaros, C.; Torensma, R.; Von den Hoff, J. W., A functional model for adult stem cells in epithelial tissues. *Wound Repair Regen* 2009, *17* (3), 296–305.
126. Dym, M.; He, Z.; Jiang, J.; Pant, D.; Kokkinaki, M., Spermatogonial stem cells: Unlimited potential. *Reprod Fertil Dev* 2009, *21* (1), 15–21.

11 Combined Therapies of Cell Transplantation and Molecular Delivery

Suk Ho Bhang and Byung-Soo Kim

CONTENTS

11.1 STEM CELL TRANSPLANTATION FOR THERAPEUTIC ANGIOGENESIS

Research involving the use of stem cells that participate in the neovascularization of adult tissues has been studied because of the potential for new cell-based angiogenic therapies. Recently, preclinical studies have shown that introducing stem cells into ischemic tissues can restore tissue vascularization in limbs, retina, and myocardium. Several studies have demonstrated that stem cells can differentiate into endothelial cells (ECs), incorporate into vessels, and promote post-ischemic neovascularization, or can secrete angiogenic and antiapoptotic factors that enhance angiogenesis.[1,2] Like embryonic stem cells, adult stem cells are self-renewing and pluripotent. Results obtained from bone marrow–derived stem cells in animal models, reiterate their potentials for regenerating myocardium and providing therapeutic benefits for ischemic heart disease. Injection of bone marrow mononuclear cells following left anterior descending coronary artery ligation decreased infarction size and increased local blood flow in a porcine model.[3] These results were an effect of increased angiogenesis in the infarcted tissue.

Another strategy to introduce bone marrow cells in infarct or ischemic tissues includes their endogenous mobilization by infusion of granulocyte colony-stimulating factor (G-CSF) and stem cell factor (SCF). This treatment increased the *in vivo* population of stem cells that were able to contribute to myocardial repair.[4,5] Cytokine treatment also has been shown to significantly improve survival following myocardial infarction and result in the formation of vascular structures and myocytes in the infarcted myocardium. Hemodynamics and left ventricle function were significantly improved and scar formation was reduced. These studies indicate that stem cells promote myocardial regeneration after infarction in animal models and suggest their potential application for the treatment of heart diseases. However, clinical studies evaluating the therapeutic potential of various stem cell therapies have reported conflicting results, generating contention amongst the research community.

Despite the angiogenic potential of stem cell therapies, low therapeutic efficacy, due to poor survival of cells transplanted into ischemic tissues, remains a large obstacle in the development of stem cell therapies for angiogenesis. Stem cells transplanted in vasculature-scarce regions (ischemic regions) are exposed to hypoxia immediately after transplantation (due to the lack of initial vasculature) and are prone to apoptosis.[6,7] A high level of cell death has been observed within a few days of cell transplantation in ischemic muscles.[6,7] The inflammatory environment of ischemic tissues also contributes to the poor survival of the transplanted cells.[8] Thus, administration of cells alone may not have a substantial effect on promoting angiogenesis. It is not known whether a single post-myocardial infarction treatment will be effective, but combination therapeutic approaches is promising. A possible adjunct to stem cell transplantation therapies may be to engineer stem cells prior to transplantation or to combine them with protein or gene delivery methods

Therapeutic angiogenesis involves the delivery of growth factors in ischemic tissues to promote the development of collateral blood vessels and to resupply circulation.[9,10] Therapeutic angiogenesis combined with pharmacotherapy eventually may eliminate the need to use angioplasty and bypass surgery for treating ischemic diseases.[9] Results from studies using animal models of ischemic diseases have been very encouraging, and numerous clinical trials are currently in progress testing proteins and genes related to vascular endothelial growth factor (VEGF) and fibroblast growth factor (FGF). Therefore, we discuss angiogenic stem cell therapies combined with protein delivery or gene delivery.

11.2 COMBINED THERAPIES OF STEM CELL TRANSPLANTATION AND PROTEIN DELIVERY FOR THERAPEUTIC ANGIOGENESIS

11.2.1 ENHANCING THE ANGIOGENIC EFFICACY OF TRANSPLANTED STEM CELLS WITH PROTEIN DELIVERY

Angiogenic protein administration may be an effective approach for improving the survival and angiogenic efficacy of stem cells transplanted in ischemic tissues. Many compounds have been identified with angiogenic activity. The administration of recombinant angiogenic growth factors (i.e., FGF,[4,11–16] G-CSF, granulocyte macrophage colony-stimulating factor (GM-CSF), VEGF[11,17]) have been shown to enhance angiogenesis in ischemic diseases and to promote neovascularization in ischemic tissues.[13,17] In 1992, Yanagisawa-Miwa et al.[18] infused FGF in the coronary arteries of dogs and improved cardiac function after myocardial infarction. Takeshita et al.[19] reported in 1994 that VEGF infused in an iliac artery promoted the development of collateral vessels in ischemic rabbit hind limbs. Moreover, in 1999, Morishita et al.[20] reported human hepatocyte growth factor (HGF) administered into an iliac artery improved collateral vessel formation in ischemic rabbit hind limbs. Several studies have shown that the delivery of angiogenic growth factors at the stem cell transplantation sites enhances their therapeutic effects in ischemic tissues.[21,22] Recently, combination therapies of stem cells and angiogenic factors have been shown to synergistically induce angiogenesis. The combination therapies were reported to produce a greater and more rapid induction of angiogenesis than administration of either stem cells or angiogenic factors alone.[23,24] The combination therapy also was shown to induce the formation of long-lasting, functionally stable vessels.[23]

Among angiogenic growth factors, FGF has been studied widely in both animal and human studies. Stem cell transplantation combined with FGF deliveries enhances the angiogenic efficacy of the transplanted stem cells, likely by improving their survival.[23] It is well known that the poor viability of stem cells transplanted in ischemic sites limits their therapeutic potential.[6,7] FGF protects transplanted stem cells from ischemia-induced apoptosis, and it prevents Fas-mediated apoptosis in fibroblasts through activation of mitogen-activated protein kinase.[25] Studies have demonstrated that the pretreatment of rodent hearts with FGF show a prominent protection against subsequent ischemia and reperfusion injury.[26,27] When administered to ischemic hearts, FGF protects against ischemic damage through the binding of FGF with tyrosine kinase FGF receptor 1, which is mediated by protein kinase C activation.[28] Therefore, FGF seems to improve cell survival by activating several survival-related signaling components. Moreover, FGF improves cell viability under hypoxic conditions by increasing expression of Bcl-2, an antiapoptotic gene,[29] and induces angiogenesis by stimulating secretion of antiapoptotic VEGF and HGF.[30,31] Coadministration of FGF with stem cells in ischemic regions resulted in a better stem cell survival and paracrine factor secretion, and consequently, FGF enhanced the therapeutic efficacy of stem cell transplantation.[23] FGF could contribute to the improvement of angiogenic efficacy of stem cell therapies by activating the hypoxia-inducible factor (HIF) families. FGF2 enhances hypoxia-induced expression of HIF-1α stem cells.[32] HIF-1α is essential for hypoxic induction of paracrine factor (VEGF, FGF, and HGF) expression and of angiogenic signaling pathways.[33,34] Therefore, exogenously adding FGF stimulates secretion of paracrine factors from transplanted stem cells by enhancing the activity of the HIF family.[35] Enhanced expression of adhesion molecules (i.e., intercellular adhesion molecule (ICAM), platelet–endothelial cell adhesion molecule [PECAM]) in stem cells by FGF might be another mechanism that is involved in the improved angiogenesis. Compared with either FGF delivery or stem cell transplantation alone, the expression of adhesion molecules was significantly increased in ischemic tissues when stem cells were transplanted with FGF delivery.[23] Enhanced paracrine actions of stem cells through FGF delivery may activate host cells to express adhesion molecules.[36] Increases in host cell adhesion molecule expression might promote the mobilization of endothe-lial progenitor cells to ischemic sites and contribute to neovascularization.[36]

11.2.2 ENHANCING THE ANGIOGENIC EFFICACY OF HOST-ORIGINATED STEM CELLS THROUGH PROTEIN DELIVERY

The administration of cytokines related to hematopoiesis could enhance neovascu-larization in ischemic tissues. G-CSF[37] and GM-CSF[38] have been shown to mobi-lize endothelial progenitor cells (EPCs) from bone marrow to circulating blood, and EPCs participate in the neovascularization of ischemic tissues. It has been reported that transplantation of autologous CD34-positive cells or mononuclear cells isolated from peripheral blood after G-CSF treatment augmented neovascularization in humans with limb ischemia or myocardial infarction.[39,40]

G-CSF and GM-CSF, which can mobilize EPCs from bone marrow into periph-eral blood, also may have the potential to promote endothelium formation in vas-cular grafts that are implanted to replace diseased vessels. Graft occlusion caused by detachment of ECs from the graft surface after implantation and/or incomplete endothelialization are the main obstacles of bioartificial vascular graft implanta-tions. A previous study reported that significantly greater endothelialization was observed on G-CSF-treated vascular grafts than on grafts that were not treated with G-CSF.[41] Combined EC transplantation and G-CSF therapy facilitated a more com-plete endothelialization than the endothelialization strategy alone. Preconditioning of EC-seeded vascular grafts with growth factors could be a method for facilitating the formation of solid endothelium.[42,43] Seeding of ECs or EPCs transfected with genes related to proliferation or migration of ECs and smooth muscle cells may also enhance endothelialization and reduce intimal hyperplasia of vascular grafts.[44]

11.3 COMBINED THERAPIES OF STEM CELL TRANSPLANTATION AND GENE DELIVERY FOR THERAPEUTIC ANGIOGENESIS

Instead of delivering recombinant proteins, which are costly and unstable, the deliv-ery of genes that encode angiogenic growth factors can induce neovascularization. Due to the short biological half-life of proteins, angiogenic protein therapies are often inefficient. One advantage of gene delivery is a relatively long-term expression of the gene of interest in ischemic regions. Angiogenic gene therapy is a rational approach for creating stable vessels for the functional improvement of ischemic tis-sue.[4] Overexpression of such genes in ischemic tissues protects the tissue against oxidative stress produced under hypoxic/reoxygenation conditions or during vari-ous inflammatory cytokine-related pathological conditions. Initially intra-arterial gene therapies were used, but systemic dilution of the gene-therapy product and diffusion in the presence of atherosclerotic occlusion limits the efficient gene trans-fer to the vicinity of the ischemic cells. For these reasons, direct intramuscular injections has become an approach for treating ischemic diseases. Gene therapies, with the goal of increasing or stimulating angiogenesis, primarily use genes encod-ing growth factors such as VEGF, FGF, or hypoxia-inducible factor 1 (HIF-1).[14–17] Plasmids encoding VEGF were injected directly into the muscle of ischemic hind limb models.[45] The results were consistent with the VEGF protein delivery, but the effects were less dramatic because of the relatively low efficiency of plasmid transfer to the tissues. Further studies confirmed that neovascularization can be induced in ischemic heart and skeletal muscles of animals by the direct injection of VEGF- or FGF-encoding DNA in plasmid or adenoviral vectors.[46,47] The genes encoding angiopoietin and HGF were shown to be as effective as genes encoding VEGF and FGF in angiogenesis-promoting animal models.[48,49]

Despite the usefulness of gene therapies, the therapeutic efficacy of gene delivery is often limited. In ischemic diseases (i.e., hind limb ischemia, myocardial infarction), partial or total occlusion of large vessels induces metabolic and contractile changes within seconds,[18] followed by apoptosis and necrosis in the ischemic tissue (Figure 11.1).[19] Thus, the therapeutic efficacy of the exogenously delivered gene is often limited by rapid apoptosis or necrosis of host cells and transplanted stem cells, which is a common occurrence in ischemic tissues.[4,18] A major drawback of DNA injections is a poor transduction efficiency. To overcome this problem, viral vectors carrying angiogenic genes have been designed and used in animal and human studies[50]; however, these studies pose several safety issues. Furthermore, angiogenic gene therapies have been shown to be associated with undesirable side effects, such as the progression of atherosclerosis or unrestricted blood vessel growth, leading to angioma formation.[51,52]

Cellular angiogenesis by implanting genetically engineered cells has advantages, including localized, regulatable gene expression or concurrent myogenesis and angiogenesis. Symes and coworkers assessed the efficacy of rat primary myoblasts transfected with the human VEGF gene for improvements in heart injuries.[53] Myocardial

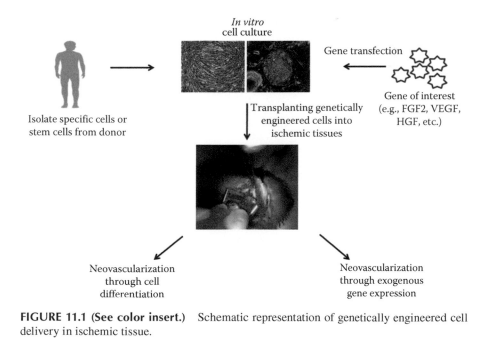

FIGURE 11.1 (See color insert.) Schematic representation of genetically engineered cell delivery in ischemic tissue.

VEGF levels in the experimental animal group increased significantly after VEGF gene-transfected cell transplantations, leading to enhanced angiogenesis and heart function. To improve the survival of bone marrow mesenchymal stem cells (MSCs) after transplantation, Song and coworkers introduced FGF-2 gene ex vivo before transplantation.[29] The FGF-2-transfected MSCs (FGF-2-MSCs) secreted increased levels of FGF-2 under hypoxic conditions and displayed a threefold increase in viability compared to normal MSCs cultured under hypoxic conditions. FGF-2-MSCs were viable until 4 weeks after implantation into infarcted myocardium. The expression of cardiac troponin T and a voltage-gated Ca^{2+} channel increased. Deramaudt and coworkers used a human heme oxygenase-1 (HO-1) plasmid to transfect rabbit coronary microvessel ECs. HO-1 participates in the regulation of EC activation, proliferation, and angiogenesis.[54] ECs transfected with human HO-1 gene demonstrated a twofold increase in HO-1 activity and maintained a similar phenotype as the non-transfected cells. ECs transfected with the human HO-1 gene exhibited a twofold increase in blood vessel formation; however, the angiogenic cascade involves complex and multiple stages, and a single angiogenic factor or specific cell type may be insufficient for appropriate angiogenic cascades. Moreover, genetically engineered cells must be able to proliferate and efficiently produce angiogenic factors after transplantation into ischemic tissues.

Cellular angiogenesis by genetically engineered cells seems useful, but many aspects of this approach need to be explored. The angiogenic cascade involves multiple stages, and a single angiogenic factor may not mimic the natural course of the angiogenic cascade. Thus, multiple types of angiogenic growth factors may need to be expressed over an appropriate time course. The choice of delivery vehicle is crucial for achieving the intended results. The type of gene delivery vehicle often affects the gene transfection efficiency, inflammation, and gene expression duration. The cells must be able to proliferate and efficiently produce the angiogenic factor after gene transduction.

11.4 CONCLUSIONS

Therapeutic angiogenesis can be induced by various methods including administration of angiogenic growth factor proteins, angiogenic gene transfections, and coadministration of angiogenic growth factor proteins or genes with stem cells. All of these approaches have shown promising results for inducing angiogenesis in experimental myocardial or limb ischemia. However, the question remains whether a single type of growth factor or gene is sufficient to produce a substantial angiogenic effect. Further studies are necessary to understand what cell type(s) and other cues (i.e., growth factor, gene combination) are necessary for satisfactory improvement in neovascularization.

REFERENCES

1. Ohnishi, S.; Ohgushi, H.; Kitamura, S.; Nagaya, N., Mesenchymal stem cells for the treatment of heart failure. *International Journal of Hematology* 2007, *86* (1), 17–21.
2. Kilroy, G.; Foster, S.; Wu, X.; Ruiz, J.; Sherwood, S.; Heifetz, A.; Ludlow, J.; Stricker, D.; Potiny, S.; Green, P., Cytokine profile of human adipose derived stem cells: Expression of angiogenic, hematopoietic, and pro inflammatory factors. *Journal of Cellular Physiology* 2007, *212* (3), 702–709.

3. Kamihata, H.; Matsubara, H.; Nishiue, T.; Fujiyama, S.; Tsutsumi, Y.; Ozono, R.; Masaki, H.; Mori, Y.; Iba, O.; Tateishi, E., Implantation of bone marrow mononuclear cells into ischemic myocardium enhances collateral perfusion and regional function via side supply of angioblasts, angiogenic ligands, and cytokines. *Circulation* 2001, *104* (9), 1046.

4. Orlic, D.; Kajstura, J.; Chimenti, S.; Limana, F.; Jakoniuk, I.; Quaini, F.; Nadal-Ginard, B.; Bodine, D.; Leri, A.; Anversa, P., Mobilized bone marrow cells repair the infarcted heart, improving function and survival. *Proceedings of the National Academy of Sciences of the United States of America* 2001, *98* (18), 10344.

5. Klocke, R.; Kuhlmann, M.; Scobioala, S.; Schabitz, W.; Nikol, S., Granulocyte colony-stimulating factor (G-CSF) for cardio-and cerebrovascular regenerative applications. *Current Medicinal Chemistry* 2008, *15* (10), 968–977.

6. Tang, Y.; Zhang, Y.; Qian, K.; Shen, L.; Phillips, M., Improved graft mesenchymal stem cell survival in ischemic heart with a hypoxia-regulated heme oxygenase-1 vector. *Journal of the American College of Cardiology* 2005, *46* (7), 1339–1350.

7. Zhu, W.; Chen, J.; Cong, X.; Hu, S.; Chen, X., Hypoxia and serum deprivation induced apoptosis in mesenchymal stem cells. *Stem Cells* 2006, *24* (2), 416–425.

8. Hill, E.; Boontheekul, T.; Mooney, D., Regulating activation of transplanted cells controls tissue regeneration. *Proceedings of the National Academy of Sciences of the United States of America* 2006, *103* (8), 2494.

9. Isner, J. M., Manipulating angiogenesis against vascular disease. *Hospital Practice* 1999, *34* (6), 69.

10. Webster, K., Therapeutic angiogenesis: A case for targeted, regulated gene delivery. *Critical Reviews in Eukaryotic Gene Expression* 2000, *10* (2), 113.

11. Asahara, T.; Bauters, C.; Zheng, L.; Takeshita, S.; Bunting, S.; Ferrara, N.; Symes, J.; Isner, J., Synergistic effect of vascular endothelial growth factor and basic fibroblast growth factor on angiogenesis in vivo. *Circulation* 1995, *92* (9), 365.

12. Unger, E.; Banai, S.; Shou, M.; Lazarous, D.; Jaklitsch, M.; Scheinowitz, M.; Correa, R.; Klingbeil, C.; Epstein, S., Basic fibroblast growth factor enhances myocardial collateral flow in a canine model. *American Journal of Physiology- Heart and Circulatory Physiology* 1994, *266* (4), H1588.

13. Tabata, Y.; Ikada, Y., Vascularization effect of basic fibroblast growth factor released from gelatin hydrogels with different biodegradabilities. *Biomaterials* 1999, *20* (22), 2169–2175.

14. Fujita, M.; Ishihara, M.; Simizu, M.; Obara, K.; Ishizuka, T.; Saito, Y.; Yura, H.; Morimoto, Y.; Takase, B.; Matsui, T., Vascularization in vivo caused by the controlled release of fibroblast growth factor-2 from an injectable chitosan/non-anticoagulant heparin hydrogel. *Biomaterials* 2004, *25* (4), 699–706.

15. Laham, R.; Rezaee, M.; Post, M.; Novicki, D.; Sellke, F.; Pearlman, J.; Simons, M.; Hung, D., Intrapericardial delivery of fibroblast growth factor-2 induces neovascularization in a porcine model of chronic myocardial ischemia. *Journal of Pharmacology and Experimental Therapeutics* 2000, *292* (2), 795.

16. Harada, K.; Grossman, W.; Friedman, M.; Edelman, E.; Prasad, P.; Keighley, C.; Manning, W.; Sellke, F.; Simons, M., Basic fibroblast growth factor improves myocardial function in chronically ischemic porcine hearts. *Journal of Clinical Investigation* 1994, *94* (2), 623.

17. Ferrara, N.; Alitalo, K., Clinical applications of angiogenic growth factors and their inhibitors. *Nature Medicine* 1999, *5* (12), 1359.

18. Yanagisawa-Miwa, A.; Uchida, Y.; Nakamura, F.; Tomaru, T.; Kido, H.; Kamijo, T.; Sugimoto, T.; Kaji, K.; Utsuyama, M.; Kurashima, C., Salvage of infarcted myocardium by angiogenic action of basic fibroblast growth factor. *Science* 1992, *257* (5075), 1401.

19. Takeshita, S.; Zheng, L.; Brogi, E.; Kearney, M.; Pu, L.; Bunting, S.; Ferrara, N.; Symes, J.; Isner, J., Therapeutic angiogenesis. A single intraarterial bolus of vascular endothelial growth factor augments revascularization in a rabbit ischemic hind limb model. *Journal of Clinical Investigation* 1994, *93* (2), 662.

20. Morishita, R.; Nakamura, S.; Hayashi, S.; Taniyama, Y.; Moriguchi, A.; Nagano, T.; Taiji, M.; Noguchi, H.; Takeshita, S.; Matsumoto, K., Therapeutic angiogenesis induced by human recombinant hepatocyte growth factor in rabbit hind limb ischemia model as cytokine supplement therapy. *Hypertension* 1999, *33* (6), 1379.

21. Sakakibara, Y.; Nishimura, K.; Tambara, K.; Yamamoto, M.; Lu, F.; Tabata, Y.; Komeda, M., Prevascularization with gelatin microspheres containing basic fibroblast growth factor enhances the benefits of cardiomyocyte transplantation. *The Journal of Thoracic and Cardiovascular Surgery* 2002, *124* (1), 50.

22. Retuerto, M.; Schalch, P.; Patejunas, G.; Carbray, J.; Liu, N.; Esser, K.; Crystal, R.; Rosengart, T., Angiogenic pretreatment improves the efficacy of cellular cardiomyoplasty performed with fetal cardiomyocyte implantation. *The Journal of Thoracic and Cardiovascular Surgery* 2004, *127* (4), 1041.

23. Bhang, S.; Cho, S.; Lim, J.; Kang, J.; Lee, T.; Yang, H.; Song, Y.; Park, M.; Kim, H.; Yoo, K., Locally delivered growth factor enhances the angiogenic efficacy of adipose derived stromal cells transplanted to ischemic limbs. *Stem Cells* 2009, *27* (8), 1976–1986.

24. Jeon, O.; Kang, S.; Lim, H.; Choi, D.; Kim, D.; Lee, S.; Chung, J.; Kim, B., Synergistic effect of sustained delivery of basic fibroblast growth factor and bone marrow mononuclear cell transplantation on angiogenesis in mouse ischemic limbs. *Biomaterials* 2006, *27* (8), 1617–1625.

25. Kazama, H.; Yonehara, S., Oncogenic K-Ras and basic fibroblast growth factor prevent Fas-mediated apoptosis in fibroblasts through activation of mitogen-activated protein kinase. *The Journal of Cell Biology* 2000, *148* (3), 557.

26. Padua, R.; Sethi, R.; Dhalla, N.; Kardami, E., Basic fibroblast growth factor is cardioprotective in ischemia-reperfusion injury. *Molecular and Cellular Biochemistry* 1995, *143* (2), 129–135.

27. Sheikh, F.; Sontag, D.; Fandrich, R.; Kardami, E.; Cattini, P., Overexpression of FGF-2 increases cardiac myocyte viability after injury in isolated mouse hearts. *American Journal of Physiology—Heart and Circulatory Physiology* 2001, *280* (3), H1039.

28. Jiang, Z.; Padua, R.; Ju, H.; Doble, B.; Jin, Y.; Hao, J.; Cattini, P.; Dixon, I.; Kardami, E., Acute protection of ischemic heart by FGF-2: Involvement of FGF-2 receptors and protein kinase C. *American Journal of Physiology—Heart and Circulatory Physiology* 2002, *282* (3), H1071.

29. Song, H.; Kwon, K.; Lim, S.; Kang, S.; Ko, Y.; Xu, Z.; Chung, J.; Kim, B.; Lee, H.; Joung, B., Transfection of mesenchymal stem cells with the FGF-2 gene improves their survival under hypoxic conditions. *Molecules and Cells* 2005, *19* (3), 402–407.

30. Wafai, R.; Tudor, E.; Angus, J.; Wright, C., Vascular effects of FGF-2 and VEGF-B in rabbits with bilateral hind limb ischemia. *Journal of Vascular Research* 2008, *46* (1), 45–54.

31. Rehman, J.; Traktuev, D.; Li, J.; Merfeld-Clauss, S.; Temm-Grove, C.; Bovenkerk, J.; Pell, C.; Johnstone, B.; Considine, R.; March, K., Secretion of angiogenic and antiapoptotic factors by human adipose stromal cells. *Circulation* 2004, *109*, 1292–1298.

32. Shi, Y.; Bingle, L.; Gong, L.; Wang, Y.; Corke, K.; Fang, W., Basic FGF augments hypoxia induced HIF-1-alpha expression and VEGF release in T47D breast cancer cells. *Pathology* 2007, *39* (4), 396–400.

33. Calvani, M.; Rapisarda, A.; Uranchimeg, B.; Shoemaker, R.; Melillo, G., Hypoxic induction of an HIF-1 alpha-dependent bFGF autocrine loop drives angiogenesis in human endothelial cells. *Blood* 2006, *107* (7), 2705.

34. Pugh, C.; Ratcliffe, P., Regulation of angiogenesis by hypoxia: Role of the HIF system. *Nature Medicine* 2003, *9* (6), 677–684.
35. Li, Q.; Wang, X.; Yang, Y.; Lin, H., Hypoxia upregulates hypoxia inducible factor (HIF)-3alpha expression in lung epithelial cells: Characterization and comparison with HIF-1alpha. *Cell Research* 2006, *16* (6), 548–558.
36. Yoon, C.; Hur, J.; Oh, I.; Park, K.; Kim, T.; Shin, J.; Kim, J.; Lee, C.; Chung, J.; Park, Y., Intercellular adhesion molecule-1 is upregulated in ischemic muscle, which mediates trafficking of endothelial progenitor cells. *Arteriosclerosis, Thrombosis, and Vascular Biology* 2006, *26* (5), 1066.
37. Ohtsuka, M.; Takano, H.; Zou, Y.; Toko, H.; Akazawa, H.; Qin, Y.; Suzuki, M.; Hasegawa, H.; Nakaya, H.; Komuro, I., Cytokine therapy prevents left ventricular remodeling and dysfunction after myocardial infarction through neovascularization. *The FASEB Journal* 2004, *18* (7), 851–853.
38. Takahashi, T.; Kalka, C.; Masuda, H.; Chen, D.; Silver, M.; Kearney, M.; Magner, M.; Isner, J.; Asahara, T., Ischemia-and cytokine-induced mobilization of bone marrow-derived endothelial progenitor cells for neovascularization. *Nature Medicine* 1999, *5* (4), 434–438.
39. Chang, S.; Kang, H.; Lee, H.; Kim, K.; Hur, J.; Han, K.; Park, Y.; Kim, H., Peripheral blood stem cell mobilisation by granulocyte-colony stimulating factor in patients with acute and old myocardial infarction for intracoronary cell infusion. *Heart* 2009, *95* (16), 1326.
40. Iwasaki, H.; Kawamoto, A.; Ishikawa, M.; Oyamada, A.; Nakamori, S.; Nishimura, H.; Sadamoto, K.; Horii, M.; Matsumoto, T.; Murasawa, S., Dose-dependent contribution of CD34-positive cell transplantation to concurrent vasculogenesis and cardiomyogenesis for functional regenerative recovery after myocardial infarction. *Circulation* 2006, *113* (10), 1311.
41. Cho, S.; Lim, J.; Chu, H.; Hyun, H.; Choi, C.; Hwang, K.; Yoo, K.; Kim, D.; Kim, B., Enhancement of in vivo endothelialization of tissue engineered vascular grafts by granulocyte colony stimulating factor. *Journal of Biomedical Materials Research Part A* 2006, *76* (2), 252–263.
42. Zemani, F.; Silvestre, J.; Fauvel-Lafeve, F.; Bruel, A.; Vilar, J.; Bieche, I.; Laurendeau, I.; Galy-Fauroux, I.; Fischer, A.; Boisson-Vidal, C., Ex vivo priming of endothelial progenitor cells with SDF-1 before transplantation could increase their proangiogenic potential. *Arteriosclerosis, Thrombosis, and Vascular Biology* 2008, *28*, 1034–1035.
43. Pasha, Z.; Wang, Y.; Sheikh, R.; Zhang, D.; Zhao, T.; Ashraf, M., Preconditioning enhances cell survival and differentiation of stem cells during transplantation in infarcted myocardium. *Cardiovascular Research* 2008, *77* (1), 134.
44. Chen, S.-Y.; Wang, F.; Yan, X.-Y.; Zhou, Q.; Ling, Q.; Ling, J.-X.; Rong, Y.-Z.; Li, Y.-G., Autologous transplantation of EPCs encoding FGF1 gene promotes neovascularization in a porcine model of chronic myocardial ischemia. *International Journal of Cardiology* 2009, *135* (2), 223–232.
45. Tsurumi, Y.; Takeshita, S.; Chen, D.; Kearney, M.; Rossow, S.; Passeri, J.; Horowitz, J.; Symes, J.; Isner, J., Direct intramuscular gene transfer of naked DNA encoding vascular endothelial growth factor augments collateral development and tissue perfusion. *Circulation* 1996, *94* (12), 3281.
46. Negishi, Y.; Matsuo, K.; Endo-Takahashi, Y.; Suzuki, K.; Matsuki, Y.; Takagi, N.; Suzuki, R.; Maruyama, K.; Aramaki Y., Delivery of an angiogenic gene into ischemic muscle by novel bubble liposomes followed by ultrasound exposure. *Pharmaceutical Research* 2011, *28* (4), 712.
47. Tio, R.; Tkebuchava, T.; Scheuermann, T.; Lebherz, C.; Magner, M.; Kearny, M.; Esakof, D.; Isner, J.; Symes, J., Intramyocardial gene therapy with naked DNA encoding vascular endothelial growth factor improves collateral flow to ischemic myocardium. *Human Gene Therapy* 1999, *10* (18), 2953–2960.

48. Van Belle, E.; Witzenbichler, B.; Chen, D.; Silver, M.; Chang, L.; Schwall, R.; Isner, J., Potentiated angiogenic effect of scatter factor/hepatocyte growth factor via induction of vascular endothelial growth factor: The case for paracrine amplification of angiogenesis. *Circulation* 1998, *97* (4), 381.

49. Aoki, M.; Morishita, R.; Taniyama, Y.; Kida, I.; Moriguchi, A.; Matsumoto, K.; Nakamura, T.; Kaneda, Y.; Higaki, J.; Ogihara, T., Angiogenesis induced by hepatocyte growth factor in non-infarcted myocardium and infarcted myocardium: Up-regulation of essential transcription factor for angiogenesis, ets. *Gene Therapy* 2000, *7* (5), 417–427.

50. Losordo, D.; Vale, P.; Hendel, R.; Milliken, C.; Fortuin, F.; Cummings, N.; Schatz, R.; Asahara, T.; Isner, J.; Kuntz, R., Phase 1/2 placebo-controlled, double-blind, dose-escalating trial of myocardial vascular endothelial growth factor 2 gene transfer by catheter delivery in patients with chronic myocardial ischemia. *Circulation* 2002, *105*, 2012–2018.

51. Lee, R.; Springer, M.; Blanco-Bose, W.; Shaw, R.; Ursell, P.; Blau, H., VEGF gene delivery to myocardium: Deleterious effects of unregulated expression. *Circulation* 2000, *102* (8), 898.

52. Schwarz, E.; Speakman, M.; Patterson, M.; Hale, S.; Isner, J.; Kedes, L.; Kloner, R., Evaluation of the effects of intramyocardial injection of DNA expressing vascular endothelial growth factor (VEGF) in a myocardial infarction model in the rat–angiogenesis and angioma formation. *Journal of the American College of Cardiology* 2000, *35* (5), 1323–1330.

53. Symes, J.; Losordo, D.; Vale, P.; Lathi, K.; Esakof, D.; Mayskiy, M.; Isner, J., Gene therapy with vascular endothelial growth factor for inoperable coronary artery disease. *The Annals of Thoracic Surgery* 1999, *68* (3), 830–836.

54. Deramaudt, B.; Braunstein, S.; Remy, P.; Abraham, N., Gene transfer of human heme oxygenase into coronary endothelial cells potentially promotes angiogenesis. *Journal of Cellular Biochemistry* 1998, *68* (1), 121–127.

Index

A

Acetylated low-density lipoprotein (acLDL), 33
Adipose stromal cells (ASCs), 38–39
Adult stem cells
  EPCs sources
    adult peripheral blood, 35–36
    HAECs, 37
    MAPCs, 37
    umbilical cord blood, 36–37
  pericyte/smooth muscle cell progenitor cell sources
    adipose tissue, 38–39
    bone marrow, 37–38
    peripheral and cord blood, 38
    vessel wall, 39
African clawed frogs, 63
Angiogenesis; *see also* Therapeutic angiogenesis
  bFGF, 166
  cell adhesion
    antiangiogenic cancer therapeutics, 168–169
    CAM model, 167–168
    cell-cell interaction, 169
    cell invasion and sprouting, 169
    cell spreading control, 167–169
    cytoskeletal tension, 166
    FAK, 167
    focal adhesions, 167
    micropatterned substrates, 167
    molecular intermediaries, 167
    proteolytic activity, 169
    Ras-Erk pathway, 167
    tubulogenesis, 169
  growth factors, 166
  mechanical regulation
    cell contractility, 170
    cytoskeletal tension, 170
    endothelial proliferation and migration, 170–171
    matrix elasticity, 171
    matrix stiffness, 172
    RhoA-ROCK-mediated contractility, 170–171
  multicellular interactions, 178
    cell-cell adhesion, 176
    dermal fibroblasts, 178
    endothelial sprouting assay, 178
    homotypic endothelial cell interactions, 177
    micropatterning technique, 177
    10T1/2 pericytes, 179
    PDGF, 166
    PlGF, 166
    VEGF, 166
Antihuman CD31-conjugated magnetic beads, 39–40
ASCs, *see* Adipose stromal cells
Atherosclerosis, 243–244

B

Basic fibroblast growth factor (bFGF), 5, 166
Blood outgrowth endothelial cells (BOECs), 32
Blood vessels
  adult stem/progenitor cells (*see* Adult stem cells)
  BOEC-expressing Factor VIII, 47
  CD34 and VEGFR-2, 47
  endothelial cells, 32–33
  isolation strategies
    human EPCs, 39–40
    human MPCs, bone marrow/cord blood, 40–41
  pericytes, 33–34
  SMCs, 34–35
  vascular networks (*see* Vascular networks)
BMPs, *see* Bone morphogenetic proteins
Bone marrow cell (BMC), 195
  CD34+ cells, 195
  chemotherapeutic agents, 194
  cytokine and growth factors, 194
  mobilizing factors, 195
  molecular therapies
    chemokines (*see* Chemokines)
    receptor-ligand bonds, 197–198
  stem cell homing, 204–205
  tissue repair, 195–196
    β-galactosidase, 206
    c-kit receptors, 206
    cross-gender cell transplantation, 206
    green fluorescent protein, 206
    Lac-Z+ cells, 205
    paracrine signaling, 206, 208
    α-smooth muscle actin, 206
Bone marrow transplantation, 219
Bone morphogenetic proteins (BMPs), 7–8, 62